THE ROLE OF BIOTECHNOLOGY IN A SUSTAINABLE FOOD SUPPLY

One of the main challenges of our generation is the creation of an efficient system that provides sustainable food, feed, fiber, and fuel from the land while also preserving biodiversity and ecosystems. The first and most immediate challenge will be to feed a human population that is expected to grow to more than nine billion by mid-century. The supply of food must grow at least as fast as the population, but the amount of land and water available for agriculture is limited; thus the need for more food must be met through higher crop yields. Agricultural biotechnology is one tool that holds promise to alleviate hunger and poverty. However, there are complex and interrelated scientific, social, political, and ethical questions regarding the widespread use of biotechnology in the food supply.

This edited volume – which includes contributions from leading scholars in many fields – discusses the numerous factors in the use of agricultural and food biotechnology as we work toward more sustainable food production systems. It includes diverse perspectives on the challenges, opportunities, success stories, barriers, and risks associated with biotechnology. It summarizes additional studies on the effects of biotechnology on the environment. The book also discusses relevant ethical and moral issues, potential changes to government policies and economics, and other social implications of agricultural biotechnology.

This comprehensive and interdisciplinary book will be of great interest to students, professionals, and researchers in various fields – from bioengineering, agriculture, and ecosystem science, to economics and political science.

JENNIE S. POPP is a Professor of Agricultural Economics and Agribusiness and Area Director of the Center for Agricultural and Rural Sustainability at the University of Arkansas. Her research areas include the economics of water and soil quality, risk management for agricultural producers, and assessment of best management practices for agriculture.

MOLLY M. JAHN is a Professor in the Laboratory of Genetics and Department of Agronomy at the University of Madison–Wisconsin. She has had a distinguished research career in plant genetics and genomics and plant breeding of vegetable crops, focusing on the molecular genetics of disease resistance and quality traits.

MARTY D. MATLOCK is a Professor of Biological and Agricultural Engineering and Area Director of the Center for Agricultural and Rural Sustainability at the University of Arkansas. He is a Certified Senior Ecologist with the Ecological Society of America and a Registered Professional Engineer. His research area is in the design and management of ecosystem services, with a focus on water resources.

NATHAN P. KEMPER is the Trade Adjustment Assistance Program Coordinator for the Southern Risk Management Education Center. He coordinates the efforts of a team made up of Cooperative Extension Service and Sea Grant faculty and staff and industry representatives in thirteen southern region states to develop curricula and deliver training to farmers and fishermen to help them adjust their business practices in response to increased import competition.

This book is dedicated to our families and to the men, women, and children who toil every day in the sun and rain to produce the food, feed, fiber, and fuel that drive our prosperity.

THE ROLE OF BIOTECHNOLOGY IN A SUSTAINABLE FOOD SUPPLY

Edited by

JENNIE S. POPP
University of Arkansas

MOLLY M. JAHN
University of Wisconsin–Madison

MARTY D. MATLOCK
University of Arkansas

NATHAN P. KEMPER
University of Arkansas

CAMBRIDGE UNIVERSITY PRESS
Cambridge, New York, Melbourne, Madrid, Cape Town,
Singapore, São Paulo, Delhi, Mexico City

Cambridge University Press
32 Avenue of the Americas, New York, NY 10013-2473, USA

www.cambridge.org
Information on this title: www.cambridge.org/9780521192347

First published 2012

Printed in the United States of America

A catalog record for this publication is available from the British Library.

Library of Congress Cataloging in Publication data

The role of biotechnology in a sustainable food supply / [edited by] Jennie Popp . . . [et al.].
p. cm.
Includes bibliographical references and index.
ISBN 978-0-521-19234-7 (hardback)
1. Food – Biotechnology. 2. Food – Supply. I. Popp, Jennie, 1966–
TP248.65.F66R64 2012
664–dc23 2011037256

ISBN 978-0-521-19234-7 Hardback

Contents

Preface

The conference that formed the foundation of this book occurred in the formative stage of the debate over the role of biotechnology in providing a sustainable food supply. In the years since the conference very significant progress has occurred in the framing of the issues presented in this book. A vibrant discourse across governmental, academic, industrial, and public stakeholders has emerged pertaining to a shared commitment toward a healthy, vibrant, prosperous future for humanity and our planet.

There is clear common agreement that making the best use of our natural resources is good. Almost everyone understands at some level that we live in a finite Earth system that is very complex and interconnected. There is also a new recognition that it is possible for our activities to exert influences on that system for better or for worse. This recognition has resulted in widespread commitments in global business communities and local governments toward efficiencies and responsibilities previously not considered part of commerce. The emerging measurements of associated co-benefits and inadvertent damages from our agricultural supply chain are resulting in detailed, holistic, cradle-to-grave or field-to-compost characterization of those supply chains. These approaches recognize the now familiar triple bottom line. Nobody wants to do things that harm the environment or communities, human or other biota, and almost everyone understands that sustainability requires optimizing across complex domains and identifying trade-offs and synergies.

The role of technology in human life has throughout history created complexity and uncertainty where before there was a certain level of comfort. We have time and time again adopted a technology to solve a problem, only to create another, sometimes much more threatening, situation. These new problems can be eminently predictable if we take the time to look. The editors of this book are cognizant of the risks of biotechnology in the food supply chain and have attempted to ensure that credible and reasonable risks are described herein.

But our progress to date has brought us to the brink of some important abysses. The number of people who are chronically malnourished is on the rise. Inputs required to produce food, feed, fiber, and fuel from the land are becoming more scarce, as is the land itself. Population pressures will generate demand from this finite Earth in excess of any experienced in history. Other life forms hang in the balance. Our global life support system is at risk.

We have experienced these brinks in the past. After a devastating Civil War, President Lincoln explicitly set our course toward a plentiful, affordable, and safe food supply by creating ingenious programs in education and technology development. In the ensuing century and a half we have made almost unimaginable progress toward his goal with consequent benefits of equally unimaginable proportions.

Our 21st-century challenge is to create a system that provides sustainable food, feed, fiber, and fuel from the land while preserving biodiversity and other ecosystem services. The editors believe that the way to approach this new brink, this 21st-Century Agricultural Challenge, is through respectful discourse so that we may better understand each other's values. This discourse should inform science-based analysis of risks and opportunities. Respect and trust come from understanding each other's values, though they differ across communities. Respectful discourse to align values with competing demands and uncertain outcomes will, in the editors' views, create our best opportunity to achieve the imaginable – a prosperous Earth for more than nine billion people.

Editors

Jennie S. Popp, Ph.D.
Professor, Department of Agricultural Economics and Agribusiness
Area Director, Center for Agricultural and Rural Sustainability
University of Arkansas–Fayetteville
Fayetteville, AR

Molly M. Jahn, Ph.D.
Professor, Departments of Agronomy and Genetics
University of Wisconsin–Madison
Madison, WI

Marty D. Matlock, Ph.D., P.E., C.S.E.
Professor, Department of Biological and Agricultural Engineering
Area Director, Center for Agricultural and Rural Sustainability
University of Arkansas–Fayetteville
Fayetteville, AR

Nathan P. Kemper
Trade Adjustment Assistance Program Coordinator
Southern Risk Management Education Center
University of Arkansas Division of Agriculture
Little Rock, AR

Contributing Authors

Caroline (Cal) Baier-Anderson, Ph.D.
Former Scientist with Environmental Defense Fund
Washington, DC

Larry Binning, Ph.D.
Professor Emeritus
Department of Horticulture
University of Wisconsin–Madison
Madison, WI

Dominique Brossard, Ph.D.
Associate Professor
Department of Life Sciences Communication
University of Wisconsin–Madison
Madison, WI

Alvin J. Bussan, Ph.D.
Associate Professor
Department of Horticulture
University of Wisconsin–Madison
Madison, WI

Anthony J. Cavalieri, Ph.D.
Consultant
International Food Policy Research Institute
Washington, DC

Jason R. Cavatorta, Ph.D.
Plant Breeder
Vegetable Seeds Division
Monsanto Company
Woodland, CA

Jed Colquhoun, Ph.D.
Associate Professor
Department of Horticulture
University of Wisconsin–Madison
Madison, WI

José Falck-Zepeda, Ph.D.
Research Fellow/Leader, Policy Team Program for Biosafety Systems
Environment and Production Technology Division
International Food Policy Research Institute
Washington, DC

Gregory D. Graff, Ph.D.
Associate Professor
Department of Agricultural and Resource Economics
Colorado State University
Fort Collins, CO

Stewart M. Gray, Ph.D.
Research Plant Pathologist, USDA/ARS
Professor, Department of Plant Pathology and Plant-Microbe Biology
Cornell University
Ithaca, NY

The Rev. Lowell E. Grisham, M. Div.
Rector
St. Paul's Episcopal Church
Fayetteville, AR

Russell Groves, Ph.D.
Assistant Professor
Department of Entomology
University of Wisconsin–Madison
Madison, WI

Michelle Mauthe Harvey, M.B.A.
Project Manager, Corporate Partnerships
Environmental Defense Fund
Bentonville, AR

Molly M. Jahn, Ph.D.
Professor and Special Advisor to Chancellor and Provost for Sustainability Sciences
University of Wisconsin–Madison
Madison, WI

Shelley Jansky, Ph.D.
Research Geneticist and Associate Professor
USDA/ARS and Department of Horticulture
University of Wisconsin–Madison
Madison, WI

Jiming Jiang, Ph.D.
Professor
Department of Horticulture
University of Wisconsin–Madison
Madison, WI

Nicholas Kalaitzandonakes, Ph.D.
MSMC Endowed Professor of Agribusiness Strategy
Director, Economics and Management of Agrobiotechnology Center
University of Missouri
Columbia, MO

Keith Kelling, Ph.D.
Professor Emeritus
Department of Soil Science
University of Wisconsin–Madison
Madison, WI

Deana Knuteson, Ph.D.
BioIPM Field Coordinator
Department of Horticulture
University of Wisconsin–Madison
Madison, WI

Peggy G. Lemaux, Ph.D.
Cooperative Extension Specialist
Department of Plant and Microbial Biology
University of California–Berkeley
Berkeley, CA

Marty D. Matlock, Ph.D., P.E., C.S.E.
Professor, Department of Biological and Agricultural Engineering
Area Director, Center for Agricultural and Rural Sustainability
University of Arkansas–Fayetteville
Fayetteville, AR

William H. Meyers, Ph.D.
Howard Cowden Professor of Agricultural and Applied Economics
Food and Agricultural Policy Research Institute
University of Missouri
Columbia, MO

Paul D. Mitchell, Ph.D.
Associate Professor
Department of Agricultural and Applied Economics
University of Wisconsin–Madison
Madison, WI

William Muir, Ph.D.
Professor
PULSe Molecular Evolutionary Genetics Program and Department of Animal Sciences
Purdue University
West Lafayette, IN

Pamela Ronald, Ph.D.
Professor
Department of Plant Pathology
University of California–Davis
Davis, CA

Matt Ruark, Ph.D.
Assistant Professor
Department of Soil Science
University of Wisconsin–Madison
Madison, WI

Eric S. Sachs, Ph.D.
Regulatory Lead, Scientific Affairs
Monsanto Company
St. Louis, MO

Mark K. Sears, Ph.D.
Professor Emeritus
Department of Environmental Biology
University of Guelph
Guelph, ON, Canada

Erin Silva, Ph.D.
Associate Scientist
Department of Agronomy
University of Wisconsin–Madison
Madison, WI

Walter R. Stevenson, Ph.D.
Professor Emeritus
Department of Plant Pathology
University of Wisconsin–Madison
Madison, WI

Alison Van Eenennaam, Ph.D.
Extension Specialist
Department of Animal Science
University of California–Davis
Davis, CA

Jeffrey D. Wolt, Ph.D.
Professor
Department of Agronomy
Biosafety Institute for Genetically Modified Agricultural Products
Iowa State University
Ames, IA

Jeff Wyman, Ph.D.
Professor Emeritus
Department of Entomology
University of Wisconsin–Madison
Madison, WI

David Zilberman, Ph.D.
Professor and Robinson Chair
Agricultural and Resource Economics
University of California–Berkeley
Berkeley, CA

Abbreviations

AATF	African Agricultural Technology Foundation
ABSPII	Agricultural Biotechnology Support Project II
AFIC	Asian Food Information Centre
APHIS	U.S. Animal and Plant Health Inspection Service
ASSOCHAM	Associated Chambers of Commerce and Industry of India
BMP(s)	Best Management Practice(s)
BSE	Bovine Spongiform Encephalopathy
Bt	*Bacillus thuringiensis*
Bt Cry toxin	The Protein Crystal Endotoxin produced by Bt
CDC	Centers for Disease Control and Prevention
CFA	UN Comprehensive Framework for Action
CFIA	Canadian Food Inspection Agency
COMESA	Common Market in East and Southern Africa
COMEST	World Commission on the Ethics of Scientific Knowledge and Technology
CPB	Cartagena Protocol on Biosafety
DALY	Disability-Adjusted Life Year
DHHS	U.S. Department of Health and Human Services
DNA	Deoxyribonucleic Acid
EC	European Commission
EDF	Environmental Defense Fund
EFSA	European Food Safety Authority
eIF4E	Eukaryotic Translation Initiation Factor
EMS	Eosinophilia-Myalgia Syndrome
EPA	U.S. Environmental Protection Agency
eQTL	Expression Quantitative Trait Loci
ERA	Environmental Risk Assessment
EU	European Union
FAO	Food and Agriculture Organization of the United Nations
FAPRI	Food and Agricultural Policy Research Institute
FDA	U.S. Food and Drug Administration

FFDCA	Federal Food Drug and Cosmetics Act
ΔG	Genetic gain
GAO	U.S. Government Accountability Office
GDP	Gross domestic product
GE	Genetic engineering; genetically engineered
GHG	Greenhouse gas
GM	Genetic modification; genetically modified
GMO	Genetically modified organism
GRAS	Generally recognized as safe
HT	Herbicide tolerant/ herbicide tolerance
HYV	High-yield variety
ICCP	Intergovernmental Committee for the Cartagena Protocol on Biosafety
IFAFS	USDA Initiative for Future Agricultural Food Systems
IFOAM	International Federation of Organic Agriculture Movements
IFPRI	International Food Policy Research Institute
ILSI	International Life Sciences Institute
IPCC	Intergovernmental Panel on Climate Change
IPM	Integrated pest management
IPPC	International Plant Protection Convention
IRM	Insect resistance management
IRRI	International Rice Research Institute
LCA	Life-cycle assessment
MBV	Molecular breeding values
MDG	Millennium Development Goals
MOU	MOU – Memorandum of Understanding signed by WPVGA and WWF
N	Nitrogen
NADA	New Animal Drug Application
NB-LRR	Nucleotide Binding Site Plus Leucine Rich Repeat
NGO	Non-governmental organization
Nitrate N	Nitrate Nitrogen
NOP	USDA National Organic Program
NOVIC	Northern Organic Vegetable Improvement Cooperative
NRC	National Research Council
NRF	Nano risk framework
OECD	Organisation for Economic Cooperation and Development
OFPA	Organic Food Production Act
OMB	The White House's Office of Management and Budget
OP	Organophosphate
PCM	Potato crop management
PDM	Potato Disease Management Software
PERV	Porcine Endogenous Retrovirus
PIP	Plant incorporated protectants
PIPRA	Public Intellectual Property Resource for Agriculture
PP	Precautionary Principle

PPB	Participatory plant breeding
PRSV	Papaya ringspot virus
PSR	Potential support ratio
PVX	Potato virus X
PVY	Potato virus Y
QTL	Quantitative Trait Loci
RABESA	The Regional Approaches to Biosafety and Biotechnology Regulations
RB	A broad-spectrum resistance gene to late blight that was isolated from *Solanum bulbocastanum*
rBST	Recombinant Bovine Somatotropin
R&D	Research and Development
rDNA	Recombinant Deoxyribonucleic Acid
RNA	Ribonucleic Acid
RR	Roundup Ready®
RTM1	Restricted TEV movement 1
RTM2	Restricted TEV movement 2
SCNT	Somatic Cell Nuclear Transfer
SEHN	Science and Environmental Health Network
shRNA	Short Hairpin RNA
siRNA	Small Interfering RNA
SNP	Single Nucleotide Polymorphisms
TEV	Tobacco etch virus
UN	United Nations
UNEP	United Nations Environment Programme
UNFCC	United Nations Framework Convention on Climate Change
USAID	U.S. Agency for International Development
USCB	U.S. Census Bureau
USDA	U.S. Department of Agriculture
UW	University of Wisconsin–Madison
VAD	Vitamin A deficiency
Vip protein	Vegetative Insecticidal Protein
VPg	Potyviral Genome-Linked Protein
WCED	World Commission on Environment and Devlopment
WFR	World Fertility Rate
WHO	World Health Organization
WISDOM	Wisconsin Decision Oriented Matrix Software
WISP	Wisconsin Irrigation Scheduling Program Software
WPVGA	Wisconsin Potato and Vegetable Growers Association
WWF	World Wildlife Fund

1

World Population Growth and Food Supply

William H. Meyers and Nicholas Kalaitzandonakes

Rapid technological change since the end of World War II has combined with the inelastic demand for food to generate declining real agricultural prices. Consumers have been the ultimate beneficiaries of agricultural innovation, while farmers have had to continually grow in size and become more efficient to offset price declines. To shelter their farmers from these price declines, governments in high-income countries have adopted various support and protective trade policies, which often contributed further to low prices. This long-term decline in real prices has periodically been interrupted by price spikes that were mostly caused by crop failures due to poor weather.

From the beginning of 2006 to the end of 2008, the world witnessed the largest surge of commodity and food prices since the early 1970s, and it seems unlikely that prices will soon return to the lower levels of the early part of this decade. This price surge has again raised the age-old Malthusian question of whether food production can keep pace with growing demand. Historically, the main driver of production has been technological progress; the drivers of consumption have been population growth, which increases the number of mouths to feed, and income growth, which increases the quality and quantity of food consumption per capita. Changing diets that accompany both increased incomes and increased urbanization generally involve more meat consumption and hence more grain consumption per person. Aging of the population, which is happening faster in higher income countries, may have the opposite effect on diet and the quantity of grain consumed.

Two new and significant factors in the growth of grain and oilseed consumption since the early 2000s have been the increase in petroleum prices and the implementation in a number of countries of policies that have stimulated biofuel production in pursuit of environmental and farm support objectives (FAO 2008b; OECD 2008). These changes have increased both the profitability of investments in biofuel capacity and the use of existing capacity, resulting in more grains and oilseeds being used as feedstock for biofuel production. More fundamentally, the growth of the biofuel

Table 1.1. *Exponential growth rates in area, yield, and production of grains and oilseeds*

	1960–70	1970–80	1980–90	1990–2000	2000–7
Grains					
Yield	2.7	1.9	2.1	1.22	1.59
Area	0.5	0.9	−0.5	−0.32	0.44
Production	3.3	2.8	1.6	0.89	2.03
Consumption	3.3	2.6	1.7	0.94	1.70
Grains and Oilseeds					
Area	1.6	1.3	−0.03	0.18	0.79
Production	4.0	3.0	2.0	1.29	2.32
Consumption	4.1	2.9	2.0	1.31	2.05

Source: USDA FAS (2010).

industry has resulted in a much stronger link between fuel and food markets that can contribute both to the higher level and the higher volatility of food prices (Meyers and Meyer 2008).

Examination of grains and oilseeds world markets indicates that the rate of production growth has been slowing since the 1970s (Table 1.1), although the new millennium saw a rebound partly in response to higher commodity prices. A comparison of growth rates in yield over each decade from 1960–2007 indicates a slowdown in yield growth rates in the 1970s, a partial recovery in the 1980s, and then a significant decline from 1990 onward.[1] According to the Intergovernmental Panel on Climate Change (IPPC), natural disasters may be becoming more frequent and extreme due to climate change (Parry et al. 2007, 299); hence, adverse weather may have contributed to a reduction in average yields. Beginning in 1980, grain acreage also declined until the late 1990s, slowing production growth to less than 1 percent per annum during the 1990–2000 period and then rebounding.

Consumption growth has generally corresponded with production growth since the 1960s (Table 1.1). This typical pattern reveals the time-series tendency of production and consumption to change synchronously. In years of bumper crops or substantial land expansion, production may temporarily outpace consumption and build buffer stocks. Conversely, declining land use or adverse weather may lead to lagging production and falling buffer stocks. Not surprisingly, since the 1980s consumption growth rates for both grains and total grains and oilseeds have declined, which suggests that weakening growth in population has not been completely offset by the consumption-boosting effects of income growth (Alexandratos 2008). Sustained reduction in buffer stocks is the hallmark of an imbalance in supply and demand in which consumption

[1] A significant share of the decline in the 1990s was due to the restructuring and reform in the former Soviet Union. Nevertheless, grain production growth would have been slower than in earlier decades.

outpaces production. Such reductions in grain stocks laid the foundation for the price shocks of 2006–8. The price surge and the emergent world food crisis captured headlines and stimulated a wide range of analytical activity and policy discourse (Meyers and Meyer 2008). The crisis caused and continues to cause hardship in many developing countries, led to social unrest in scores of these countries, added 75 million people to the number of undernourished, and reversed progress toward the Millennium Development Goals (MDG) hunger target (FAO 2008a). It also raised questions about the capacity of the global food system to sustain its past success in expanding food and fuel security for a growing part of the world population.

This chapter begins the quest for answers to these food and fuel security questions by first studying population growth projections, then the implications for food demand growth, and finally grain and oilseed supply projections from several sources. Obviously, growth in demand and supply for foodstuffs cannot be viewed in isolation from food prices and income growth. Higher incomes both stimulate demand and change demand patterns, and prices influence the supply as well as the level and composition of consumption. Thus, supply and demand growth is analyzed relative to the expectation of the effect of supply–demand pressures on prices and potentially necessary developments on the supply side to keep pace with likely demand growth.

Population Growth Dynamics and Projections to 2050

World population is projected by the U.S. Census Bureau (USCB) to reach 9.5 billion persons by 2050, which is roughly a 42 percent increase over the population in 2008 (USCB 2008), although the medium variant[2] update issued by the United Nations (UN) in early 2009 was 9.15 billion (UN 2009) and revised by the USCB to 9.28 billion in June 2010 (USCB 2010). This 9.5 billion would not yet be a steady-state population, because world population is expected to peak and begin to decline in about 2070. However, it has been the pattern of population projections that rates of growth are adjusted downward in nearly every projection update, so this peak population could occur sooner than current projections suggest.

Although world population continues to increase, rates of population growth have been declining for decades as incomes grow and education becomes more universal (Table 1.2). Both income growth and education are well-established factors that increase the marriage age of women and reduce the number of children per family, which more than offset the increased life expectancy at birth and reduce population growth rates. Only in Africa, the poorest continent, was the population growth rate still increasing until 1983, when it reached 3.09 percent year over year before starting

[2] The medium variant assumption is that the total fertility rate (TFR) will decline from 2.56 children per women in 2005–10 to 2.02 in 2045–50.

Table 1.2. *Population growth rates, historical and projections in 10-year increments*

Region	1960–70	1970–80	1980–90	1990–2000	2000–10	2010–20	2020–30	2030–40	2040–50
World	2.06	1.82	1.73	1.42	1.21	1.10	0.89	0.73	0.58
Developing	2.50	2.20	2.08	1.68	1.44	1.30	1.05	0.85	0.68
Developed	0.97	0.74	0.55	0.42	0.25	0.15	0.05	−0.11	−0.08
Africa	2.55	2.70	2.84	2.48	2.29	2.11	1.88	1.69	1.52
North America	1.73	1.43	1.33	1.37	1.05	0.95	0.84	0.71	0.59
South America	2.70	2.38	2.06	1.64	1.33	1.09	0.84	0.60	0.37
Europe	0.82	0.55	0.38	0.11	−0.03	−0.12	−0.25	−0.34	−0.45
Oceana	2.10	1.58	1.55	1.43	1.43	1.21	0.98	0.75	0.57
Asia	2.39	2.04	1.87	1.47	1.21	1.06	0.80	0.58	0.39
China	2.59	1.81	1.57	1.00	0.60	0.61	0.21	−0.05	−0.21
India	2.21	2.14	2.05	1.83	1.66	1.41	1.19	0.95	0.71

Source: USCB (2008).

to decline (USCB 2008). In the aggregate of developed countries, population is forecasted to decline after 2030; Europe already began this decline in 1999. In China, which has mandated a reduced family size with its one-child policy, the population growth rate has been declining from the developing-country level in 1960 and is projected to become negative in 2033 and reach the developed-country level by 2050.

Another important aspect of global population growth is its regional distribution and that distribution's change over time. From 1970 to 2010, 63 percent of world population growth was in Asia and 21 percent in Africa (Table 1.3). From 2010–50 the growth in Asia will be about 51 percent of the world total and Africa 40 percent. In 1970, Africa accounted for about 10 percent of world population, but because it has the fastest population growth rate, by 2010 it is projected to be 15 percent and by 2050 almost 22 percent of the total. Meanwhile, the European population, which has been declining (negative population growth) since 1999, dropped from about 18 percent of world population in 1970 to 10 percent in 2010 and is forecasted to be less than 7 percent in 2050. The share of Asia remains close to 60 percent throughout this period, whereas North America and South America remain relatively steady at 8 and 6 percent, respectively (Table 1.3).[3]

In summary, between 2010 and 2030 there are expected to be 1.5 billion more mouths to feed in the world and an additional 1.16 billion by 2050. The result is 2.67 billion additional persons, or approximately two more Chinas over the next 40 years. Additionally, the fastest growing parts of the world are the lower income

[3] Figures assume no population change due to immigration or other displacement.

Table 1.3. *Regional shares in population and population changes, in millions*

	Asia	Africa	N. America	S. America	Europe	Oceana	World
Change 1970–2010	1992.5	650.6	217.5	207.2	71.2	16.0	3155.1
% of total	63%	21%	7%	7%	2%	1%	100%
Change 2010–50	1353.9	1056.5	193.8	132.9	–79.9	14.8	2672.1
% of total	51%	40%	7%	5%	–3%	1%	100%
Share of World							
% in 1970	58.1	9.9	7.7	5.6	17.7	0.5	100
% in 2010	60.4	14.8	7.8	5.8	10.6	0.5	100
% in 2050	57.7	21.7	8.7	5.6	6.8	0.5	100

Source: USCB (2008).

regions, where per capita consumption will also be growing and dietary patterns changing as incomes increase. As is explored further in many of the chapters, this is the challenge for agriculture in the decades ahead.

Potential Income and Demand Growth

As a first approximation to the growth in demand for food, we can theorize that, if there were no change in per capita consumption in any country, food demand would grow at a slower rate than population simply because the populations with lower per capita consumption levels also tend to have higher population growth rates (Alexandratos 1999). In reality, per capita consumption does grow with income, which is especially true in low-income populations for which calorie intake and dietary quality are inadequate. At higher income levels, when people have obtained an adequate diet, the income effect becomes very small. The percent growth in demand for food for a 1 percent growth in income is the income elasticity of demand. We can illustrate this concept by comparing the calculated growth in demand for food with no change in per capita consumption (or no income effect) compared to scenarios where the income elasticity of demand for food is 0.2 or 0.4, meaning that food consumption would grow by 2 or 4 percent for every 10 percent growth in income, assuming no other changes.[4]

This calculation is:

Food demand growth = population growth + (income elasticity * income per capita growth)

The calculation requires a projection of the growth in income per capita for which we use the projected growth in real gross domestic product (GDP) per capita. By applying the different income elasticities and using the world population growth rates from the

[4] "No other changes" means there are no changes in prices or other factors that may influence food demand.

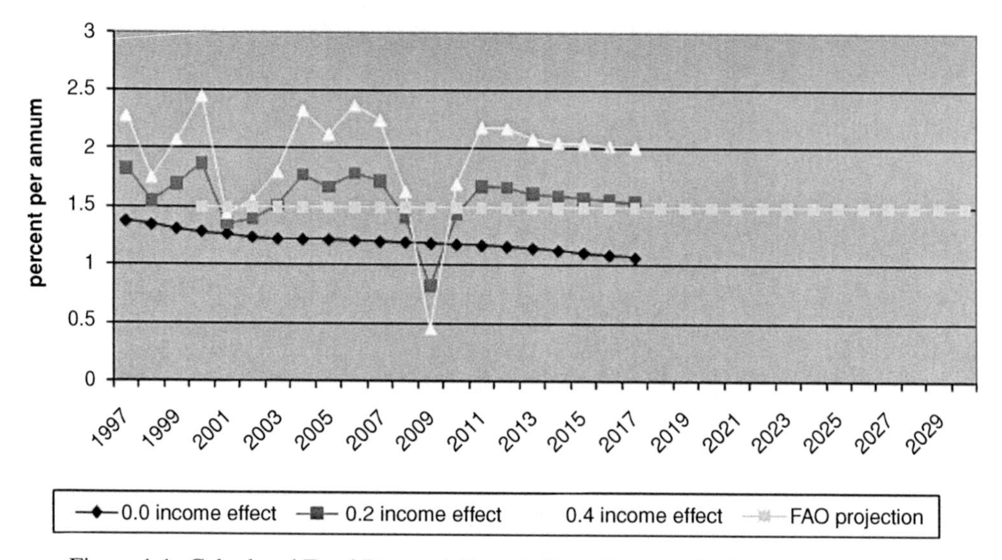

Figure 1.1. Calculated Food Demand Growth Rate Compared with FAO Long-Term Projections. *Source:* GDP growth rates are IHS Global Insight projections used in the FAPRI (2009) baseline.

previous section, we can obtain food demand growth rates that range from about 2 percent per annum for the higher income elasticity to about 1 percent per annum when there is no income effect or no change in per capita consumption (Figure 1.1). To determine which of these rates may be closest to a recent long-run projection of food demand growth, we use the Food and Agriculture Organization (FAO) projection for growth rate of demand for all commodities and all uses. For the period 2000–30 they estimate an average growth rate of 1.5 percent per annum, which is virtually identical to the middle calculation from 2017 onward (Figure 1.1). Because population growth rates are declining, it makes sense that this food demand growth is higher than 1.5 percent per annum before 2017 and likely to decline further in the remaining years.[5]

We do not want to dwell too much on food demand growth without speaking of supply and price conditions, but a reference point is useful when discussing potential growth in supply. When looking at supply growth, we focus not on food in general but on grains and oilseeds, because they are the basic commodities from which most foods derive. When incomes grow, individuals tend to shift from direct consumption of grains to meats, which consume grain as a feedstuff. This shift comprises much of what is reflected in the income effect represented by income elasticity, because more grain is needed per capita with the addition or increase of meat in the diet.

[5] This global illustration greatly oversimplifies the process actually used in doing such projections. Generally, projections of demand growth must be done in more disaggregated ways and preferably country by country (Alexandratos 1997).

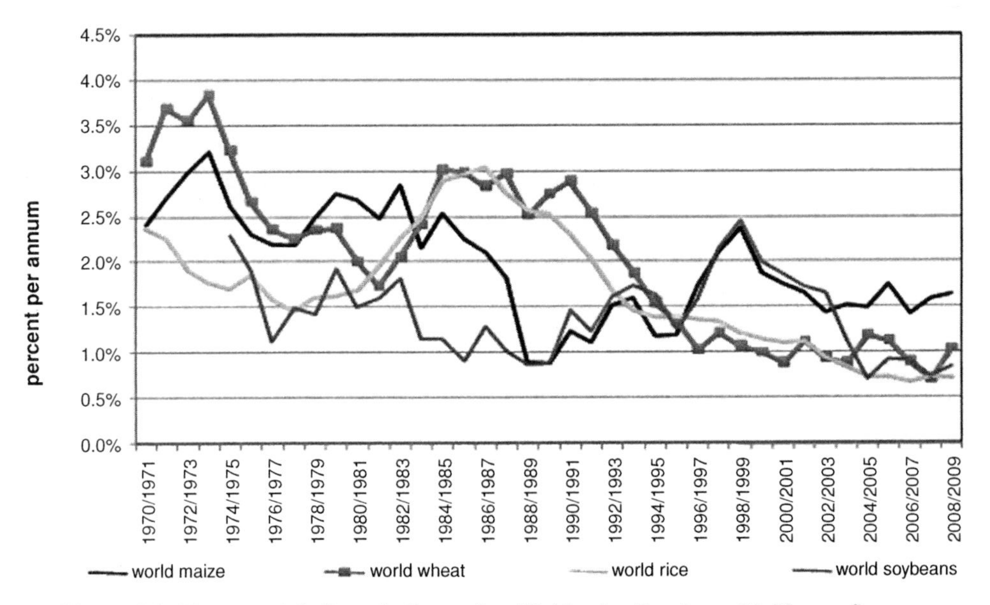

Figure 1.2. Exponential Growth Rates for Yields the Previous 10 Years. *Source:* Calculated from the USDA FAS (2010).

Supply Projections for Cereals from Different Sources

We have discussed the declining growth rates in production that have characterized the last two decades. Several factors contributed to this relatively slow production growth, but the key market factor was declining real prices for an extended period that reduced market incentives to invest and produce. The decline was interrupted only by short-lived price surges in short crop years in 1988/89 and 1995/96 (Meyers and Meyer 2008). In short, grain acreage declined (Table 1.1), and yield growth rates slowed simultaneously.

In addition to low prices, another important contributing factor in slowing yield growth rates, especially for rice and wheat (Figure 1.2), is the diminishing national and international public investment in agricultural research and development (R&D). Such investment has slowed in both developing and developed countries since the 1990s. The international research investments of the 1960s were deliberate policies to enhance agricultural productivity in developing countries and resulted in the high-yielding Green Revolution wheat and rice varieties that spurred yield growth and enhanced multiple cropping opportunities with shorter growing seasons. Along with continuing public and private agricultural R&D in industrial countries, these improved technologies supported grain yield growth of 2.4 percent and production growth of 3.1 percent annually from 1960–80. Although yield growth remained relatively high in the 1980s, grain acreage declined. Grain acreage peaked in 1981, and some land moved into oilseeds. As a result, total land in grains and oilseeds grew by less than 0.2 percent per annum from 1981–2007. Since 1990, yield and production growth have been slowing for both grains and oilseeds.

It has been well established in numerous documents of the World Bank, FAO, and International Food Policy Research Institute (IFPRI) that investment in agricultural development has been lagging, especially in developing countries. The World Bank *World Development Report 2008* (IBRD 2007) shows that developing countries have "suffered from neglect and underinvestment over the past 20 years. While 75 percent of the world's poor live in rural areas, a mere 4 percent of official development assistance goes to agriculture in developing countries. In Sub–Saharan Africa, a region heavily reliant on agriculture for overall growth, public spending for farming is also only 4 percent of total government spending" (World Bank 2007). Pardey et al. (2006) found that growth in public agricultural R&D spending, which was critical to the Green Revolution, declined by more than 50 percent in most developing countries from 1980 onward and even turned negative in high-income countries beginning in 1991. There were important exceptions in China and India (IBRD 2007), but national governments and international organizations have neglected these investments despite the high rates of return that were demonstrated in past R&D projects.

It is noteworthy that corn and soybean yield growth rates have been significantly higher than for wheat and rice (Figure 1.2) since the Green Revolution's effects diminished in the mid-1980s. This trend reflects the role of private investments in corn and soybean germplasm research and biotechnology development, as well as the commercialization of that research.

In response to the commodity price surge since 2006, grain area and production have been increasing. The most recent 10-year projections of future supply have been conducted in this context. These projections were also completed within the context of the public R&D deficiency, because even if urgent action were taken to reverse the investment path for agriculture, it is a long-term endeavor.

The first comparison of grain and oilseed production projections is from three of the best known annual global market assessments. One is by the Food and Agricultural Policy Research Institute (FAPRI), which is a collaboration of the University of Missouri–Columbia and Iowa State University and the University of Arkansas's global rice market analysis. The second assessment is by the U.S. Department of Agriculture (USDA), and the third is conducted jointly by the Organisation for Economic Cooperation and Development (OECD) and FAO. These were completed at somewhat different times of the year: USDA was in December 2007 (USDA 2008), FAPRI was in March 2008 (FAPRI 2008), and OECD was in May 2008 (OECD/FAO 2008). Some of their differences are due to different release dates and the information available at publication, but in any case the implications of these three projections are quite similar.

Grains and oilseeds experienced significant growth in production in the middle part of this decade, primarily in response to rapidly rising prices. The rates of growth in grain (except for rice) and oilseed production from 2000–7 were significantly higher than those of the previous decade, with coarse grains growing nearly three times as

Table 1.4. *Comparison of growth rates for grains and oilseeds production, percent per annum*

Crop	1990–2000	2000–7	FAPRI 2007–17	USDA 2007–17	OECD 2007–17
Rice	1.61	1.02	0.88	0.73	0.89
Wheat	0.78	1.08	0.99	0.92	0.97
Coarse grains	1.04	2.96	1.46	1.53	1.46
Total grains	1.07	1.96	1.21	1.16	1.17
Oilseeds	4.01	4.29	2.30	2.89	2.22

Sources: FAPRI (2008); OECD/FAO 2008; USDA (2008) historical figures. Used 3-year average of production.

fast and total grains growing nearly twice as fast (Table 1.4). From 2007–17, projected production growth rates are not as high as in the recovery period 2000–7, but all grain production growth rates remain somewhat higher than in the 1990s.

Finally, we compare these 10-year projections to long-term projections conducted by FAO (FAO 2006) and by IFPRI (Rosegrant et al. 2009). For FAO, the long-term projections use different types of modeling systems than the 10-year projections, and they essentially assume constant real prices such as we illustrated in the demand growth exercise of the previous section. Their long-term projections were completed without including the use of grains and oilseeds for biofuel feedstock. In contrast, IFPRI does estimate prices, includes biofuel feedstock demands, and analyzes alternatives under different assumptions of investment in technology to enhance productivity. Their long-term estimates also take into account water and land resource constraints in the analysis. In any case, looking at the longer term is helpful to gauge how it differs from the 10-year outlook. We use the FAPRI results for comparison and find a rather consistent outlook. In the 2000–30 period, production growth rates are projected to be lower than in the 2000–17 period of the FAPRI analysis (Table 1.5). When the

Table 1.5. *Comparison of growth rates for grains and oilseeds production, percent per annum*

Crop	FAPRI 2000–17	FAPRI 2007–17	FAO 2000–30	FAO 2030–50	FAO 2000–50	IFPRI 2000–50
Rice	1.02	0.88	0.8	0.2	0.50	0.44
Wheat	1.16	0.99	1.1	0.5	0.90	0.79
Coarse grains	2.21	1.46	1.4	0.8	1.10	1.16
Total grains	1.53	1.21	1.2	0.6	0.90	0.92
Oilseeds	3.06	2.30	2.2	1.6	1.96	Na
Population	1.18	1.15	1.0	0.5	0.80	0.87

Sources: FAPRI (2008); FAO (2006); Rosegrant et al. (2009).

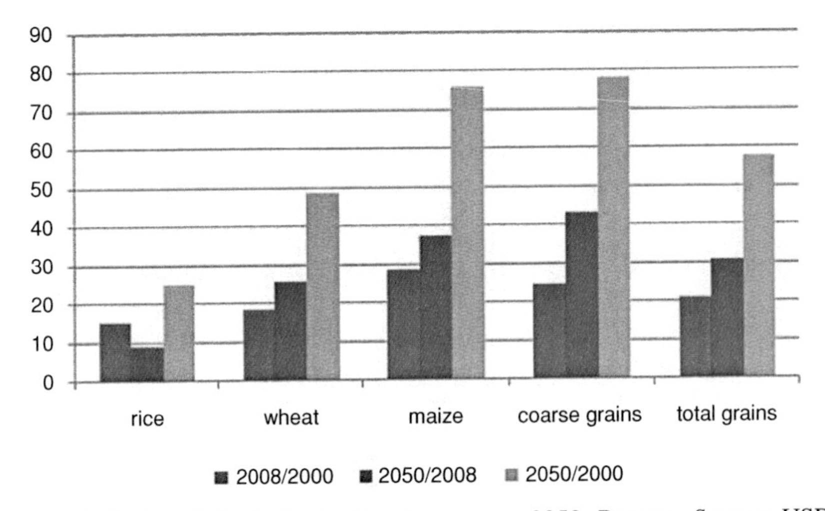

Figure 1.3. Projected Grain Production Increase to 2050, Percent. *Source:* USDA 2000–8 estimates (USDA 2008), IFPRI 2000–50 projection (Rosegrant et al. 2009).

production recovery period up to 2007 is removed and we look at the 2007–17 period only, the differences in growth rates are much smaller. FAO projects that the rates of growth in the 10-year FAPRI projection would continue to 2030. Longer term production growth is slightly lower.

Comparing FAO and IFPRI and looking only at the 2000 to 2050 summaries, there is sufficient evidence to conclude that the assessments are very similar. IFPRI analysis projects slightly higher growth rates in grain production, especially coarse grains, which reflect the growth of biofuel feedstock demand at least through 2020 and a slightly higher rate of population growth. Using IFPRI's growth rate results, which are slightly higher, these figures would result in an increase in total grain production of 58 percent from 2000 to 2050 (Figure 1.3). However, world grain production was quite flat during the late 1990s and early 2000s and then grew by more than 20 percent from 2003 to 2008 as prices increased and policy incentives became more favorable. So, from the current 2008 levels to 2050, the projected total production growth is 30 percent for total grains, closer to 40 percent for coarse grains and maize, and somewhat lower for wheat and rice (Figure 1.3).

Regional patterns of production are important as well. Production growth in areas closer to population and consumption growth centers results in lower transportation costs and reduced losses of quality and quantity associated with transportation. The historical path shows again the declining rates of growth up to 2001, but regional differences are revealing. The rate of production growth increased in industrial countries and Latin America, slowed most in developing regions, and was increasingly negative in transition countries (former Soviet Union and East European countries). As in the past, future cereal production growth is projected to be higher in developing than industrial or transition countries (Table 1.6). In the projection period 2000–30, the small increase in the world cereal production growth rate is due to major

Table 1.6. *Regional pattern of cereal production growth rates, historical and FAO projection, percent per annum*

	1971–2001	1981–2001	1991–2001	2000–30	2030–50	2000–50
World	1.7	1.2	1.1	1.2	0.6	0.9
Developing	2.7	2.2	1.6	1.4	0.7	1.1
Industrial countries	1.2	0.8	1.4	0.9	0.5	0.7
Transition	−0.6	−1.4	−2.2	0.8	0.4	0.6
Sub-Saharan Africa	2.7	3.0	2.3	2.5	1.8	2.3
Middle East and North Africa	2.2	2.0	−0.4	1.9	1.0	1.6
South Asia	2.8	2.6	2.5	1.3	0.7	1.0
East Asia	2.8	2.0	1.1	1.0	0.2	0.7
Latin America and Caribbean	2.2	1.7	3.0	2.1	0.9	1.6

Source: FAO (2006).

changes in transition countries, which have already begun. Although all production growth rates are lower in the later 2030–50 period, these growth rates in developing countries continue to be higher than in industrial and transition countries; the highest projected growth rates are in Sub-Saharan Africa and Latin America, where growth in area planted is still feasible. These cereal production growth rates imply that production will more than triple in Sub-Saharan Africa, more than double in Latin America/Caribbean and Middle East/North Africa, and increase more than 50 percent for the world as a whole (Figure 1.4).

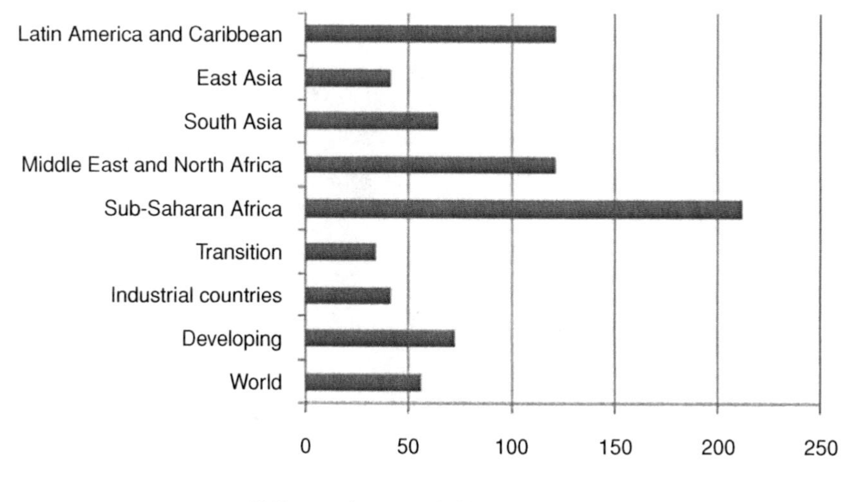

Figure 1.4. Projected Regional Growth in Cereal Production to 2050, Percent. *Source:* FAO (2006).

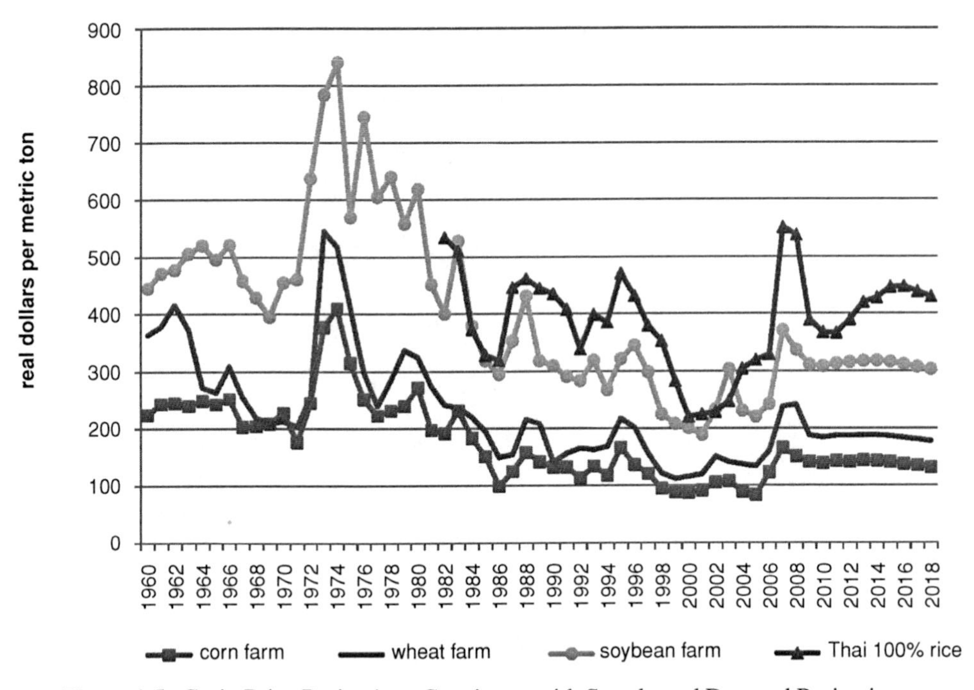

Figure 1.5. Grain Price Projections Consistent with Supply and Demand Projections. *Source:* History from USDA FAS (2010), projections from FAPRI (2009) baseline.

Prices Matter Too

While reviewing these projections, we have not yet discussed the importance of commodity and food prices. The FAPRI, OECD, and USDA projections were conducted with models that included a complete range of crop, livestock, and dairy products; their production, consumption, and trade; and their prices. Returning to the initial purpose of this chapter, these facts suggest that we cannot achieve the necessary higher production growth rates unless prices remain relatively higher than the trend since the late 1990s and the early 2000s (Figure 1.5). The new demand growth for grains and oilseeds as feedstock for biofuels adds significantly to the traditional demands for food grains and animal feeds. Future demand depends on the trend of petroleum prices and the path of economic recovery that follow the current economic crisis. The current financial crisis and declining petroleum prices have diminished these price pressures in the near term, but during economic recovery we are unlikely to see a return to the patterns of low and declining real prices that existed in the late 1990s and early 2000s.

In fact, the IFPRI projection has analyzed alternative technological investment and productivity assumptions that generate different growth rates of production and different price and consumption forecasts. The analysis is instructive because it addresses the tradeoffs between the status quo of relatively slow productivity growth and a more rapid growth path that could be generated with increased investment. In the baseline case, the growth of cereals yield over the period 2000–50 is assumed to be 1.02 percent annually. This baseline generates the production growth rates already reported, which are close to those of FAO. Together with the income and population growth rate assumptions and the economic relationships in the model, the baseline projects that real grain prices would need to increase by 67 percent in the 2000–25 period and slightly more to a total of 74 percent by 2050. This percentage seems like a large increase in real prices, but recall that real prices were extremely low in 2000 and real prices of wheat and maize had already risen by 60 to 70 percent and rice by more than 100 percent from 2000 to 2008. Summarily, with current rates of productivity growth, real prices in 2025 must be at least as high as the current year (2008/9) and increase further by 2050. These results imply a very sluggish production response to continually rising real prices and deterioration of the well-being of low-income consumers, who have been hardest hit by recent food price increases.

The higher R&D investment scenario is designed to reverse the low investment and slow productivity growth path of recent decades that were projected to continue in the baseline scenario. It assumes that increased investment leads to crop yield growth rates that are 40 percent higher, or approximately 1.43 percent per year. Given no change in assumptions on the demand side, the higher growth of production leads to cereal production 70 percent higher in 2050 compared to 2000 and to grain prices that continue to decline in real terms. Grain production grows at a rate of 1.06, rather than 0.92, percent per annum in this scenario. Under these conditions, real grain prices in 2025 would regress to the level of 2000 and, from 2025 to 2050, would decline 25 percent further (Figure 1.6). The fruition of these conditions would imply a real price decline from 2008 to 2025 similar to the real wheat and maize price declines from 1985 to 2000 and a decline from 2008 to 2050 similar to the real wheat and maize price declines from 1981 to 2000; therefore these projections are well within the recent patterns of market behavior in even shorter time periods. The increased production and lower real prices would significantly increase the calorie availability and reduce the number of malnourished children, according to the IFPRI analysis. The high-technology growth scenario would also increase grain production by nearly 70 percent from 2000 to 2050 and by 40 percent from 2008 to 2050 (Figure 1.7).

Conclusions and Implications

During the last two decades, production and consumption growth rates have been slowing. Parallel declines in real food prices over a long period of time suggest that, until recently, demand pressure was not building up in the food system from slowing production. However, declining stocks in the recent decade and a series of recent market shocks have led to markedly higher prices, which in turn have stimulated land use and more rapid growth of yields and production. Nevertheless, the possibility for

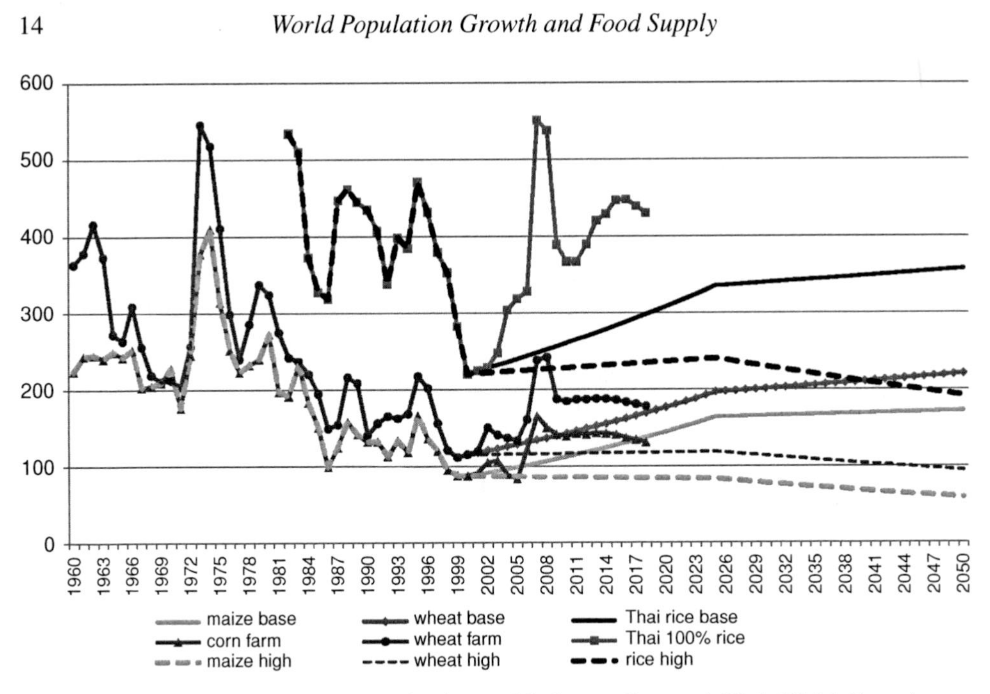

Figure 1.6. Alternative Price Projections with Status Quo and High-Yield Growth. *Source:* 2025 and 2050 alternatives from Rosegrant et al. (2009).

land use expansion is physically limited, and other factors, such as water and natural resource constraints, also limit the rate of growth in production. To meet the growth in demand for food and biofuels in the years ahead, prices must remain higher than they have been in the early part of this decade so as to increase production growth rates,

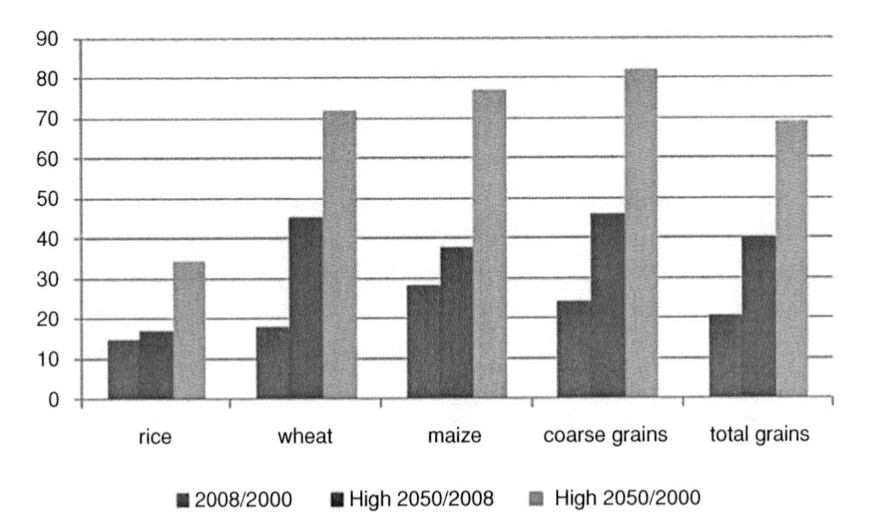

Figure 1.7. High-Technology Projected Grain Production Increase to 2050, Percent. *Source:* USDA 2000–8 estimates (USDA 2008), IFPRI 2000–50 projection with high-technology growth (Rosegrant et al. 2009).

primarily through productivity enhancements and increased purchased inputs. Thus as a consequence of an extended period of higher prices, we can expect greater land use pressure as well as greater pressure on food import costs and food security among lower income households, especially in developing countries. These higher prices may also provide an incentive for increased agricultural R&D and productivity growth.

Most of these baseline projections assume a continuation of technology growth patterns of the recent past, which have seen a slowing of yield growth rates compared with the pre-1990 period. An alternative future would increase the rate of technological advance through higher private and public investment in agricultural R&D, which would create a higher yield growth path, permit substitution of technology for cropland, and benefit farmers through higher productivity and consumers through lower food prices. In other words, it could realign the future path of real prices with their projected long-run path before the recent price surge and market conditions, instead of keeping prices on a new and higher path. Such an alternative higher technology future would improve the well-being of consumers, especially in developing countries, and contribute to the long-term sustainability of agricultural resources by substituting technology for land and thereby reducing pressure on cropland.

References

Alexandratos, N. 1997. "The World Food Outlook: A Review Essay." *Population and Development Review* 23 (4): 877–88.

Alexandratos, N. 1999. "World Food and Agriculture: Outlook for the Medium and Longer Term." *Proceedings of the National Academy of Sciences of the United States of America* 96: 5908–14.

Alexandratos, N. 2008. "Food Price Surges: Possible Causes, Past Experiences, Relevance for Exploring Long-Term Prospects." *Population and Development Review* 34 (4): 663–97.

FAO (Food and Agriculture Organization). 2006. *World Agriculture: Towards 2030/2050: Prospects for Food, Nutrition, Agriculture and Major Commodity Groups*. Rome: FAO Global Perspective Studies Unit. http://www.fao.org/fileadmin/user_upload/esag/docs/Interim_report_AT2050web.pdf.

———. 2008a. "Hunger on the Rise: Soaring Prices Add 75 Million People to Global Hunger Rolls." September 18. Rome: FAO Media Office. http://www.fao.org/newsroom/en/news/2008/1000923/.

———. 2008b. *The State of Food and Agriculture 2008: Biofuels: Prospects, Risks and Opportunities*. Rome: FAO Electronic Publishing Policy and Support Branch Communication Division.

FAPRI (Food and Agricultural Policy Research Institute). 2008. *FAPRI 2008 U.S. and International Agricultural Outlook*. FAPRI Staff Report 08-FSR 1. Ames, IA: FAPRI.

———. 2009. *FAPRI 2009 U.S. and International Agricultural Outlook*. FAPRI Staff Report 09-FSR 1. Ames, IA: FAPRI.

IBRD (The International Bank for Reconstruction and Development). 2007. *World Development Report 2008: Agriculture for Development*. Washington, DC: World Bank. doi:10.1596/978-0-8213-7235-7.

Meyers, W. H., and S. Meyer. 2008. *Causes and Implications of the Food Price Surge*. FAPRI–MU Report 12-08. Columbia, MO: FAPRI.

OECD (Organisation for Economic Cooperation and Development). 2008. *Biofuel Support Policies: An Economic Assessment*. Paris: OECD. http://www.oecd.org/document/0,3343,en_2649_33785_41211998_1_1_1_1,00.html.

OECD/FAO (Organisation for Economic Cooperation and Development/ Food and Agriculture Organization of the United Nations). 2008. *OECD–FAO Agricultural Outlook 2008–2017*. Paris: OECD. http://www.fao.org/es/esc/common/ecg/550/en/AgOut2017E.pdf.

Pardey, P. G., N. Beintema, S. Dehmer, and S. Wood. 2006. *Agricultural Research: A Growing Global Divide?* Food Policy Report 17. Washington, DC: IFPRI.

Parry, M. L., O. F. Canziani, J. P. Palutikof, P. J. Van der Linden, and C. E. Hanson, eds. 2007. *Contribution of Working Group II to the Fourth Assessment Report of the Intergovernmental Panel on Climate Change*. Cambridge: Cambridge University Press.

Rosegrant, M. W., C. Ringler, A. Sinha, J. Huang, H. Ahammad, T. Zhu, T. S. Msangi, B. Sulser, and M. Batka. 2009. *Exploring Alternative Futures for Agricultural Knowledge, Science, and Technology (AKST)*. ACIAR (Australian Center for International Agricultural Research) Project Report ADP/2004/045. Canberra: ACIAR.

UN (United Nations). 2009. *World Population Prospects: The 2008 Revision Population Database* (accessed December 2). http://esa.un.org/unpp/.

USCB (U.S. Census Bureau). 2008. *International Data Base* (accessed November 5). http://www.census.gov/ipc/www/idb/.

______. 2010. *International Data Base* (accessed June 21). http://www.census.gov/ipc/www/idb/.

USDA (U.S. Department of Agriculture). 2008. *International Long–Term Projections to 2017*. OCE-2008-1. Washington, DC: USDA. http://www.ers.usda.gov/Publications/OCE081/OCE20081fm.pdf.

USDA FAS (U.S. Department of Agriculture Foreign Agricultural Service). 2010. *Production, Supply, and Distribution Database* (yield, world totals; accessed December 12). http://www.fas.usda.gov/psdonline/psdQuery.aspx.

World Bank. 2007. "World Bank Calls for Renewed Emphasis on Agriculture for Development." Press Release No: 2008/080/DEC, October 19. http://go.worldbank.org/IUIGDTF9M0.

2

Social Challenges

Public Opinion and Agricultural Biotechnology

Dominique Brossard

Agricultural biotechnology has come to the forefront as a tool that can help solve the food crisis the world is facing, as described in the first chapter. Many have argued that genetic engineering (GE), also known as genetic modification (GM; both terms are used throughout this book), can be part of that solution, because it can help increase the yields of staple crops by making them more resistant to climatic stress, diseases, and pests or can improve their nutritional value. Although biotechnology offers numerous opportunities for solving the food crisis, several social and political challenges raised by its use should not be dismissed.

The purpose of this chapter is to outline some of the major sociocultural and political challenges that need to be taken into account when considering the role of agricultural biotechnology in a sustainable food supply, with special emphasis on the issues of acceptance and rejection. Experts might agree on the opportunities offered by biotechnology as related to food production. However, intended beneficiaries of technological innovations do not always adopt these technologies, even though the innovations seem to have clear benefits to the scientists who created them. As is discussed later, the use of GE in food production has evoked significant resistance in different parts of the world. The social dynamics producing acceptance (or rejection) of this technology are therefore a crucial determinant of the success of biotechnology in feeding the world in a sustainable way and also provide an important lesson for future technologies that will likely encounter similar responses.

The first section of this chapter summarizes the issues and arguments put forward by proponents and opponents of the use of GE in food production, focusing on the challenges posed by GE crops. The following sections briefly outline some of the major political and social challenges facing GE foods and review the state of public opinion around the world as it relates to the technology's level of adoption and to the hunger map discussed in Chapter 1. Finally, the chapter discusses the factors that are likely to affect public attitudes toward agricultural biotechnology and the consumption of GE foods.

Using Genetically Engineered Crops in Food Production: Pro and Con Arguments

The debates that have surrounded the use of GE in food production are a good illustration of discourses around controversial issues, during which proponents bring forward arguments that are rejected by opponents and vice versa, with each side having difficulty accepting the opposing arguments. In particular, the impact of GE on the environment, human health, economics, and ethics is the main point of disagreement between proponents and opponents; opponents emphasize the risks, proponents the benefits.

For instance, when proponents argue that the environment will benefit from agricultural biotechnology because it enables a decreased use of polluting pesticides (Phipps and Park 2002), opponents point out that pollen drift from modified to unmodified species may cause unwanted genetic exchanges that may lead to the creation of "super weeds" and to the introduction of genetic traits into neighboring crops. Opponents also point out that GE crops may hurt beneficial insects because of the toxins introduced to create genetic resistance (Nelson 2001).

Environmental and health-related safety issues are at the core of the debate around GE crops, with related concerns regarding the adequacy of the safety testing mechanisms (and associated biosafety regulations) that are or should be put into place. For instance, supporters argue that GE can help increase the nutritious content of food and that no risks for human health are associated with the consumption of genetically modified organisms (GMOs; Gura 1999). Some also argue that agricultural biotechnology is one of the means of feeding the world's growing population. However, others wonder whether enriching soils that could increase the nutritional value of all plants fit for human consumption is more prudent than trying to produce plants with higher nutritional value through biotechnology (Kaufman 2008). Opponents have also pointed out that unknown allergic reactions to GE foods may occur (Franck-Oberaspach and Keller 1997).

Economic arguments have also generated great controversy. Agricultural biotechnology has been presented as a cost-saving and consumer-oriented technology. By producing GE plants, which are more resistant to pests, have higher yields, and are less labor intensive, industrialized farmers incur fewer expenses related to production, pesticides, and herbicides. The consumer also can potentially benefit economically from this technology by getting high-valued product attributes at a reasonable price. However, opponents argue that (1) agricultural biotechnology exemplifies monopoly capital and big corporations' interests, (2) it will ultimately hurt small farmers by destroying their way of life, (3) third world countries may not be able to produce or buy costly GM seeds, and (4) patenting issues are far from being resolved (Bender and Westgren 2001).

Finally, the moral implications of biotechnology have also generated controversy. Some assert that not using our technological knowledge to feed the planet would be unethical and that helping the developing world gain access to biotechnology at affordable prices is a moral imperative (Nuffield Council on Bioethics 1999). Others have argued that the technology itself poses a moral dilemma: tampering with nature. As Isserman (2001) asserted, human beings have the ethical obligation to help conserve the earth's diversity, but their ability to create new species through genetic engineering raises new ethical issues that need to be debated.

To add another controversial dimension to those just summarized, some have argued that consumers have the right to know when they consume GE products. Following this line of argument, all products containing GE organisms should be labeled as such, something that current Food and Drug Administration (FDA) policy in the United States does not require (Greger 2000; McCallum 2000). Other countries mandate GE labeling. Labeling has been an issue central to the debate in Europe, with 94.6 percent of a 2001 representative sample of Europeans wanting to have "the right to choose" and demanding labeling (Gaskell, Allansdottir, et al. 2006).

In sum "the balance of evidence suggests that GM organisms have the potential to both degrade and improve the functioning of agro ecosystems" (Peterson et al. 2000, 1), and proper risk assessment should rely on a multidisciplinary approach taking into account all relevant dimensions (ethics, health, environment, and economics). A cost/benefit calculation that would quantitatively assess the pro and con arguments briefly outlined earlier is not the goal of this chapter. Yet it is clear that agricultural biotechnology, despite presenting opportunities for feeding a growing world population, poses a number of serious social and political challenges because it is rejected by some individuals and groups, often on the grounds outlined earlier. Often these discussions are not informed by the scientific validity of the claims at stake. Rather, attitudes are based on complex sociopsychological mechanisms, which are discussed later in the chapter.

Therefore it is important to keep in mind that the issues that have been associated with GE crops in the past are still relevant in a sustainable food supply context and that these issues are likely to continue to be discussed in a partisan manner. The brief summary of these issues in this chapter is by no means exhaustive, and many of these topics are explored in later chapters in detail. The point of this overview is to illustrate that biotechnology might help resolve the food crisis if the viewpoints and arguments of all relevant partners are taken into account and not dismissed as irrelevant and if the sociopolitical contexts of the countries under the spotlight are well understood.

Macro-Level Challenges: The Sociopolitical Context

As described in Chapter 1, the countries with the most pressing malnourishment problems are on the African and Asian continents, yet as Figure 2.1 shows, these are the parts of the world in which the least amount of GE crops were grown, as of 2010. At that time, more than half of the GE crops acreage planted was located in North America. Apart from some Latin American countries that grew mostly soybean, maize,

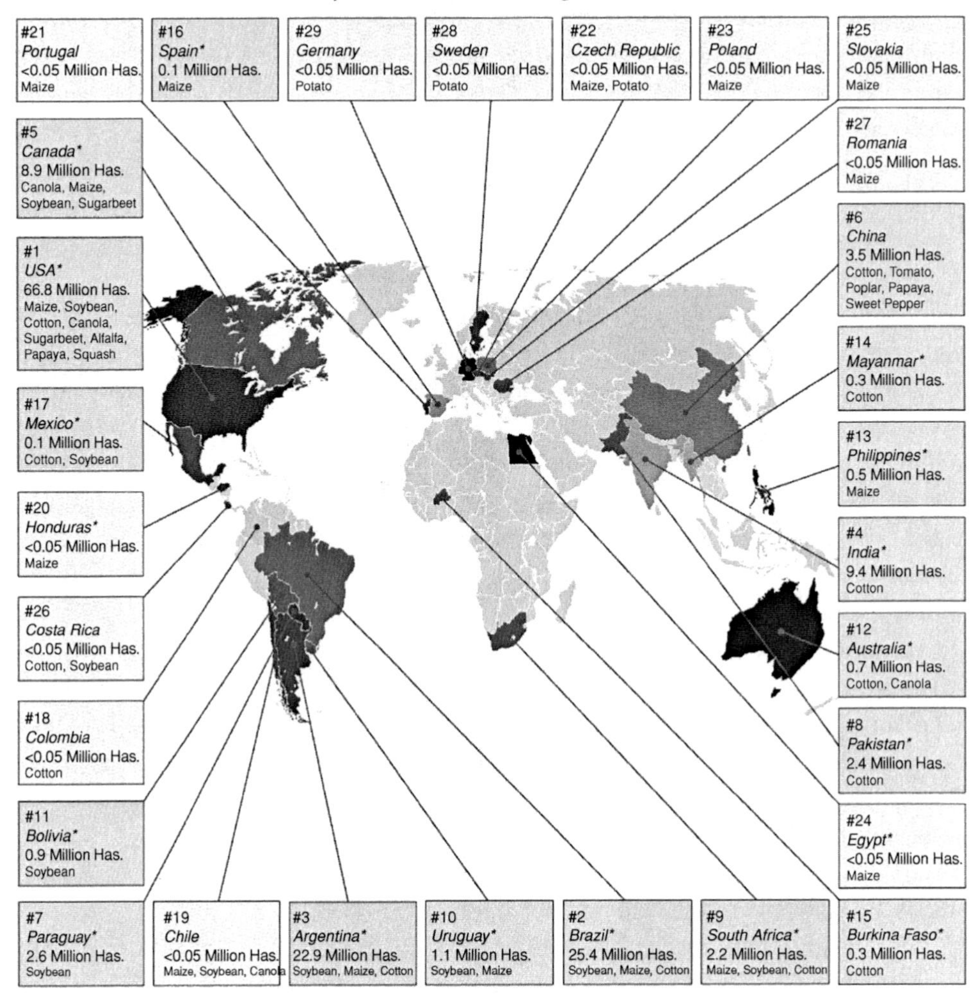

Figure 2.1. Levels of Adoption of GE Crops Worldwide. *Source:* James (2010).

and cotton and India's growing *Bacillus thuringiensis* (*Bt*) cotton, the developing world had a fraction of its total area planted in GE crops. Most acreage there is still not devoted to edible crops, but the distribution is likely to change in the near future. India's first GE food crop (GM brinjal, i.e., eggplant) received regulatory approval for commercial cultivation at the end of 2009. Concurrently, China approved its first strain of GE rice for commercial production. The uneven adoption levels beg the question: Why have GE crops been slow to gain acceptance when other technologies, such as the cell phone, have been rapidly and widely adopted worldwide? The reason is largely that developing countries do not have the sophisticated infrastructures in place necessary for the regulation and commercialization of GE crops (more specifically,

GE crops for human consumption) and the price of the technology is still prohibitively high for most farmers in the developing world. In fact, detailed regulatory processes regarding agricultural biotechnology have yet to be clearly defined in most developing countries. When these regulations are in place, they vary substantially from country to country, as do the ethical, political, and social climates that shaped their final forms. A brief discussion follows of the Cartagena Protocol on Biosafety (CPB) of 2000, as one of the key sociopolitical dimensions influencing regulatory safety processes related to GE crops – and, by extension, their adoption worldwide and ultimately human consumption.

The CPB was the first international treaty categorizing GMOs as organisms requiring a specific regulatory framework. It was established in 2000 to protect the world's biodiversity. The protocol ascertains the framework for international legislation on the international movement of GE crops using a "precautionary approach." It states that, in the absence of full scientific certainty, a country can limit imports of GE crops on the grounds of their potentially damaging effects. One should note that, although most European countries believe in the precautionary principle, the United States and Canada rely on a different approach. In Canada and the United States, a GE food is not required to undergo mandatory premarket approval if it is considered substantially equivalent to other foods (McCallum 2000).

The CPB came into effect in September 2003. Since then, some regional agreements have been put into place to supplement its clauses. For example, the African Biosafety Model Law includes mandatory labeling and traceability requirements for GMOs and liability clauses for harm caused by GMOs to human health or the environment and the ensuing economic damages (M. Mayet, pers. comm., January 29, 2004). As of January 2009, 103 countries had signed the CPB. By May 2011, 160 countries and the European Union had ratified or agreed to the Protocol (Secretariat of the CBD, 2011).

Some wonder whether the CPB has hindered the adoption of GE crops in the developing world. It would be beyond the scope of this chapter to discuss this point further.[1] However, the way that risks are conceptualized at the societal level (as exemplified by the European precautionary principle or the American concept of substantial equivalence) reflects cultural choices that will affect the adoption of a specific technology and, ergo, market dynamics and consumer choices.

As Bender and Westgren (2001) pointed out when discussing biotechnology in the marketplace, buyers and sellers socially construct markets, which are set in a broader sociopolitical environment. International market forces are indeed another important determinant of GE adoption worldwide. For instance, a number of African countries have in place major trade exchanges with European countries. The fear of losing one of their major export markets if GMOs were to be introduced into their agricultural

[1] For a discussion on the topic, visit the Science and Development Network, an online information site about science, technology, and the developing world (SciDev.Net 2006).

supply may partially explain the reluctance of some countries to embrace GE crops. Some have argued that the Zambian rejection of food aid containing GE organisms in 2002 was in part fueled by such a concern.[2]

However, the lack of a regulatory framework is not always a deterrent to the adoption of GE crops. In fact, illegal planting has, in some instances, fueled and accelerated the regulatory process. Brazil is the exemplar of such a situation. The South American country was the site of an intense controversy in 2003 when it was formally announced that a major proportion of Brazilian soybean crops in the southern state of Rio Grande do Sul were transgenic due to illegal planting. Rio Grande do Sul shares a border with Argentina, where GE crops are widely planted. Shortly after an announcement maintaining the general ban on GE crops despite the situation, the government decided to allow the sale of GE soybeans for animal and human consumption, sparking protest within the government and from environmental groups. Three years of intense controversy followed. Despite initial public opposition, Brazilian parliamentarians finally approved a "Biosafety Law" that established the regulatory process for the approval of biotechnology crops in 2005 after 17 months of intense debate in Congress.[3] The first crops received approval in 2008.

Brazilian farmers were an important force pushing for a regulatory framework allowing the planting of GE crops in Brazil (Brossard, Massaroni, et al. 2008). Worldwide, the impact of stakeholders in the GE issue (which include not only the farmers but also policy makers, research institutions, farmer groups, consumer groups, industry groups, the mass media, and ultimately the consumers themselves) in the decision-making process has differed significantly across countries. However, it is clear that public opinion has been a driving force behind the slow adoption of GE crops in the European Union.

Public Opinion about GE Crops Worldwide

As illustrated by the Brazilian case, the level of adoption by farmers does not always reflect public opinion about a technology. Biotechnology could play a major role in a sustainable food supply if individuals would agree to consume GE foods. Therefore, it is of utmost importance to monitor levels of acceptance and rejection of the technology at the consumer level, and not only political or producer opinions.

It can be safely assumed that a large proportion of the world population is unaware of GE and hence would be unable to form any attitudes or judgments about it without first being provided educational information. I discuss later the processes by which such judgments may be informed, but I first provide a summary of worldwide attitudes toward GE. This summary comes with certain caveats. First, reliable public

[2] For a discussion of the Zambian case, see Mumba (2007).

[3] For details on the legislative debate, see Dolabella, Araujo, and Faria (2005); for a discussion of media coverage of the issue, see Brossard, Massaroni, et al. (2008).

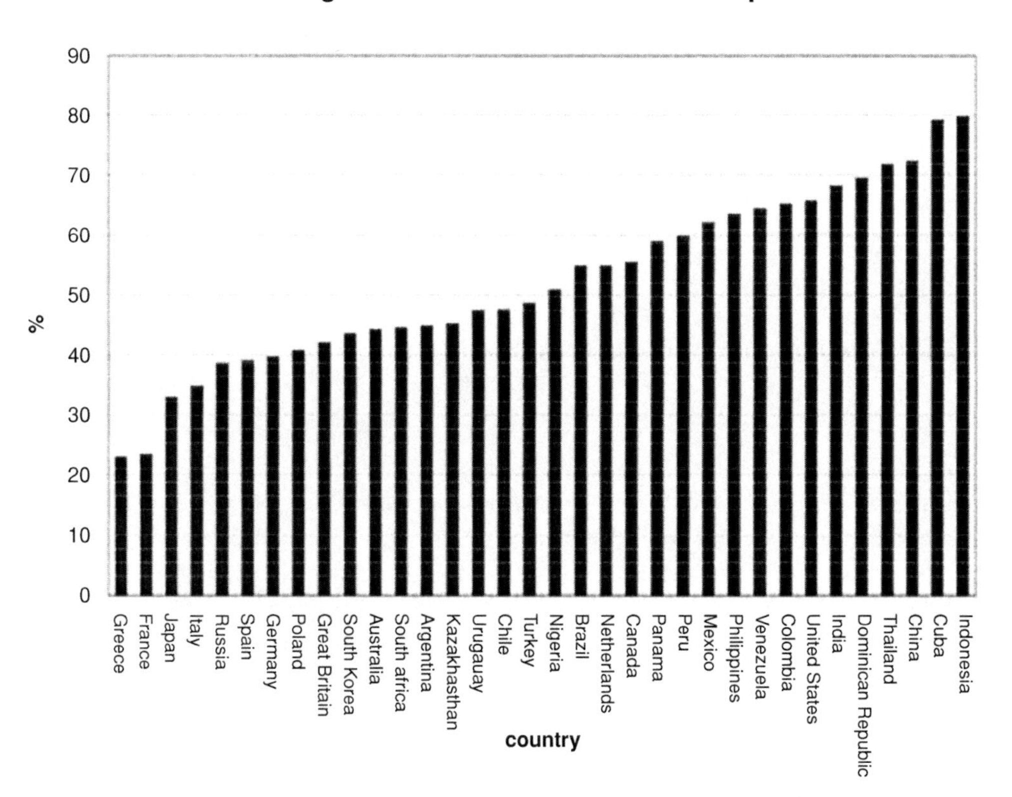

Figure 2.2. Worldwide Public Attitudes toward GM Plants. *Source:* Reproduced from Brossard and Shanahan (2007), based on data from Environics International (2000).

opinion data for Africa have not been published, and Asian data yield conflicting results. Second, because most surveys do not use the same questions or methodology, comparison of public opinion across countries is difficult.

One study, however, provides some interesting insights. An Environics International survey (2000) is one of the rare studies that compared worldwide attitudes toward GE. Figure 2.2 presents aggregated percentages of respondents who agreed that GE crops were useful if they helped decrease the use of pesticides. Although the numbers may have changed since then, the 2000 percentages are nonetheless useful because the same question matrix and methodology were used in all countries under study. Figure 2.2 clearly illustrates that public attitudes toward GE varied extremely in 2000 (from around 20 percent agreement on the usefulness of GE crops in Greece to 80 percent agreement in Indonesia), even for a topic traditionally assumed to be one of the least controversial aspects of GE crops.

It is clear that worldwide public opinion about biotechnology comprises a large spectrum of attitudes and that opinions vary with the type of use of the technology. In fact, the use of GE in food production (as opposed to medicine and other applications) is one of the least favored applications of the technology worldwide (Brossard and Shanahan 2007; Environics 2000). As stressed earlier, European resistance to GE crops is a driving factor of resistance by exporters to Europe, even though the continents that would benefit most from the technology are Africa and Asia, net exporters to Europe. Therefore, data on public opinion toward GE crops and food in Europe, Africa, and Asia are presented next.[4]

European Attitudes

Public opinion about GE crops is still largely negative in Europe. According to an extensive 2005 study of public opinion based on a representative sample of 25,000 respondents (roughly 1,000 in each European Union [EU] member state), a majority of Europeans perceived GE food as being unsuitable, morally unacceptable, and a risk for society (Gaskell, Allansdottir, et al. 2006). Even in Spain, where GE crops have been planted widely, support for using GE in food crops was only 7 percent higher than the European average of 27 percent. Evidently the passage of new EU regulations on the commercialization and labeling of GE crops and food in March 2001 has had little impact on European reluctance to adopt agri-food biotechnology.

It remains to be seen how European public opinion will evolve and if sustainability will be an argument in future debates. The British government is currently organizing new public debates (following the GM Nation consultation, an extensive public consultation in 2003) and is asking the public to ponder whether GE has the potential to resolve global food problems and counter rising food prices.

African Attitudes

Although peer-reviewed studies reporting African opinions and attitudes toward GE crops and food are rare, some well-conducted but unpublished studies give some insights on the state of public opinion. South Africa is the only African country with a system in place for the regulation, planting, and commercialization of GE crops, and it is growing GE cotton, soybeans, and corn. An unpublished 2001 study conducted in South Africa revealed overall positive attitudes with some mixed feelings about the technology among the various South African stakeholders. Although biotechnology was perceived as an aid in solving important agronomic issues that have not been resolved by conventional breeding, respondents did not seem to trust the biosafety regulations in place. They were also strongly in favor of the labeling of GM food (Aerni 2002). Interestingly, the South African case is not representative of public opinion in neighboring countries. Zambian resistance to GE crops and food (Mumba 2007) was mentioned earlier. Opinions in other African countries vary as well.

[4] For more detailed accounts, see Brossard, Nesbitt, and Shanahan (2007).

Unpublished reports and anecdotal evidence seem also to suggest that Eastern Africa (where a number of field trials are underway and biosafety regulations have been established) is overall more tolerant of the technology than Western Africa. A study conducted in 2004 in Nairobi, Kenya, among supermarket shoppers showed that the vast majority (80 percent) of respondents thought that GE could help solve the world food problems, once the basic tenets of the technology were explained to them. However, significant proportions of respondents were also worried about the potential environmental consequences (50 percent) and health risks (40 percent) of the technology (Kimeju et al. 2005).

The climate of opinion is less favorable in West Africa. Burkina Faso, one of the poorest countries in the world, began field trials in early 2000 before establishing biosafety laws, but Mali, also among the poorest countries, has been reluctant to pursue the technology. In January 2006, a Malian citizen jury comprising small farmers and medium-sized producers rejected the introduction of GE crops (particularly GE cotton) in Mali, a decision that was widely broadcast on radio, the main communication medium. The foremost concerns were to prevent dependence on multinational corporations and to preserve local crop varieties and traditional agricultural practices. The jury's decision is likely to have important implications because Mali is a signatory of the CPB. Public consultation at a national level would be needed before the introduction of any GE crops in Mali, even for testing (Gaillard 2006).

Asian Attitudes

Asia also provides a mixed landscape as far as public opinion and acceptance of GE food are concerned, according to 2008 data collected by Nielsen Company Research on behalf of the Asian Food Information Centre (AFIC). Nielsen conducted an online survey of 1,007 adult respondents in Beijing, New Delhi, Manila, Seoul, and Tokyo in August 2008. Quotas were set to best reflect the demographic diversity in the cities (AFIC 2008).[5]

Results were mixed, although it seemed clear that specific benefits or risks perceived in GE foods were linked to the dietary habits and agricultural practices of each country and that no country made a labeling request for biotechnology-derived ingredients in food. As expected, levels of awareness of biotechnology were low in all the countries surveyed (AFIC 2008).

Chinese, Indian, and Filipino respondents were more positive than their Japanese and South Korean counterparts about the benefits that GE crops could bring. This may be explained by the fact that China, India, and the Philippines produce most of their food, whereas Japan and South Korea import the majority of their food (AFIC 2008).

[5] Readers should note that AFIC is a lobbying group, which might render its survey results suspicious to some. In particular, some of the survey questions may have used leading wording and could have biased the results toward support for GE.

Figure 2.3. Factors Influencing Individual Attitudes toward GE Food.

Whereas 71 percent of the Japanese respondents were unsure about biotechnology-related benefits, the landscape was more positive in China, the Philippines, and India. In the Philippines (where *Bt* corn has been grown commercially since 2003), roughly 55 percent of the respondents believed that biotechnology would bring benefits to them or their families (almost 50 percent believed so in China). When participants were educated on sustainability, however, 85 percent of the respondents in each country said they would support biotechnology if it contributed to sustainable food production.

In India, the picture was universally more positive for GE food. Seventy percent of respondents were favorable toward the technology, with a majority believing that food biotechnology would bring benefits in the next few years, such as improved food quality (39 percent), better health (26 percent), and food sustainability (18 percent; AFIC 2008). Yet not to be overlooked is the very active and vocal opposition to GE in India as illustrated by the controversy surrounding the regulatory approval of GE eggplant at the end of 2009.

Understanding Public Opinion Dynamics

Clearly, the majority of individuals in the countries most in need of biotechnology may not yet have formed stable attitudes about it because levels of awareness of the technology are relatively low worldwide. Therefore opinions could change rapidly and could be influenced by a range of factors (see Figure 2.3).

Sociocultural and political contexts can affect levels of adoption of GE crops and public opinion dynamics related to GE products. In addition, negative media coverage of GE (portrayals in entertainment media, in news stories, etc.) is often mentioned as an important determinant of public attitudes. However, media coverage is only one factor among many influencing attitudes toward GE food and support of the technology (Brossard and Nisbet 2007).

Media can make some aspects of an issue particularly salient (such as those highlighted in the first section of this chapter) and therefore indirectly foster the agenda of certain interest groups. In fact, public interest groups that have strongly opposed the introduction of GE crops have been active in disseminating their points of view and garnering media attention. For instance, the mass media routinely cover events around public opposition to GE. Through processes called "agenda building" and "framing" in communication and public opinion research, interest groups can influence media coverage of an issue and the weight given to specific aspects of it.[6] Individuals often rely on cognitive shortcuts to make sense of complex issues; the framing by the media often provides these shortcuts (Scheufele 2006a, 2006b). However, research has also shown that the processes by which people make sense of mediated information are complex. Steadfast values, such as religious beliefs or deference to scientific authority, sometimes act as perceptual filters that individuals use to process unfamiliar information related to complex scientific concepts. As a result, two individuals may interpret the same media information very differently according to their perceptual filters and reach conflicting conclusions (Brossard et al. 2009; Scheufele 2006a).

Often, the level of understanding and knowledge about an issue will determine levels of support. However, from a standpoint of communication effectiveness, knowledge has not been shown to be a major factor predicting support for biotechnology. In fact, there is virtually no scientific basis for assuming that increased knowledge of technical facts leads to greater support for biotechnology. Instead, other variables, such as trust in institutions, groups, or individuals providing information, are more predictive of the type of support the technology encounters. Furthermore, moral values are the most important predictors of levels of support in some cases (Brossard and Nisbet 2007).

Of course, the perception of the risks and benefits associated with the technology also influences how people feel about it (Brossard and Nisbet 2007). Because risk is a social and cultural construction, the cultural meanings of the terms "risk" and "benefit" may differ significantly among societies with dissimilar political and economic contexts (Beck 1992). What may seem trivial to someone living in an industrialized society may be perceived as a risk by others, and perceived benefits in one cultural context may not be relevant in another. In some instances, the weight of benefits may be far more important than that of risk, particularly related to new technologies and products. Research has shown that although some individuals may recognize the tradeoff between risks and benefits of GE food and use this information when

[6] See Nisbet and Scheufele (2007) and Scheufele (2006a) for a discussion on framing.

determining their levels of support for the technology, perceptions of GE benefits can dominate others' judgment heuristically (Gaskell et al. 2004).

In sum, this chapter has illustrated the host of social, cultural, individual-level and scientific factors that can influence an individual's perception of GE food, which have implications for communication-related endeavors. Advancing specific benefits or dismissing certain risks related to the technology may result in failed communication efforts if these benefits/risks are not relevant to the targeted audiences. Education efforts focused on explaining the science behind the technology are also likely to fail. Successful communication efforts will need to rely on a thorough and, more importantly, scientific understanding of the sociocultural and political context, on an acknowledgment of audiences' underlying beliefs and attitudes, and on the involvement of trusted spokespersons.

Conclusion

Individuals and social groups are plagued by preconceived notions of GE foods' potential negative or unknown societal consequences. In this context, the role of expertise is especially tricky, and providing "right" or "wrong" answers to these broader questions from a purely scientific point of view is especially difficult.

As discussed, the main challenge facing biotechnology in a sustainable food supply system is the availability of GE products, either grown or consumed, in the most critical parts of the world. Acceptance or rejection of these products is contingent on a wide variety of factors. Strongly held beliefs, perceptions of benefits versus risks, and trust in the institutions involved in the process are some of the important determinants of technology acceptance. Lord Mandelson (former European Commissioner) summarized the situation very well in June 2007:

> While technology determines what is possible, consumer demand determines what is economically viable. Public fears may be misplaced, but they cannot and should not be dismissed.... We – and by that I mean you the industry and we, public authorities and governments – need to do a better job of setting out the issues. So that people are aware of the potential benefits of GM food; and – crucially – so they have confidence in our testing and approval regime (EC 2007).

The implication is that biotechnology will be integrated into the world food supply without strong opposition if it focuses on the improvement of products that fit local sociocultural contexts, answer local needs, and have been chosen through a mechanism that is perceived by all involved parties to be fair and equitable. These products may not be the traditionally bioengineered crops, such as corn or soybeans. For instance, roots and tubers are dietary staples in Africa, where cassava is the third most important source of calories overall (Gregory et al. 2008). These foodstuffs would likely be perceived as priorities in the context of food security at the local or regional level.

As Gregory et al. (2008) emphasized, communication efforts that engage all public and private stakeholders at the local level (scientists, policy makers, producers, retailers, processors, consumers, etc.) early in the process of choosing which foods to develop with biotechnology are likely to increase their acceptance, adoption, and consumption. This type of priority-setting exercise has successfully been used in India, Bangladesh, and the Philippines by the Agricultural Biotechnology Support Project II (ABSPII)[7] and led to the identification of *Bt* eggplant as a high-priority product. Gregory et al. (2008, 286) noted, "This was not only because of its verified technology and potential economic, health and environmental benefits, but also because of the absence of road blocks due to international property rights, favorable prospects for regulatory approval, strong local partnership organizations and the high likelihood of gaining public acceptance for the product." Of course, transparency of the policy decision-making process involving different institutional actors once the bioengineered development is launched will also foster trust in the safety mechanisms and will ultimately be a major determinant of acceptance of GE.

Finally, one should keep in mind that long-term sustainability can be promoted only by fostering development and capacity-building in the countries most in need of food resources. Providing access to bioengineered food alone cannot guarantee sustainability.

References

Aerni, P. 2002. *Public Attitudes towards Agricultural Biotechnology in South Africa*. Cambridge, MA: Harvard University Press.

AFIC (Asian Food Information Centre). 2008. *Executive Summary Food Biotechnology: Consumer Perceptions of Food Biotechnology in Asia*. Singapore: AFIC. http://www.whybiotech.com/resources/tps/AsiaConsumerPerceptions.pdf.

Beck, U. 1992. *Risk Society: Towards a New Modernity*. London: Sage.

Bender, K. L., and R. E. Westgren. 2001. "Social Construction of the Markets for Genetically Modified and Nonmodified Crops." *American Behavioral Scientist* 44 (8): 1350–70.

Brossard, D., E. Kim, D. A. Scheufele, and B. V. Lewenstein. 2009. "Religiosity as a Perceptual Filter: Examining Processes of Opinion Formation about Nanotechnology." *Public Understanding of Science* 18 (5): 546–58.

Brossard, D., L. E. Massarani, B. Buys, E. Acosta, and C. Almeida. 2008. "Media Frame Building and Culture: Genetically Modified Organisms in Brazilian Print Media." Paper presented at the Mass Communication Division of the 2008 International Communication Association (ICA) Convention, Montreal, QC, May 22–6.

Brossard, D., and M. C. Nisbet. 2007. "Deference to Scientific Authority among a Low Information Public: Understanding U.S. Opinion on Agricultural Biotechnology." *International Journal of Public Opinion Research* 19 (1): 24–52.

Brossard, D., T. C. Nesbitt and J. Shanahan. (Eds.) 2007. *The Public, the Media, and Agricultural Biotechnology*. Cambridge, MA: CABI Press.

[7] ABSPII is funded by the U.S. Agency for International Development (USAID) and is a multipartner project led by Cornell University.

Brossard, D., and J. Shanahan. 2007. "Perspectives on Communication about Agricultural Biotechnology." In *The Public, the Media, and Agricultural Biotechnology*, edited by D. Brossard, T. C. Nesbitt, and J. Shanahan, 3–20. Cambridge, MA: CABI Press.

Dolabella R., J. Araujo, and C. Faria. 2005. "A Lei De Biossegurança E Seu Processo De Construções No Congresso Nacional." *Cadernos ASLEGIS* 25: 63–75.

EC (European Commission). 2007. "Biotechnology and the EU." Summary of a Speech by P. Mandelson, European Biotechnology Info Day. Brussels: Bavarian Representation. http://trade.ec.europa.eu/doclib/docs/2007/june/tradoc_134910.pdf.

Environics International. 2000. *International Environmental Monitor 2000. Public Opinion Survey*. Toronto: Environics International.

Franck-Oberaspach, S. L., and B., Keller. 1997. "Review: Consequences Of Classical and Biotechnological Resistance Breeding for Food Toxicology and Allergenicity." *Plant Breeding* 116: 1–17.

Gaillard, R. 2006. "Resistance Continues to GM Crops – Mali: Not on My Farm," *Le Monde diplomatique*, April. http://mondediplo.com/2006/04/11gmomali.

Gaskell, G., A. Allansdottir, N. Allum, C. Corchero, C. Fischler, J. Hampel, J. Jackson, et al. 2006. *Europeans and Biotechnology in 2005: Patterns and Trends: Eurobarometer 64.3*. A Report to the European Commission's Directorate-General for Research. Brussels: DG Research.

Gaskell, G., N. Allum, W. Wagner, N. Kronberger, H. Torgersen, J. Hampel, and J. Bardes. 2004. "GM Foods and the Misperception of Risk Perception." *Risk Analysis* 24 (1): 185–94.

Greger, J. L. 2000. "Biotechnology: Mobilizing Dietitians to Be a Resource." *Journal of the American Dietetic Association* 100 (11): 1306–8.

Gregory, P., R. H. Potter, F. A. Shotkoski, D. Hautea, K. V. Raman, V. Vijayaraghavan, W. H. Lesser, G. Norton, and W. R. Coffman. 2008. "Bioengineered Crops as Tools for International Development: Opportunities and Strategic Considerations." *Experimental Agriculture* 44: 277–99.

Gura, T. 1999. "New Genes Boost Rice Nutrients." *Science* 285: 994–5.

Isserman, A. M. 2001. "Genetically Modified Foods: Understanding the Societal Dilemma." *American Behavioral Scientist* 44 (8): 1225–32.

James, C. 2010. *Executive Summary, Global Status of Commercialized Biotech/GM Crops*. ISAAA Brief 42. International Service for the Acquisition of Agri-Biotech Applications (ISAAA). http://www.isaaa.org/resources/publications/briefs/42/executivesummary/default.asp.

Kaufman, M. 2008. "Extra-Nutritious Bioengineered Foods Still Years Away." *Washington Post*, November 3, A12. http://www.washingtonpost.com/wp-dyn/content/article/2008/11/02/AR2008110201939.html.

Kimenju, S. C., H. De Groote, J. Karugia, S. Mbogoh, and D. Poland. 2005. "Consumer Awareness and Attitudes toward GM Foods in Kenya." *African Journal of Biotechnology* 4 (10): 1066–75.

McCallum, C. 2000. "Food Biotechnology in the New Millennium: Promises, Realities, and Challenges." *Journal of the American Dietetic Association* 100 (11): 1311–15.

Mumba, L. E. 2007. "Food Aid Crisis and Communication about GM Foods: Experiences from Southern Africa." In *The Public, the Media, and Agricultural Biotechnology*, edited by D. Brossard, T. C. Nesbitt, and J. Shanahan, 338–64. Cambridge: CABI Press.

Nelson, C. H. 2001. "Risk Perception, Behavior, and Consumer Response to Genetically Modified Organisms." *American Behavioral Scientist* 44 (8): 1371–88.

Nisbet, M. C., and D. A. Scheufele. 2007. "The Future of Public Engagement." *The Scientist* 21 (10): 38–44.

Nuffield Council on Bioethics. 1999. *Genetically Modified Crops: Ethical and Social Issues*. London: Nuffield Council on Bioethics.
Peterson, G., S. Cunningham, L. Deutsch, J. Erickson, A. Quinlan, E. Raez-Luna, R. Tinch, M. Troell, P. Woodbury, and S. Zens. 2000. "The Risks and Benefits of Genetically Modified Crops: A Multidisciplinary Perspective." *Conservation Ecology* 4 (1): 13. http://www.consecol.org/vol4/iss1/art13/.
Phipps, R. H., and J. R. Park. 2002. "Environmental Benefits of Genetically Modified Crops: Global and European Perspectives on Their Ability to Reduce Pesticide Use." *Journal of Animal and Feed Sciences* 11: 1–18.
Scheufele, D. A. 2006a. "Five Lessons in Nano Outreach." *Materials Today* 9 (5): 64.
______. 2006b. "Messages and Heuristics: How Audiences Form Attitudes about Emerging Technologies." In *Engaging Science: Thoughts, Deeds, Analysis and Action*, edited by J. Turney, 20–25. London: Wellcome Trust.
SciDev.Net Opinions. 2006. "The Cartagena Protocol: The Debate Goes On." May 12. http://www.scidev.net/en/opinions/the-cartagena-protocol-the-debate-goes-on.html.
Secretariat of the CBD (Convention on Biological Diversity). 2011. *The Cartagena Protocol on Biosafety, and its Nagoya-Kuala Lumpur Supplementary Protocol on Liability and Redress*. Montreal: United Nations. http://www.cbd.int/undb/media/factsheets/en/undb-factsheet-biosafety-en.pdf.

3

Loving Biotechnology

Ethical Considerations

The Rev. Lowell E. Grisham

> The greatest service a citizen can do for his country is to add a new crop for his countrymen.
>
> *Thomas Jefferson*

> It is not simply wrong, it is a piece of stupidity on the grandest scale for us to assume that we can simply take over the earth as though it were part farm, part park, part zoo, and domesticate it, and still survive as a species.
>
> *Thomas (1992, 122)*

The creation myths of many religious traditions describe earth as a living entity. One myth of the Hebrew Bible imagines God's Spirit hovering over the void and speaking creation into being. The other myth tells of God's molding humanity from the earth and planting a garden to sustain this new human life. The Maasai tribe of Kenya relates the story of the Creator's splitting a single tree into three pieces, giving the original tribes a stick to herd animals, a hoe to cultivate the ground, and a bow and arrow to hunt. Tales from the ancient Finns say that the world arose from an egg that was broken. Some Eastern traditions believe the universe has always existed, passing through endless cycles. The Cherokee tell how the animals brought the mud from below the water and attached the sky with four strings. Anthropologists track the wondrous interrelationships that have allowed life to evolve into consciousness, and ecologists explore the interrelatedness of complex living systems.

Earth is a living organism – evolving, always in the act of becoming. The history of life from single-cell organisms to the mystery of human consciousness is a story of continual response to challenge. In response to challenge, growth, learning, and adaptation occur.

Among the challenges for our generation is the struggle for survival facing more than a billion of our sisters and brothers. How do we produce and distribute food for a growing world population while preserving the limited resources of earth's water and soil? The question calls for more than the technical answers that will be provided in later chapters. If technology is to be exercised in service to humanity, we will need an ethical – spiritual – framework for our approach to our research and our applications. We will need to use knowledge and power with wisdom.

The faith traditions offer some maps through the complicated territory of power, knowledge, and finitude. Religion seeks to connect humanity to our deepest longings and to enlighten our encounter with mystery and being itself. Faith invites us into a path of hope as we face the unknown or the threatening.

The drama of planetary evolution has been a story of biological yearning for new and fearless adaptations in the presence of challenge. Our calling is to respond with hope and fearless realism to the planetary challenges of our generation. We do so as a species with a stunning history of survival and adaptation.

The Christian Ethic Celebrates Discovery

In monotheistic religious tradition three things are reserved for deity alone: omniscience, omnipotence, and immortality. To be healthy and sane, human beings must accept the limitations of our knowledge, power, and time. It is hubristic for us to claim or to imagine, even subconsciously, that given enough knowledge, power, and time, we could solve anything to which we put our minds. Life is more than a series of engineering problems to be solved.

Researchers at the edge of scientific inquiry experience the endless struggle against ever receding horizons. Discovering more answers uncovers more questions. One definition of humility is to keep growing while accepting our limitations. The healthy scientific quest is a humble journey of fearless hope.

I contend that the pursuit of knowledge is a spiritual quest. The Greeks extolled the virtue of humanity's search for truth, beauty, and goodness. Christian religious tradition ascribes to God the qualities of perfect truth, perfect beauty, and perfect goodness. Whenever science discovers new truth, science adds to humanity's understanding of reality. Those of us from the Abrahamic faiths hold that God is Reality Itself, so there can never be a conflict between the broadening of scientific truth and the exercise of religious faith. From this perspective every new discovery reveals more about reality and therefore reveals more about God. The influential fifth-century Christian theologian Augustine of Hippo argued passionately that nature and Scripture are complementary books of understanding. As our understanding of creation expands, our understanding of the nature of God expands as well. Whenever we participate in new discovery, it is an encounter with God.

One of the challenges that science faces is the exercise of its pursuit of ever expanding truth, coupled with a corresponding commitment to using the new knowledge for good. Goodness must expand as truth expands. Whenever we find new discoveries, will we exercise enough goodness not to abuse the earth or its creatures? With new power comes new responsibility.

Biotechnology's search for a new order of reality is a spiritual search. God has endowed us with intellect and curiosity. It is in the evolutionary nature of humanity to continue to probe deeper and deeper into our understanding of reality. Quashing the instinct for adventure and learning is a violation of the human spirit. The search for knowledge is endorsed by the spiritual traditions.

The use of science is an ethical question. Is science an end in itself, or is science always a means to some other end? In earlier centuries, science was considered the handmaiden of theology. Arguably, science has become the handmaiden of colonization, war, consumerism, globalism, and a host of other purposes since the industrial era. Nuclear technology can be used to produce efficient electricity or devastating bombs. How will the discoveries of biotechnology be used? To what end will we use our knowledge?

It is significant for the ethics of biotechnology that the pioneer of gene theory was an Augustinian monk, Gregor Mendel (1822–84). Mendel has been called the father of genetics, because his work opened inquiry in the early part of the 20th century into the molecular processes of inheritance (Henig 2000). Some have argued that technologies such as genetic engineering (GE) are nothing more than an acceleration of Mendel's basic processes of cross-breeding.

It is not uncommon for new discoveries to meet resistance. As explained in Chapter 2, many people fear the new and unexplored. New insight often challenges embedded paradigms. For centuries there has been opposition to biological exploration, and researchers were accused of tampering with nature. One school of thought alleges that God made things the way they should be, and we should not try to improve on God's handiwork. "If we were meant to fly, God would have given us wings," said the skeptics to the Wright brothers. When artificial insemination was introduced, it provoked a firestorm of opposition and controversy. Today, air travel and livestock production via insemination are widely accepted.

The preponderance of religious tradition invites believers into a humble spirit of fearless adventure on a journey toward enlightenment and salvation. All of the founders of the great spiritual traditions were pioneers on a fearless pilgrimage into ultimate reality. The Hebrew and Christian Bibles have hundreds of verses reminding readers to "fear not" or "be not afraid." The scientific spirit of inquiry into the mystery of reality is consistent with the spiritual journey into ultimate reality, which we call God.

Responsibility toward Our Neighbor

Christianity: "Do unto others as you would have them do unto you."
Judaism: "What you hate, do not do to anyone."
Islam: "No one of you is a believer until he loves for his brother what he loves for himself."
Hinduism: "Do nothing to thy neighbor which thou wouldst not have him do to thee."

Buddhism:	"Hurt not others with that which pains thyself."
Sikhism:	"Treat others as you would be treated yourself."
Confucianism:	"What you do not want done to yourself, do not do to others."
Aristotle:	"We should behave to our friends as we wish our friends to behave to us."
Plato:	"May I do to others as I would that they should do unto me."

Covey (2006, 145)

All enduring religious traditions enshrine in their teachings a responsibility toward our neighbors and an ethic of compassion. It is unacceptable for some to eat while others starve. The quest to improve our agriculture and enhance its sustainability is a good quest with good intentions. The potential for creating crops that need fewer chemicals and use less water is exciting. The promise of enriched edibles containing increased levels of nutrition is hopeful. The search for bio-based pharmaceuticals and fuels holds potential for an improved and more wholesome earth. The quest for sustainable forms of agriculture is good stewardship. Modern biotechnology researchers are continuing the heritage of Mendel, and it is therefore appropriate that they also adopt his attitude to work with fearless, hopeful humility. *That* we discover and *what* we discover are ethically neutral. It involves simply the pursuit of reality. It is in the nature of humanity that the curiosity gene/genie is out of the bottle. The real question is, What *shall* we do with what we discover? That question is usually a power question, and knowledge is power. Who or what powers will knowledge serve?

Responsibility for Creation

"The Earth is God's and all that is in it," proclaims Psalm 24 (also Exodus 9:29 and 1 Corinthians 10:28 NRSV; O'Day and Peterson 1999). The earth has its own intrinsic value and integrity, and it is not ours to exploit or abuse. Humanity is a fellow creature on earth, along with the birds and the fish, the microbes, and the animals. We share the resources of earth and water and vegetation. Humankind does not have a superior or privileged place among the other creatures except insofar as we are given a certain level of consciousness to use to care for creation and exercise responsibility for it.

In the ancient Hebrew myth of the seven days of creation, God gives exactly the same blessing and value to each part of creation. The vegetation and living creatures are all declared to be good: "The earth brought forth vegetation: plants yielding seed of every kind, and trees of every kind bearing fruit with the seed in it. And God saw that it was good." On successive days the myth narrates God's creating the "swarms of living creatures," birds, fish, cattle, creeping things, wild animals, and even the "great sea monsters." Repeated on each day is the refrain: "And God saw that it was good" (Exod 1). All of creation is God's creation. All of creation is good. Everything carries God's blessing.

When God creates humankind in God's image, God gives humankind dominion over the other animals. However, in the Genesis creation account, God does not give us dominion over animals in order to eat them. Our original sustenance was a vegetarian diet, according to Genesis. God gives to the animals exactly the same food source as humankind: "everything that has the breath of life, I have given every green plant for food" (Gen 1:30).

In the biblical story, it is only after the flood – after humanity has corrupted God's creation and God must make a new start – that God gives humans permission to use animals for food. Even with that gift, the life of the animal is to be taken with respect, releasing its life-blood back to the earth and to God: "Every moving thing that lives shall be food for you; and just as I gave you the green plants, I give you everything. Only, you shall not eat flesh with its life, that is its blood" (Gen 9:3–4).

We experience this ancient sensitivity today particularly in regard to biotechnology techniques involving animals. Higher up the evolutionary chain – the more like us the animal seems – there is greater public sensitivity to the treatment of animal life. In our culture we do not eat dogs or horses, but we do eat cows, which are considered sacred by some cultures. Many notable theologians – many of the Desert Fathers, Basil the Great, Tertullian, Origen, and John Wesley – were vegetarians, believing that practice to be truer to God's original intentions. Many people today abstain from animal products to varying degrees as a spiritual practice.

It is also instructive that the Genesis story contains an account of an earthly destruction of universal proportions, which was caused by our pride and violence. One can easily see the potential for such a universal catastrophe in our own day.

Although humankind is given a place of dominion within the order of creation, the rest of the planet – vegetable and animal – holds a place of honor in God's economy as well. To each order God has given blessing and pronounced that it is good. The dominion that human beings are to exercise over the other life forms is a dominion of stewardship. The earth is God's, and we are called to care for it as such.

The other creation myth in the Bible elaborates on humanity's relationship with the plants and animals: "The Lord God took the man and put him in the garden of Eden to till it and keep it" (Gen 2:15). One may call Adam the first ecologist. Caring for the garden through ecology and sustainable development is a form of stewardship of the earth.

Whereas the first creation account emphasizes God's creation of men and women in the image of God, and thus as equals, the second creation story suggests a valuable relationship between humans and the rest of the natural world. God formed out of the ground "every animal of the field and every bird of the air, and brought them to the man to see what he would call them; and whatever the man called every living creature, that was its name" (Gen 2:19). There is an intimacy between the man and the living creatures. The human being knows each creature by name, personally, as a God-given potential partner. We humans, male and female, are to live in intimate partnership with the other animals.

All of creation is good and is imbued with God's creativity and care. God speaks, breathes, and forms all things into being; therefore all parts of creation share in God's spirit and participate at some level in God's consciousness. All creation is holy because of this relationship. We live in a creation that is an interconnected whole, filled with divine life. And it is good. Ours is an interdependent existence. We participate in the web of life. Everything is connected to everything else.

Other Faith Traditions' Attitudes toward Earth

All enduring religious traditions have made significant contributions to our contemporary conversations about the environment, sustainability, and ecology.[1] Eastern religion has traditions of respect and honor toward the earth. Writing in his seminal study, *The World's Religions*, Smith says, "Taoism seeks attunement with nature, not dominance. Its approach is ecological. . . . Taoist temples do not stand out from their surroundings. They nestle against the hills, back under the trees, blending in with the environment. At best, human beings do likewise" (Smith 1991, 213). "Nature is to be befriended. When the British scaled earth's highest peak, the exploit was widely hailed as 'the conquest of Everest.' D.T. Suzuki remarked, 'We orientals would have spoken of befriending Everest'" (Smith 1991, 212).

Stewardship, Trust, and Responsibility

From a Christian perspective, humanity's role is to befriend creation and be stewards of God's earth. We are to approach our work with holy awe, as servants. Ecology is the service of stewardship. I believe that the greatest challenge facing the future development of biotechnology is the public's level of trust and that the future of biotechnology is limited more by trust issues than its scientific challenges. Where public trust is low, biotechnology faces its strongest resistance. If biotechnology is to thrive, the science has to earn and build trust among the populace.

Trust is a function of character and competence. Although researchers have spent much energy on perfecting the performance and safety of genetically modified organisms (GMOs), the public perception seems to be that biotechnology development has given less attention to promotion of a trust-inspiring culture of integrity and openness. Until the public trusts the character of those who are developing GMOs, there will be a corresponding degree of mistrust and resistance.

1 Harvard University Press has published the nine-volume *Religions of the World and Ecology*, which assesses the attitudes of nine religious traditions toward ecology. The Forum on Religion and Ecology at Yale is a robust international multi-religious project that explores environmental concerns.

The spiritual traditions have much to teach about the work of building trust. Trust is a synonym for faith. Will the public have faith in the science, industry, and economics of the biotechnology revolution? The answer to that question will be highly dependent on the character of the work and its workers. Can we do our work as trustees of the earth in a humble spirit of stewardship and service?

The invitation to stewardship can be contrasted to an attitude of exploitation for possession and profit. The Hindu leader Mahatma Gandhi proclaimed, "The earth has enough for mankind's needs, but not for its greeds" (quoted in Dwivedi and Vajpeyi 1995, 47). The Hebrew tradition of the Jubilee Year, when all property reverted to the original tribes to ensure its balanced distribution, stands as a challenge to the presumption of the ownership of land conveying an absolute control over it. The Eastern Woodland tribes of North America embodied their faith traditions into their strategy for producing crops. They grew corn, beans, and squash in fertile river valleys and moved their fields with the rhythms of flood and drought. They understood at a cultural level the footprint of their communities and defended that footprint as territory, but never presumed to own the land. For most native peoples, the land is a part of their self-understanding and identity. When native peoples are removed from their land, that removal is a form of cultural genocide. Western notions of rights of land ownership become critical in this discussion. Air and water travel beyond boundaries. So do genes, apparently. Life is remarkably creative in spreading genetic material with profound fecundity.

Giving Voice to Stakeholders

The conversation about the use of earth's resources must be an open one. Multinational companies that can collect under one umbrella earth and water, lab and scientist, seed and sower, consumer and producer, accountant and stockholder must also be accountable to the whole. Consumers and even nonconsuming neighbors must have a substantial stake in the conversation. Someone must speak for the interests of the animal and plant kingdoms. If industry cannot give them voice and vote, we must create appropriate instruments of regulatory accountability.

Beginning in 2008, the United States faced a credit crisis that some called the greatest economic challenge since the Depression. The banking and credit industries experienced the consequences of deregulation of an economic system that had lost a portion of its accountability in the credit market. The effect spiraled into all sectors of the economy. When any entity loses its ability to self-regulate it will create systemic damage – a cancer. The recession of 2008 was essentially a failure of trust and of trustworthiness.

If the emerging biotechnology revolution is to be healthy, now is the time to create trustworthy pathways of communication and cooperation. The organic nature of all living things offers us a model. In a healthy, living body all of the cells are interconnected. Redundant systems of neurological and chemical transmitters and receptors ensure that the whole is completely informed by its parts. Each cell, organ, and system in the body is self-defined, knowing its role and staying within the boundaries of its effectiveness. The flow of communication is open and uninhibited. The body's mechanisms for growth and protection pay respectful attention to threats even at the molecular level.

Can we pioneers of the new technology create an interdependent collaborative norm that will foster health and trust on behalf of the whole planet? Or will we choose to be independent pioneers, focused on the exercise of domination that is manifested in power and control? If we choose the latter path, it would be wise to establish constraints and regulation preemptively.

In the United States, there are at least three federal agencies responsible for review and oversight of agricultural products involving biotechnology: the Department of Agriculture's Animal and Plant Health Inspection Service (APHIS), the Environmental Protection Agency (EPA), and the Food and Drug Administration of the Department of Health and Human Services (DHHS). These agencies have discrete and overlapping jurisdiction, resulting in a fragmented and often discordant process for reviewing and regulating crop and animal biotechnology. The voices of other interests are often limited to federal comment periods or political action through advocacy campaigns. These approaches result in an inevitable confrontational culture, pitting biotechnology companies against stakeholders rather than engaging parties to discuss, understand, and perhaps learn from each other's perspectives. This scenario is a picture of a body that is in conflict with itself – researchers, producers, accountants, environmentalists, consumers, and stockholders each assert their claim for priority and power. Wisdom is lost in the fray.

The international community through the United Nations Environment Programme (UNEP) established the Intergovernmental Committee for the Cartagena Protocol on Biosafety (ICCP) in 2000. The purpose of the ICCP was to "contribute to ensuring an adequate level of protection in the field of the safe transfer, handling and use of living modified organisms resulting from modern biotechnology that may have adverse effects on the conservation and sustainable use of biological diversity, taking also into account risks to human health, and specifically focusing on transboundary movements" (CPB 2000, Article 1). The United States has refused to formally adopt the CPB because of conflicts over labeling, intellectual property rights, and risk assessment protocols, among others.

In the passage 1 Cor 12:16f, Paul addresses a similar conflict within the congregation in Corinth:

> If the ear would say, "Because I am not an eye, I do not belong to the body," that would not make it any less a part of the body. If the whole body were an eye, where would the hearing be? . . . As it is, there are many members, yet one body. The eye cannot say to the hand, "I have no need of you." . . . On the contrary, the members of the body that seem to be weaker are

indispensable . . . and our less respectable members are treated with greater respect; whereas our more respectable members do not need this.

As were the members of the church at Corinth, all of the stakeholders in the development of agricultural biotechnology are members of one body, and we are joined to the very life of the planet. We must regard the smallest of genes and the most remote impoverished village as worthy of greater respect. For the most part, the multinational companies have sufficient power and money to generate their own abundant supply of "respect."

Paul closes his essay on cooperation and interdependence by advocating a more superior way than respect of interacting with others: the way of love. 1 Corinthians 13, Paul's "love chapter," may offer an ethical map for scientific practices and the organization of our agricultural industries. First, Paul calls love "patient" and "kind." Good science and sustainable agricultural practices look patiently toward the long-term benefit. Native American wisdom calls on leaders to make every decision with regard to its consequences for the whole group, now and far into the future – in one tradition, even to the seventh generation. James Oglethorpe, colonial governor of Georgia in 1764, was famously frustrated by the lack of autocratic rule in the Muscogee Creek Nation. He described the Native Americans' decision-making process with distaste: "There is no coercive power. (Their leaders) can do no more than persuade. They reason together with great temper and modesty till they have brought each other into some unanimous resolution" (O'Brien 1989). When patience and kindness supersede the narrow kind of efficiency that looks primarily to the immediate gain, we are more likely to create systems that will bring long-term and wider benefit.

There is important work to be done to enable the many different stakeholders to come to a common understanding of what sustainability means, maybe along the lines of this definition from the UN Food and Agriculture Organization (FAO): "Sustainable development is the management and conservation of the natural resource base and the orientation of technological and institutional change in such a manner as to ensure the attainment and continued satisfaction of human needs for the present and future generations" (FAO 1994). That definition could be strengthened by taking into account not only "human needs" but also the needs of the nonhuman elements that share this planet with us.

Richard Crossman has offered two recommendations as a touchstone for biotechnology development that parallel Paul's admonition for patience and kindness. First is the *precautionary principle*, articulated by the 1992 UN Conference on Environment and Development: "to act in such a way as to not make the planet a laboratory in a trial and error experiment" (Evangelical Lutheran Church in America 2001, 50). He argues that the burden of proof falls on proponents of the activity or those who are financially benefiting from it. Paul closes his essay on cooperation and interdependence by advocating what he calls "a more excellent way": the way of love (1 Cor 12:31).

Although the precautionary principle (PP) seems commendable, it must be balanced against the compelling need for sustainable development in a hungry world. How long can the hungry wait? How long does it take to prove the safety of a GMO? The PP defines the criteria for acceptance of GM technology in food as demonstration of substantial equivalence, based on chemical analysis of GM and non-GM foods. A GM food that is characterized as substantially equivalent to the non-GM food item is appropriate for commercial use (Myhr and Traavic 2002). Some products have been in use for more than a decade with no evidence of harm. Has the practice of allowing GMOs that meet a standard of substantial equivalence produced negative externalities? Should the bar be lower for products that address critical, life-saving needs like famine? Should the bar be higher for products that offer mostly economic benefits for producers?

Crossman's second recommendation is the *involvement principle.* He argues that biotechnology requires close monitoring, transparency, and public accountability. To support these principles, Crossman recommends the following:

1. Support the labeling of genetically modified food.
2. Give keen attention to monitoring economic and political activity regarding the development and approval of biotech processes and products (both short- and long-range).
3. Encourage public participation in and awareness of public debate on biotech concerns.
4. Press government and corporations to pursue activities that benefit the whole of creation (including those who are marginalized) rather than only those activities that will generate a large profit.
5. Advocate that there be a period of time in all biotech processes that gives space for ethical reflection as part of any biotech development activity. This would be a requisite time in which research, education, and global monitoring would allow large numbers of people to understand the problems they face and offer them the means to [ethically] address these problems.

Evangelical Lutheran Church in America (2001, 50)

Openness and Trust as Process

> I am not an Athenian or a Greek but a citizen of the world.
>
> *Diogenes*

> Love is not envious or boastful or arrogant or rude. It does not insist on its own way; it is not irritable or resentful.
>
> *1 Cor 13:4–5*

How open can our biotechnical research be? Part of the effectiveness of the 20th-century revolution in mathematics and physics was due to the remarkable sharing of information across continents and academic boundaries. The scientists who discovered the quantum universe were part of an international community of scientists exercising healthy competition while maintaining a vibrant conversation with openness and transparency.

Creating a culture of integrity, openness, and care can inspire confidence that builds trust. A firm commitment to quality control and safety can create a track record that also builds trust. Take, for example, those who created the Mississippi Delta farm-raised catfish industry. Knowing that they were marketing a product that had a compromised image as an ugly, bottom-feeding scavenger, they made an absolute commitment to quality control from their inception. Their goal was that no consumer would ever experience a single bite of catfish whose taste was off-color or fishy. They established strict policies of testing fish from a pond three days before harvest, on the day of harvest, and from the tankard before delivery for processing. Any sign of flaw at any stage aborted the process. Their adherence to such a strict policy of quality control gained the trust of *New York Times* food editor Craig Claiborne and led to the acceptance of Mississippi farm-raised catfish on the menus of fine restaurants.

It will be important for the political and industrial players in the biotechnology endeavor to be able to build trust – to listen humbly and compromise. Committing to an ethic of love would be beneficial in negotiating complicated interests in the pursuit of a safe, abundant supply of food that protects the earth and deals justly with all. This journey requires a firm commitment to truth, wherever it may lead us. "Love does not rejoice in wrongdoing, but rejoices in the truth" (1 Cor 13:6). One of the qualities that led to the opening of our understanding of the micro-universe was the willingness of mathematicians and physicists to follow the strange and counterintuitive results of their computations and experiments. Inconvenient truths must be embraced as truth nonetheless. Yet truth is not always self-evident; it is often revealed only partially, over time. "We see through a glass darkly," Paul says later (1 Cor 13:12 KJV). Complex systems usually involve ambiguous exchanges of risk and benefit. Love embraces the wide, patient view: "Love bears all things, believes all things, hopes all things, endures all things" (1 Cor 13:7 NRSV). Interestingly, it seems Paul may also have been describing the qualities of a good scientist, politician, or executive.

This conversation about love is not out of place in biotechnology. Love is at the core of every biological and evolutionary process. All matter and life are in a relationship. Love is the relationship that seeks union without destroying. The relationship of nucleus to electron or of planet to sun is a dance of attraction and integrity. Dante ends the *Divine Comedy* with a final vision of the central circle of Paradise, spinning like a balanced wheel rotated "by the Love that moves the Sun and the other stars" (Birk and Sanders 2005, 204). It is not too far-fetched to describe the fundamental yearning of cells to multiply and divide as the call of love. The whole history of evolution is a history of love transcending itself into consciousness. Love is our highest consciousness, so why cannot we bring love into our highest biotechnology pursuits on behalf of the health and sustainability of our planet and the human race?

Deep in the human heart is a desire for a peaceful earth, where hunger and thirst are no more and there is neither violence nor oppression. The dream of a fulfilled earth is a vision present in most spiritual traditions. We long for a world where justice reigns and all may live lives of wholeness and peace. The spiritual traditions invite us to embrace the dream of a fulfilled earth as a present reality coming to life, in utero, waiting to be born, to awaken. Like a mother eagerly preparing for the birth of her child, so aware of the life inside her, we are to live in the expectant reality of this coming hope. We are to prepare for it as if the day had already arrived. We are to live out of a vision for the end.

Our hope is for a healed world, where no one is hungry, no one is sick, and each person is equally valued. We hope for a clean, whole, and balanced world. The vision is close, and we strive toward it. Out of that vision and energy, let us live and work.

Living with such an end in sight, we are better prepared to make decisions for action in the present. We are moved with hope to address those places where the vision is not yet realized. It may take only a bit of trust and integrity, a little self-restraint and courage to frame this endeavor in order to reshape and support life in a new and loving way.

Toward the end of Graham Greene's (1990) novel, *The Power and the Glory*, the Whisky Priest sits in his prison cell, looking outside his window at the gallows that will hang him in the morning. He has lived a morally ambiguous life. Although he stayed to provide the sacrament to the people after the army arrived, he had fathered an illegitimate child and drowned much of his fear in liquor. As he waited to die,

> He felt only an immense disappointment because he had to go to God empty-handed, with nothing done at all. It seemed to him at that moment, that it would have been quite easy to have been a saint. It would only have needed a little self-restraint and a little courage. He felt like someone who had missed happiness by seconds at an appointed place. He knew now that at the end there was only one thing that counted – to be a saint.
>
> *Greene (1990, 211)*

While some of us see signs of potential planetary catastrophe outside our window, we also lay claim to possibilities for hope. To use the capacities of science and technology in a saving way may only take "a little self-restraint and a little courage" on our part. If we are to expand happiness and security for the earth during our lifetime, we are going to have to do this work of stewardship and do it well. At the end there will be only one thing that counts – to care for all life in a responsible and compassionate manner, to prove ourselves trustworthy.

The Ethic of Development

> Live simply so others may simply live.
>
> *Gandhi*

The world faces a challenge as population increases strain our capacity to produce and deliver affordable food in an ecologically responsible manner, particularly in developing areas. As we seek to love our neighbor as ourselves, we strive to find ways to alleviate hunger, poverty, malnutrition, and disease.

It can be tempting for those of us in the developed world to research potential solutions and seek to apply them without giving adequate voice and power to those whom we intend to help. It can be an act of oppression to solve another's problem for them, especially if we do so without their input. Can we give equal voice to the communities that are the potential beneficiaries of our planning? If we fail to hear them and engage their active leadership, it is likely that we will overlook important social and economic factors that might affect our planning, and we may find resistance to a solution that appears to be offered from a position of power *over* rather than power *with* or power *for*.

It can be presumptuous for an outsider to attempt to solve a problem for others. How can we accurately understand the problems of the developing world until we have developed trusting and reciprocal relationships with them? Our motives make a difference. Are we trying to fix others or to love them? St. Vincent de Paul said, "It is because of our love alone that the poor will forgive us for the bread we give them."

The developing world can teach the developed world many lessons in community and survival. The vibrant liberation theology movement that energized the Roman Catholic Church in the second half of the 20th century emerged from the immersion of priests, theologians, and lay leaders living in community with peasants in parts of Latin America and elsewhere in the developing world.

It is important for agricultural developers to live with and listen to the poor and hungry. With deep and humble listening we will learn from them things that will augment our growth and integrity. Any plans for relief and development will need to be appropriate to the culture and context of those whom we identify as needy. We need to hear from them what are their real needs. What power and control do they need over the processes that might bring them new resources?

We need to ask them to help us anticipate how change might affect the balance of their societies, economies, and ecology. How can we empower servant leaders within their cultures while avoiding empowering tyrants who will use seed or food oppressively? We need to approach our relationship with the poor as a pilgrimage in which we ask them to bless us. In the Beatitudes, Jesus said, "Blessed are the poor," not those who care for them.

For a number of years, Roman Catholic theologian and priest Henri Nouwen lived in the L'Arche community, which was founded by Jean Vanier. L'Arche is a community where people with and without disabilities live together in a remarkably interdependent way. Nouwen reflected on his own service as he bathed and clothed and fed Adam, a severely disabled man who could not speak and experienced violent seizures. Nouwen called Adam "my friend, my teacher, and my guide."

I want to help. I want to do something for people in need. I want to offer consolation to those who are in grief and alleviate the suffering of those who are in pain. There is obviously nothing wrong with that desire. It is a noble and grace-filled desire. But unless I realize that God's

> blessing is coming to me from those I want to serve, my help will be short-lived, and soon I will be "burned out."
>
> How is it possible to keep caring for the poor when the poor only get poorer? How is it possible to keep nursing the sick when they are not getting better? How can I keep consoling the dying when their deaths only bring me more grief? The answer is that they all hold a blessing for me, a blessing that I need to receive. Ministry is, first of all, receiving God's blessing from those to whom we minister. What is this blessing? It is a glimpse of the face of God. Seeing God is what heaven is all about! We can see God in the face of Jesus, and we can see the face of Jesus in all those who need our care.
>
> Once I asked Jean Vanier: "How do you find the strength to see so many people each day and listen to their many problems and pains?" He gently smiled and said: "They show me Jesus and give me life."
>
> *Nouwen (2006, 69–70)*

Nouwen speaks from a Christian framework, but we can all commit ourselves to respect the dignity of every human being, especially those whom we seek to feed and to help. We can expect to see the face of human dignity in their faces, and we shall be blessed. Humble, fearless hope will give us eyes to see the dignity and grace in all people.

A similar story is told in the Islamic faith. "When one of Muhammad's followers ran up to him crying, 'My Mother is dead; what is the best alms I can give away for the good of her soul?' the Prophet, thinking of the heat of the desert, answered instantly, 'Water! Dig a well for her, and give water to the thirsty'" (Smith 1991, 249).

Some pharmaceutical companies have used profits from the first-world sale of their human immunodeficiency virus (HIV) drugs to help underwrite the delivery of low-cost medicines in response to the epidemic in Africa. Could seed companies do the same to provide access to improved GM varieties in areas of the world hit by hunger and famine?

Might not justice be served, or karma, if pharmaceutical and seed companies would acknowledge and generously compensate native peoples for products and profits developed from local knowledge? Whenever pharmaceutical and seed companies have benefited from native peoples' knowledge of indigenous plants and their uses, the companies owe a special obligation of respect and generosity.

Model: *Three Cups of Tea*

Greg Mortenson's story narrated in his book *Three Cups of Tea* offers a parable about compassionate development in a challenging environment. Although disputes about his account have yet to be resolved, the story is compelling as an illustration of how compassionate development might proceed. According to Mortenson, he has built more than 50 schools, mostly for girls, in the Taliban-inhabited Karakoram Mountains. He says that as he worked feverishly on his first school in his pragmatic, linear Western style, the village chief Haji Ali told Morenson, "Sit down. And shut your mouth. . . . You're making everyone crazy," (Mortenson and Relin 2006, 150). Then Haji Ali had his wife make them a cup of tea. It was a slow, half-hour brewing process.

It was only when the porcelain bowls of scalding butter tea were steaming in their hands that Haji Ali spoke. "If you want to thrive in Baltistan, you must respect our ways," Haji Ali said, blowing on his bowl. "The first time you share tea with a Balti, you are a stranger. The second time you take tea, you are an honored guest. The third time you share a cup of tea, you become family, and for our family, we are prepared to do anything, even die," he said, laying his hand warmly on Mortenson's own. "Doctor Greg, you must make time to share three cups of tea. We may be uneducated, but we are not stupid. We have lived and survived here for a long time" (Mortenson and Relin 2006, 150).

Mortenson writes,

> That day, Haji Ali taught me the most important lesson I've ever learned in my life. We Americans think you have to accomplish everything quickly. We're the country of thirty-minute power lunches and two-minute football drills. Our leaders thought their "shock and awe" campaign could end the war in Iraq before it even started. Haji Ali taught me to share three cups of tea, to slow down and make building relationship as important as building projects. He taught me that I have more to learn from the people I work with than I could ever hope to teach them.
>
> *Mortenson and Relin (2006, 150)*

A great deal of the energy that motivates this book and its collection of essays is the desire to analyze the challenges that face our planet and to ask questions related to the potential for new technologies to address those challenges. We are trying to anticipate possible futures. What potential for good does biotechnology hold? Who or what is good? Who will gain from the proliferation of new biotechnology products and techniques? What are the potential risks? Who is at risk? Whose potential loss?

These questions have to be framed from various points of view. The potential for gain or loss is different for a multinational corporation than for a community experiencing famine. The loss and gain perspective is different for a scientist hoping to develop a new strain of seed, a cautious environmentalist, and an economist analyzing the market impacts on both the producer and consumer.

This is the work that I would urge us to do with humility and love. This is the conversation that needs to avoid love's opposites: hubris and fear. Beware of attachment. When we become attached to a particular way of seeing things or a particular outcome, we tend to become deaf and blind. Being patient and intentional about examining the impact of change from the perspective of every stakeholder is hard work. It is easier to retreat into technical solutions. The poet T. S. Eliot (1934) asks, "Where is the knowledge that is lost in information? Where is the wisdom that is lost in knowledge?"

Just because we can do something does not mean we should do it. Embrace the truth, the whole truth, and nothing but the truth. Let the data and perspectives be real, realistic, and grounded. Buddhism offers a tradition of pragmatic, egalitarian compassion. Buddhists invite us to judge our actions according to the Noble Eightfold Path: right view, right intention, right speech, right action, right livelihood, right effort, right mindfulness, right concentration.

As we do our biotechnology research and development, where there are competing truths, we must hold onto the tension between them. Theologians and most scientists know that some of the deepest truths are discovered in paradox. Light displays the properties of a wave and of a particle. Matter is almost entirely nonmaterial space. God is imminent and transcendent. So much wisdom is discovered in the dialectic of thesis, antithesis, synthesis. Do not give in to an available answer too easily. A heretic is someone who grasps part of the truth and beats the rest of the truth to death with it.

When there are competing interests at stake, remember Paul's advice to give greater respect to the weaker parts. Liberation theology states that the Scripture consistently reveals God's preferential treatment of the poor. What would it mean to place the interests of the poor as the highest interest in this endeavor? How might our conversations change if the motivations of profit and pride took a back seat to the needs of the poor and the health of the planet? Give voice to the voiceless. Help the hungry to feed themselves.

Ultimately the most important part of the biotechnological revolution will not be about technique and data. It will be about relationships. How can we use our discoveries responsibly to honor the land, water, and living beings who share our planet? The religious communities invite our fellow explorers to embrace the mystery and ambiguity of the journey. Life is complex – always evolving, changing, growing. We can do our best and then humbly trust, being open to surprise. Sometimes intuition opens doors that hours of research cannot unlock. There is playfulness at the core of creation. Let the fun of digging around create an atmosphere of openness and nonpossessiveness that honors science as its own end, something good in and of itself. The border-crossing friendly competition of research is like a game of skill, concentration, and openness. We join the long history of life's evolution as we respond to challenge with hope-filled adaptation seasoned with human wisdom. May the wonder of the lure of mystery be at the center of our work. "Faith, hope, and love abide, these three; and the greatest of these is love" (1 Cor 13:13).

> The choice is always ours. Then let me choose
> The longest art, the hard Promethean way
> Cherishingly to tend and feed and fan
> That inward fire, whose small precarious flame,
> Kindled or quenched, creates
> The noble or the ignoble folk we are,
> The worlds we live in and the very fates,
> Our bright or muddy star.
>
> *Aldous Huxley (1971, 152, "Orion," stanza 10)*

References

Birk, S., and M. Sanders. 2005. *Dante's Paradiso*. San Francisco: Chronicle Books.
Covey, S. M. R., with R. R. Merrill. 2006. *The Speed of Trust: The One Thing That Changes Everything*. New York: Free Press.
CPB (Cartagena Protocol on Biosafety). 2000. *Cartagena Protocol on Biosafety to the Convention on Biological Diversity: Texts and Annexes*. Montreal: Secretariat of the Convention on Biological Diversity.
Dwivedi, O. P., and D. K. Vajpeyi. 1995. *Environmental Policies in the Third World: A Comparative Analysis*. Westport, CT: Greenwood Press.
Eliot, T. S. 1934. *Choruses from "The Rock."* New York: Harcourt Press.
Evangelical Lutheran Church in America. 2001. *Genetics! Where Do We Stand as Christians?* Chicago: Augsburg Fortress Publishers.
FAO (Food and Agriculture Organization). 1994. *New Directions for Agriculture, Forestry and Fisheries, Strategies for Sustainable Agriculture and Rural Development*. Rome: FAO.
Greene, G. 1990. *The Power and the Glory*. New York: Penguin Books.
Henig, R. M. 2000. *The Monk in the Garden: The Lost and Found Genius of Gregor Mendel, the Father of Genetics*. Boston: Houghton Mifflin.
Huxley, A. 1971. *The Collected Poetry of Aldous Huxley*. Edited by D. Watt. New York: Harper & Row.
Mortenson, G., and D. O. Relin. 2006. *Three Cups of Tea: One Man's Mission to Promote Peace . . . One School at a Time*. New York: Penguin Books.
Myhr, A. I., and T. Traavik, 2002. "The Precautionary Principle: Scientific Uncertainty and Omitted Research in the Context of GMO Use and Release." *Journal of Agricultural and Environmental Ethics* 15: 73–86.
Nouwen, H. J. M. 2006. *Here and Now: Living in the Spirit*. New York: Crossroad.
O'Brien, S. 1989. *American Indian Tribal Governments*. Norman: University of Oklahoma Press.
O'Day, G. R., and D. Peterson, eds. 1999. *The Access Bible, New Revised Standard Version*. Boston: Oxford University Press.
Smith, H. 1991. *The World's Religions: Our Great Wisdom Traditions*. New York: HarperCollins.
Thomas, L. 1992. *The Fragile Species*. New York: Touchstone Books.

4

Biotechnology in Crop Production

Eric S. Sachs

> It took some 10,000 years to expand food production to the current level of about 5 billion tons per year. . . . By 2025, we will have to nearly double current production again. This cannot be done unless farmers across the world have access to current high-yielding crop-production methods as well as new biotechnological breakthroughs that can increase the yields, dependability, and nutritional quality of our basic food crops.
>
> *Borlaug (2000, 21)*

Crop improvement methods developed during the twentieth century have been essential for improving food quality and abundance. However, increasing productivity through intensification of agricultural systems has contributed to a degradation of natural resources and loss of biodiversity across agricultural landscapes (Evans 1998). Twenty-first-century agriculture has the potential to facilitate more sustainable development as modern technologies are applied to help meet the food security needs of a growing population and as farmers implement improved agronomic practices with decreasing impacts on the environment. Breakthroughs in agricultural science, modern breeding, and agricultural biotechnology promise to provide farmers with an array of new solutions for reducing crop yield losses caused by weed competition, pest damage, disease, and abiotic stressors, such as drought, heat, and salinity (the role of biotechnology in controlling viral diseases in crops is explored in the next chapter). These breakthroughs also have increased yield gains by improving the efficiency of inputs, such as energy, fertilizer, and water. Plant scientists are leveraging traditional and modern approaches in tandem to increase crop yields, quality, and economic returns, while reducing the environmental consequences associated with the consumption of natural resources for producing agricultural commodities.

The need to accelerate agricultural productivity on a global scale has never been greater or more urgent. Importantly, the tools for implementing more sustainable approaches for conserving natural resources and preserving native habitats have never been more attainable. The challenge for the agricultural sector is to develop and adopt more rational approaches for introducing new agricultural and food technologies that will lead to more widespread use and broad societal acceptance. This chapter discusses the key drivers for increasing agricultural productivity, the importance of addressing environmental challenges, and the role of advancing technologies for meeting the future food, feed, fuel, and fiber needs of a global society.

The Demand for More Food, Feed, Fuel, and Fiber

In Chapter 1 it was noted that, as the global population increases, so does the demand for more food, feed, fuel, and fiber. Historically, agriculture has kept pace with global demands for agricultural commodities, despite localized shortages in areas that have lagged in adopting available agricultural tools and agronomic practices, suffered from severe losses caused by drought or other climate stressors, experienced wars or civil conflicts, or lacked roads and other infrastructure needed for a functional agricultural production system. However, the growth in meat consumption in emerging economies and the use of food crops for biofuel production in developed countries have led to a reduction in grain carryover stocks and growing concerns about the adequacy of global grain supplies. At the same time, the rate of yield gain in the core commodity crops has been limited by a variety of factors, including inadequate funding of public breeding programs, insufficient knowledge of crop genomes, slow adoption of marker-assisted breeding methods in some crops, and the economic barriers to meeting the regulatory requirements of GM crops developed using the tools of agricultural biotechnology.

The global demand for corn and wheat is accelerating because socioeconomic development in developing countries has resulted in more disposable income and an increase in meat consumption. Currently global consumption levels in developing countries remain far below the meat consumption rate of developed countries. FAO compared the meat consumption rates of developed countries that are part of OECD with non-OECD developing countries (Figure 4.1; OECD–FAO 2008). By 2017, the OECD–FAO estimates that meat consumption will grow by 55 Tg to 310 Tg per year, (1 Tg = 1 million tonnes) with 88 percent of the increase in consumption occurring in non-OECD countries. The production of more meat will significantly increase the demand for animal feed grains. (This topic is further explored in Chapter 6.) The Food and Agricultural Policy Research Institute (FAPRI) estimated that feed grain production of corn and wheat will need to increase by about 50 Tg by 2018 (Table 4.1; FAPRI 2008).

The global demand for production of biofuels is also increasing as countries implement policies to develop alternative energy sources to petroleum-based fuel. FAPRI (2008) estimated that global grain-based biofuel production will increase by about

Table 4.1. *Global corn and wheat production and consumption estimates from FAPRI's 2008 world outlook (Tg per year)*

Crop	Crop year	Production	Feed	Fuel	Food/other
Corn	07/08	767	492	84	191
	17/18	896	528	143	225
Increase		129	36	58	35
Wheat	07/08	603	98	1	503
	17/18	688	113	3	572
Increase		85	14	2	68
Combined	07/08	1370	590	85	694
	17/18	1584	641	146	797
Increase		214	50	60	103

Source: FAPRI (2008).

72 percent, from 0.039 to 0.067 km^3/year by 2018 (Figure 4.2). Producing the additional 0.028 km^3 of biofuels needed per year in 2018 will require that an additional 60 Tg of grains be used for fuel rather than feed and food each year (Table 4.1). In total, over the next decade, the amount of corn and wheat needed to meet estimated food, feed, and fuel demand will exceed 1.5 Pg per year (1 Pg = 1 billion tonnes). Agricultural production will need to increase by about 15 percent, or about 200 Tg per year, to a total of approximately 1.5 Pg per year.

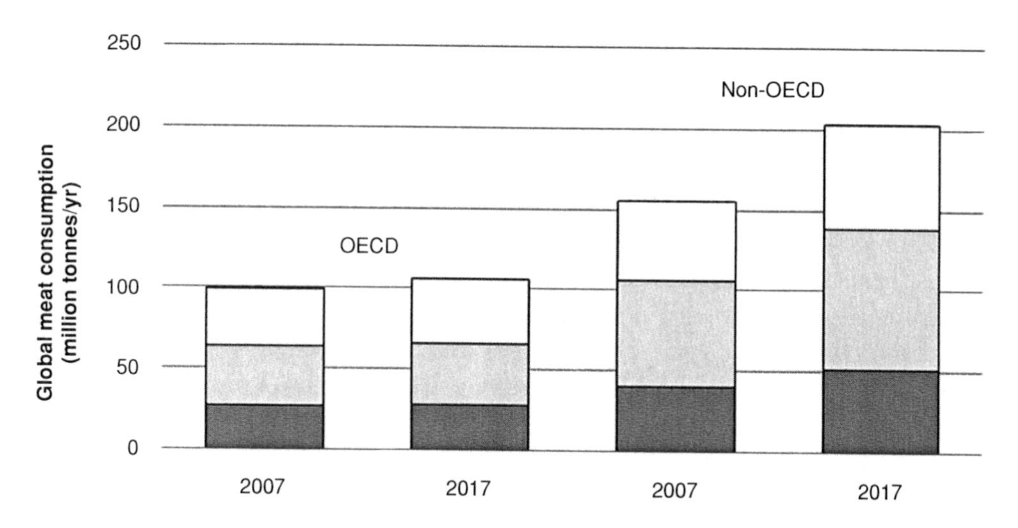

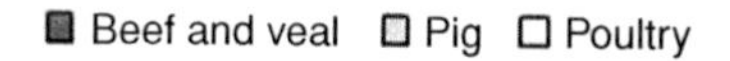

Figure 4.1. Estimates of Global Meat Consumption. Meat consumption outside of the OECD is expected to increase by 48 million tonnes per year in the next decade. *Source:* Edgerton (2009).

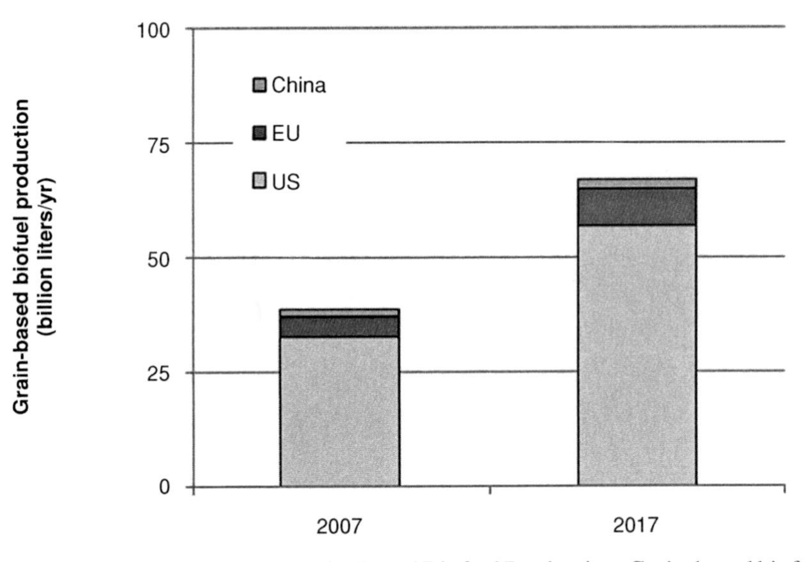

Figure 4.2. Estimates of Global Grain-Based Biofuel Production. Grain-based biofuel production is expected to increase by 28 billion liters per year in the next decade. *Source:* Edgerton (2009).

Population Growth, Food Prices, Hunger, and Environmental Degradation

> Despite the successes of the Green Revolution, the battle to ensure food security for hundreds of millions of miserably poor people is far from won. Mushrooming populations, changing demographics and inadequate poverty intervention programs have eaten up many of the food production gains. This is not to say that the Green Revolution is over. Improvements in crop management productivity can be made all along the line – in tillage, water use, fertilization, weed and pest control, and harvesting.
>
> *Borlaug (2001)*

Demographers predict that the global population will grow from about 6.6 billion in 2008 to more than 9 billion in 2050 (USCB 2009). At the same time, with continued economic growth in Asia, the world will need twice as much food in 2050 as is produced today. Indeed, a UN forum on food demand predicted that agriculture will be called on to produce more food over the next 50 years than has been produced over the course of the past 10,000 years combined (Tacio 2009).

Food prices and availability are of major concern. A variety of factors have led to increased food-price inflation and food shortages in some world areas during the first decade of the 21st century (Thompson 2008). These factors include a loss of production due to drought in some world areas, such as Australia and Africa; increasing consumption of meat protein in Asia in response to rising incomes in urban centers; rising commodity prices in response to increased demand and limited supply; spiking energy prices; and, to a lesser extent, increasing use of corn and soybeans for biofuel production. The top ten warmest years on record have all occurred since 1998, with 2010 being tied with 2005 as warmest. In the 2000–10 decade, only 2008 was not in the top ten (NOAA National Climactic Data Center 2010), advancing fears of global warming and impacts on food production, particularly in the Southern Hemisphere. Even as food security concerns become more acute, there remain more than 850 million malnourished and hungry people, and nearly 2 billion people suffer from micronutrient malnutrition caused by a lack of diversity in their diets (FAO 2002).

Raven (2008) describes several factors that are contributing to environmental degradation, reduced biodiversity across agricultural landscapes, and a reduced potential for agricultural productivity. Even as the demand for food continues to increase, the production of food is degrading agricultural land, making less sustainable and productive the approximately one-third of the world's land surface that is used for agriculture and grazing. The land available for food production is shrinking to 11 percent of the world's land surface as a result of salinization, desertification, erosion, and urban expansion.

Environment advocates are critical of modern intensive agricultural methods because they result in environmental degradation and reduce biodiversity across agricultural landscapes. However, some of the environmental consequences are the result of voluntary actions, as farmer practices designed to increase agricultural productivity also may reduce agro-ecosystem biodiversity. In contrast, more diverse multiple crop production systems seek to increase biodiversity and soil health, but produce lower yields than intensely managed, mono-crop production systems. There remains a wide gap in perspective between advocates for more traditional methods of agriculture that maintain biodiversity on farms and producers favoring modern agricultural methods intended to maximize yield and productivity.

The rapidly increasing global population demands that agriculture become more productive. The elimination of hunger and food insecurity is a widely accepted mandate that requires collaboration among all stakeholders and the development and implementation of creative solutions. The amount of arable land is limited and shrinking, and low-yield, traditional methods of agriculture cannot produce enough food to feed the global population (Raven 2008). Global agriculture uses 70 percent of the world's freshwater (Rosegrant et al. 2002) and is responsible for production of 13.5 percent of atmospheric greenhouse gases (IPCC 2007). The majority of the world's nations are involved in international treaties aimed at protecting biodiversity (Secretariat of the Convention on Biological Diversity 2000) and reducing greenhouse gas emissions (United Nations Framework Convention on Climate Change 2011). In the 21st century, agriculture must develop strategies to reduce environmental degradation and conserve available water resources, or else the amount of arable land will continue to shrink and further constrain productivity. The focus must be on making the existing agricultural lands more productive, so that native lands remain intact to support biodiversity. Degraded lands also must be rehabilitated to preserve biodiversity and to support agricultural productivity and ecosystem sustainability.

The Role of Technology and Innovation

> We must be more rational about our approach to risks. We need to think in broader terms, recognizing, for example, that the world cannot feed all its 6.3 billion people from organic farms or power all its cities and industries by wind and solar energy.
>
> *Borlaug (2004, xii)*

Innovation in agriculture is essential for increasing productivity and preserving ecosystem sustainability. Agriculture must adapt quickly in the decades ahead to meet the food and energy needs of a growing population, while preserving and rehabilitating the environment. The agricultural sector must reduce environmental impacts by getting more from each unit of land, water, and energy committed to crop production (Keystone Center 2009). Increased production and conservation will lower greenhouse gas emissions and reduce the amount of irrigation water per unit of crop yield produced.

Sustainable Development in Agriculture

Sustainable development is a primary objective for policy makers at international, national, and local levels. The UN (1987) World Commission on Environment and Development (the Bruntland Commission) acknowledged the role of development in society and defined sustainable development as meeting "the needs of the present without compromising the ability of future generations to meet their own needs." Inherent in this definition is the concept that sustainable development must meet the food security needs of the developing nations while preserving the environment. Because supplying food for a growing world population and the environment are dynamic systems, sustainability is a course, not a destination.

Sustainable development in agriculture must focus on three arenas simultaneously and equally; any imbalance would impede success. The key areas of focus are (1) increasing productivity to meet future food, feed, fuel, and fiber needs, while decreasing impacts on the environment; (2) improving human health through access to safe, nutritious food; and (3) improving the social and economic well-being of agricultural communities and people. Sustainable development is an ambitious objective and requires future policy considerations to fully support all three aspects: environment, yield and productivity, and community economic viability.

To support these comprehensive sustainable development objectives, 21st-century agriculture must focus on three key sustainability imperatives: (1) reduce environmental impacts by getting more from each unit of land, water, and energy devoted to crop production; (2) deliver twice as much food in 2050 as is produced today; and (3) deliver economic benefits for all farmers, small and large. However, the pursuit of sustainable development continues to face significant social, economic, and environmental challenges. As previously discussed, ongoing population growth, food insecurity, poverty, exploitation of natural resources, and maintenance of biodiversity all require urgent attention (Scherr and McNeely 2007). Furthermore, there are increasing concerns about the rate and severity of climate change and the development of critical infrastructure, such as roads, storage, and agricultural channels, which is absent or nascent in critical areas of the developing world.

The Principal Tools

Over the past decade, we have witnessed the success of plant biotechnology in helping farmers throughout the world produce higher yields, while reducing pesticide use and soil erosion. The benefits and safety of biotechnology have been proven over the past decade in countries in which more than half of the world's population lives. What we need is courage by the leaders of those countries where farmers still have no choice but to use older and less effective methods. The Green Revolution and now plant biotechnology are helping meet the growing demand for food production, while preserving our environment for future generations (Borlaug 2001).

To deliver on sustainability imperatives it is essential that 21st-century agriculture use all available agronomic strategies, including approaches that are being used successfully today and evolving approaches that will contribute in the future. Importantly, policy makers should become knowledgeable about the range of agricultural technologies becoming available and seek to implement promising approaches as rapidly as possible. Yet it would be naïve to underestimate the barriers to change that exist within society. In an earlier paper I discuss (Sachs 2007) how knowledge and understanding must extend beyond the scientific realm to society at large, else society may ignore or reject important technological advancements as a result of ignorance or poor quality information.

To meet the food, feed, fuel, and fiber needs of the 21st century, farmers will need an array of tools with the potential to increase crop yields on their farms through advanced agronomic practices, modern breeding and germplasm development, and genetic engineering of crops. Given the magnitude of the social, economic, and environmental challenges, all three development areas must be priorities for the public and private sector, and innovations in all three areas will be needed to support the sustainable development objectives.

The rate of yield gain has been variable among the different commodity crops grown in the United States and reflects different research priorities and levels of funding by public and private breeding programs. In addition, genome knowledge and application of marker-assisted breeding have lagged in some crops, whereas regulatory costs and uncertainty about consumer acceptance in export markets have preempted the use of transgenic improvements in most crops.

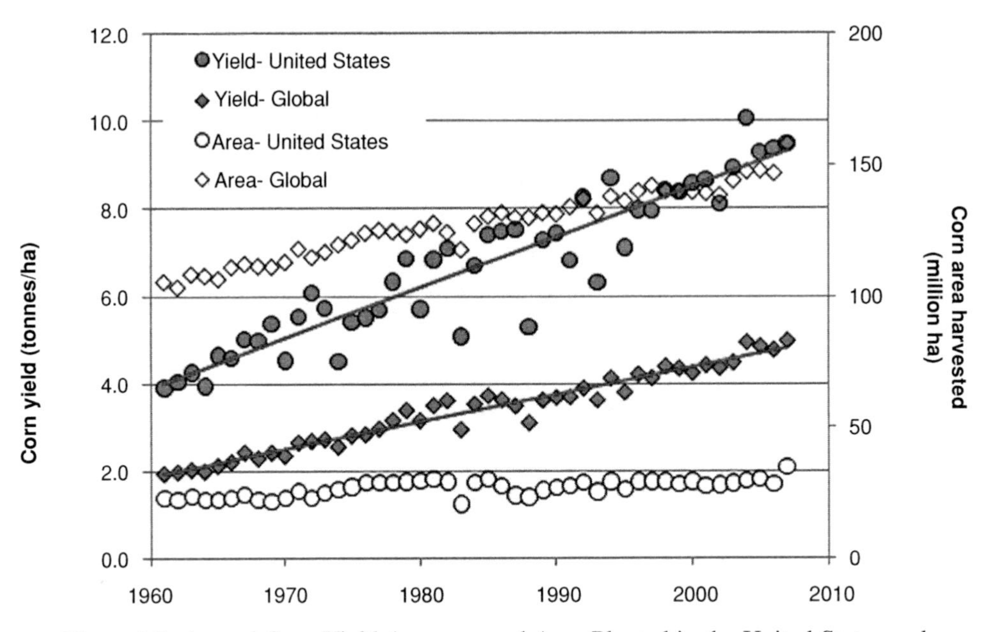

Figure 4.3. Annual Corn Yield Averages and Area Planted in the United States and World. Lines indicate yield trend line. *Source:* Edgerton (2009).

Advanced Agronomic Practices

The development and adoption of new farming technologies, such as hybridization, synthetic fertilizers, and farm machinery, all have contributed to dramatic increases in corn yield in the United States and to a lesser extent globally (Edgerton 2009). Before the 1930s, average annual corn yields in the United States were about 1.5 Mg per ha, rising to about 4.0 Mg per ha with the introduction of single-cross hybrids in the 1950s, and then doubling to roughly 8.0 Mg per ha in the 1990s, as improvements in agronomic practices were implemented over several decades. Several agronomic practices were responsible for about 50 percent of the yield gain (Duvick 2005), including increased fertilizer application, use of more efficient farm machinery, use of earlier planting dates to lengthen the growing season (Kucharik 2008), and increasing plant populations to improve the efficiency of energy capture and conversion into yield (Tollenaar and Lee 2002).

The average rate of yield gain in the United States since the introduction of hybrid corn has been nearly double the rate of corn yield gain globally (Figure 4.3; OECD-FAO 2008). From 1961 to 2007, U.S. farmers have increased corn yield by 0.11 Mg per ha per year, whereas farmers globally have increased yields by only 0.06 Mg per ha per year. The greater yield gains in the United States almost certainly reflect the adoption of improved agronomic practices, as well as use of modern breeding, germplasm development, and GE crops (discussed in the next two sections).

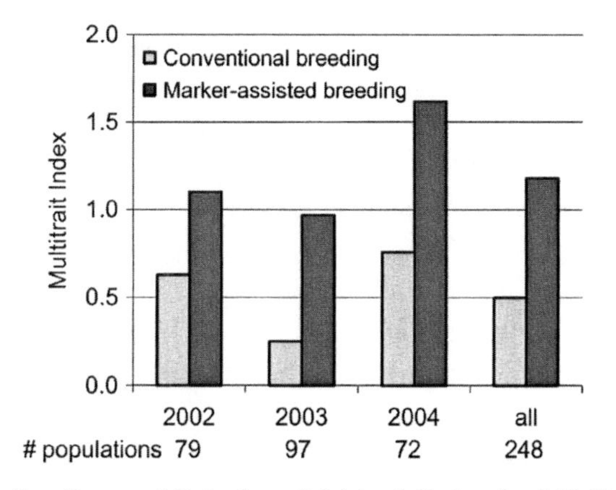

Figure 4.4. Breeding Rates of Gain for a Multitrait Index for 248 Corn Populations Initiated across Three Years. The multitrait index is weighted toward yield, but also incorporates other agronomic traits such as grain moisture and stalk strength. *Source:* Edgerton (2009).

Over the same period, from 1961–2007, the harvested corn area globally has increased at the rate of 9,300 km^2 per year, whereas the harvested corn area in the United States has remained flat at 1,500 km^2 per year. There was an increase of about 50,000 km^2 in 2007 and 2008, but this reflects a spike in the corn commodity price, driven partially by the production of corn-based ethanol mandated by the U.S. renewable fuels standard (EPA 2007). The increase in planted area globally helped meet the demand for more corn production; however, the strategy of increasing productivity on existing agricultural land is preferable because it avoids an increase in greenhouse gas emissions and the large-scale disruption of existing ecosystems associated with bringing new land into production.

Modern Breeding and Germplasm Improvement

Selection breeding, use of quantitative genetics, and advanced statistical procedures continue to be key components of plant breeding programs to discover genetic and environmental factors responsible for improving plant performance (Crosbie et al. 2006). Marker-assisted breeding (also known as marker-assisted selection) is a relatively new technology for improving the pace of trait introgression and selecting for traits to improve productivity. The first molecular markers were identified in corn more than 20 years ago (Helentjaris et al. 1985; Paterson et al. 1988), but the first hybrids developed using molecular markers have only recently been commercialized. Use of this approach is expected to increase the rate of gain beyond that seen in recent decades (Edgerton 2009), as molecular marker technology becomes more widely applied to hybrid improvement. A large study of commercial corn populations showed that the use of markers can increase the rate of gain in yield, as well as decrease grain moisture and stalk lodging (Figure 4.4; Eathington et al. 2007).

Another modern breeding technology for advancing genetic selection uses robotics, computer automation, deoxyribonucleic acid (DNA) analysis, and massive processing capacity to improve the pace of genetic improvement and breeding efficiency. Used in tandem, the seed chipper, automated genetic fingerprinter, and seed sorter shave off a tiny section of a seed, test and analyze that section for desirable genetic traits, and then automatically sort the most desirable seeds into labeled packets for immediate shipping to various field locations for multisite field evaluation. The majority of seeds, which in the past would have been planted in the field and later culled and discarded, now can be screened in the laboratory, enabling a much larger number of improved seeds to be tested each year.

Genetic Engineering

Breakthroughs in molecular genetics, protein chemistry, genomics, and bioinformatics have enabled rapid growth in agricultural biotechnology. Already GE of agronomic crops, including corn, soybean, cotton, canola, alfalfa, and sugar beet, as well as papaya and squash, has produced commercial varieties with advantages for farmers, the environment, and society. The first traits developed were single-gene inserts into the crop genomes, introducing insect and virus resistance, herbicide tolerance, and improved nutrition. For example, the addition of one or more insecticidal protein genes from *Bt* to the crop provides protection against susceptible target insect pests but does not affect beneficial predatory insects and other nontarget species. Another approach is to add one or more herbicide-tolerant (HT) genes, usually from bacteria, to the crop to provide tolerance to specific herbicides that would otherwise kill conventional varieties of the same crop. These traits were accomplished by the expression of a given bacterial gene in the crops. In the case of insect tolerance, expression of an insecticidal protein gene from *Bt* in plants resulted in protection of the plants from damage caused by insect feeding (Perlak et al. 1991). Similarly, expression of a glyphosate-resistant form of the gene cp4 epsps resulted in plants being tolerant to glyphosate (Padgette et al. 1995). Transgenic crops with pest protection and herbicide-tolerant traits enable farmers to more effectively manage insect and weed pests compared to traditional management practices.

By providing protection from pest feeding, *Bt* crops help reduce both direct yield losses caused by pest injury and indirect yield losses associated with environmental stresses and secondary disease infections. Technology providers have developed *Bt* traits based on the expression of different and complementary *Bt* proteins that target stalk-boring and foliage-feeding pests, such as the European corn borer, Southwestern corn borer, Fall armyworm, and corn earworm, as well as root-feeding pests, such as the Western, Northern, and Southern corn rootworm. As a result, *Bt* crops on average have higher yields than comparable conventionally managed crops and realize more of the potential yield of the crop's genetic makeup. Similarly, by providing more

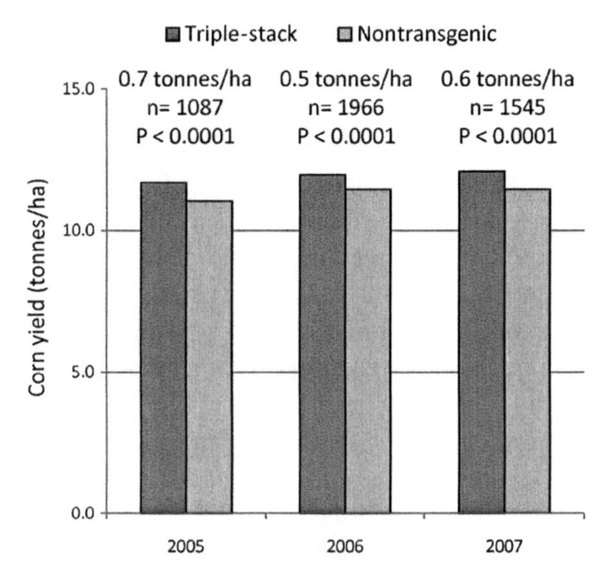

Figure 4.5. Yield Advantage of "Triple-Stack" Corn Hybrids. Average yield values are shown in the bars, and the yield difference between "triple-stack" and conventional corn is indicated in the text above the bars. These are average values from yield trials run across corn-growing regions in the United States. Values can be significantly higher in regions with more insect pressure. *Source:* Edgerton (2009).

herbicide choices and weed management options, HT crops enable farmers to better control weeds that can compete with the crop for sun, water, and nutrients; on average, they produce equivalent or higher yields compared to conventional cropping practices that rely on combinations of various selective herbicides.

So-called triple-stack corn hybrids with *Bt* traits for control of above- and below-ground target pests and an HT trait with tolerance to the glyphosate herbicide family increased yield on average from 0.5 to 0.7 Mg per ha compared to similar conventional corn hybrids treated with standard weed and insect management programs (Figure 4.5). These data provide strong evidence that using GE to improve crops can significantly increase yields compared to conventionally bred crops employing traditional weed and pest management.

In addition to delivering higher average yields, *Bt*/HT corn reduces the risk of significant yield loss from weeds and pests. The reduction in production risk has been studied and acknowledged by the Federal Crop Insurance Corporation's Biotech Yield Endorsement, a risk management instrument that offers an insurance premium rate reduction for farmers using crops with qualified *Bt* and HT traits (USDA FCIC 2010). The discount on crop insurance premiums for U.S. farmers using corn hybrids with a combination of *Bt* and HT traits was $7.41/ha in 2008 (Brookes and Barfoot 2009a).

Pest management strategies that lower the risk that pest populations will develop resistance to control tactics are potentially much more sustainable than higher risk approaches and also may help farmers be more productive and profitable. The use of multiple *Bt* traits with complementary modes of action against key target pests, versus the use of single *Bt* traits, significantly reduces the risk of pest resistance (Roush 1998). Securing approval for the commercial use of *Bt* crops requires proactive resistance management that focuses on preserving susceptibility of the target pest populations. Toward this goal, regulators have authorized planting a *Bt* crop in association with a conventional crop refuge. Refuges provide the foundation on which insect resistance management (IRM) is based for crops containing *Bt* technology by functioning as sources of susceptible target insects in the vicinity of *Bt* crop fields. The susceptible insects produced in refuges mate with rare resistant insects that potentially emerge from *Bt* crops. This ensures that the offspring of resistant individuals remain susceptible to *Bt* crops, thus preserving the long-term effectiveness of *Bt* technology, as well as the benefits it provides to growers. The EPA's approval of crops with multiple *Bt* traits with complementary modes of action has allowed farmers who are planting Bollgard II or Widestrike varieties the option to plant 100 percent of the planted area in *Bt* cotton. In this case, other crops and weeds serve as important natural refuges for susceptible insects. As a result of reducing the size of the refuge, cotton growers realize the yield protection benefit afforded by *Bt* traits on more acres, producing higher yields and profits.

The first generation of corn pest control traits included genes for protection from root-feeding pests (e.g., Cry3Bb1, Cry34/Cry35) and stalk-boring pests (e.g., Cry1Ab, Cry1F); this protection lessens the effects of drought stress (Figure 4.6). With their roots protected, plants absorb more water and nutrients than unprotected plants. Similarly, with their stalks protected, the plants are able to distribute more water and nutrients throughout.

Researchers in the public and private sectors already are developing the next generation of transgenic traits that promise to help increase agricultural yield and productivity in the 21st century. The use of large-scale, high-throughput screening methods (Creelman et al. 2008; Riechmann et al. 2000; Van Camp 2005) has produced a large number of gene candidates that are now being evaluated in laboratories, greenhouses, and field research sites. These new genes include an array of new approaches for reducing crop yield losses and increasing yield gains (Edgerton 2009). These products most likely will involve regulation of key endogenous plant pathways that can help mitigate the effects of abiotic stressors, such as drought, heat, and salinity. In addition, genes have been identified with the potential to further increase yield gains by improving the efficient use of inputs, such as energy, fertilizer, and water.

Researchers are focused in the near term on developing crops with improved drought tolerance. GM crops that include genes for drought tolerance promise to (1) increase yields in regions experiencing frequent drought; (2) increase average yields in areas that experience occasional drought, particularly during flowering and grain fill (Campos et al. 2006); and (3) lessen the amount of water required in irrigated systems. One candidate gene, a ribonucleic acid (RNA) chaperone, functions through

Figure 4.6. Transgenic Stalk and Root Protection from Pests Lessens Effects of Drought Stress Compared to Soil Insecticides. Corn with protection from corn rootworms and corn borers is able to access more moisture and support better plant development compared to conventional corn protected with corn rootworm soil insecticides. *Source:* The photo was taken by Monsanto Technology Development in Cambridge, Illinois, in 2006, during a period of low rainfall and low soil moisture.

an association with nucleic acid cellular processes in rapidly growing tissues to mitigate water stress and stabilize yield under water-limiting conditions. The presence of this gene in corn plants reduces the effects of periodic water stress and has achieved the goal of 8 to 10 percent average yield advantage over a range of yield performance under water-stress conditions (Figure 4.7; Castiglioni et al. 2008). A second candidate gene, nfb2, is a transcription factor that regulates the activity of several genes present in corn plants involved in responding to drought stress. The nfb2 transcription factor activates key genes that help the corn plant tolerate drought conditions, leading to improved seed set and yield (Nelson et al. 2007). The nfb2 and RNA chaperone gene leads are two examples of drought-tolerant traits that are intended to be combined with *Bt* and HT traits, as well as other new traits in corn, to protect yield potential or push potential yield to higher and higher levels. Drought tolerance also is under development in rice (Hu et al. 2006), as researchers strive to improve rice productivity in areas of China subject to drought.

Researchers also are working to develop crops that improve resource use efficiency. For example, a variety of canola has been developed that uses nitrogen more efficiently, and it is anticipated that nitrogen use efficiency in other major crops could be increased by 20 to 30 percent (Strange et al. 2008). Such products would be beneficial because

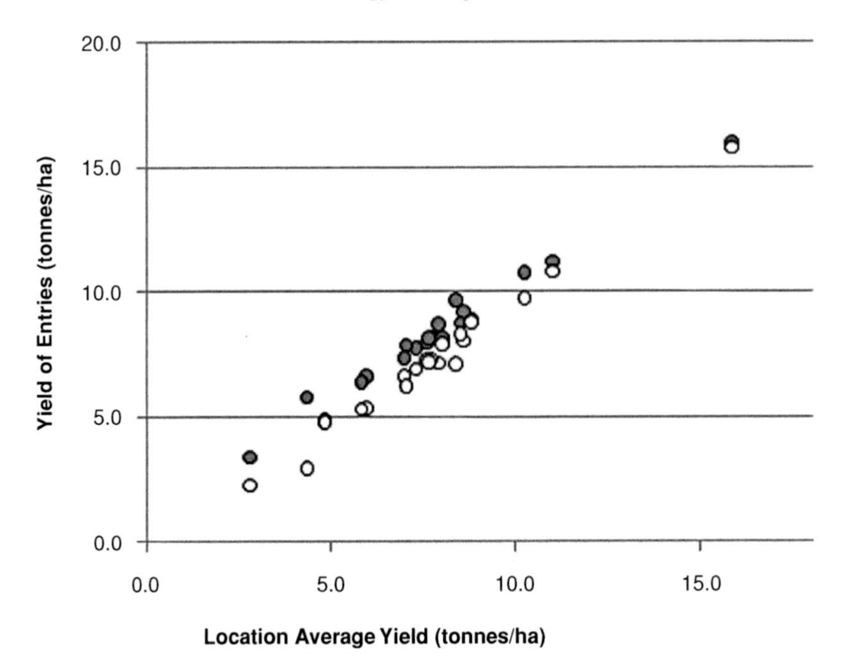

Figure 4.7. Yield Increase in Corn Plants That Express cspB, an RNA Chaperone from *B. subtilis*. Hybrids from a single transgenic event were tested in yield trials over three years at managed stress locations. Yield of the transgenic hybrid (green circles) and nontransgenic isogenic hybrids (open circles) at individual locations are plotted against the yield of all entries tested at that location. *Source:* Castiglioni et al. (2008).

nitrogen is costly for farmers and requires large amounts of energy (usually derived from fossil fuels) in its manufacture and transportation, which in turn release large amounts of carbon dioxide into the atmosphere. In addition, overuse and off-site movement of nitrogen fertilizers can lead to pollution and degradation of waterways.

There are additional promising biotech traits expected to help alleviate biotic and abiotic stressors and improve the yield of other crops. Biotech soybeans with improved yield (Lundry et al. 2008) or oil concentration (Lardizabal et al. 2008) are examples of new approaches for improving global supplies of vegetable oil and protein meal. Additional biotech traits with the potential to further improve soybean yield and nitrogen use efficiency are expected after 2015 (Padgette 2008). Nitrogen use efficiency (the amount of crop produced per unit of input) has steadily increased in the United States since the 1980s (Frink et al. 1999). Continued improvements in nitrogen use efficiency are expected as new genes are identified and evaluated, but there is a limit to how much nitrogen requirements can be reduced. All crops require nitrogen for protein synthesis, and adequate nitrogen must be available to support growth and development. A corn crop producing 10 Mg per ha contains about 100 kg nitrogen per ha as protein; consequently at least this amount of nitrogen must be added back to the field to maintain fertility (Edgerton 2009). Collectively, the next generation of transgenic traits should contribute significantly to productivity on

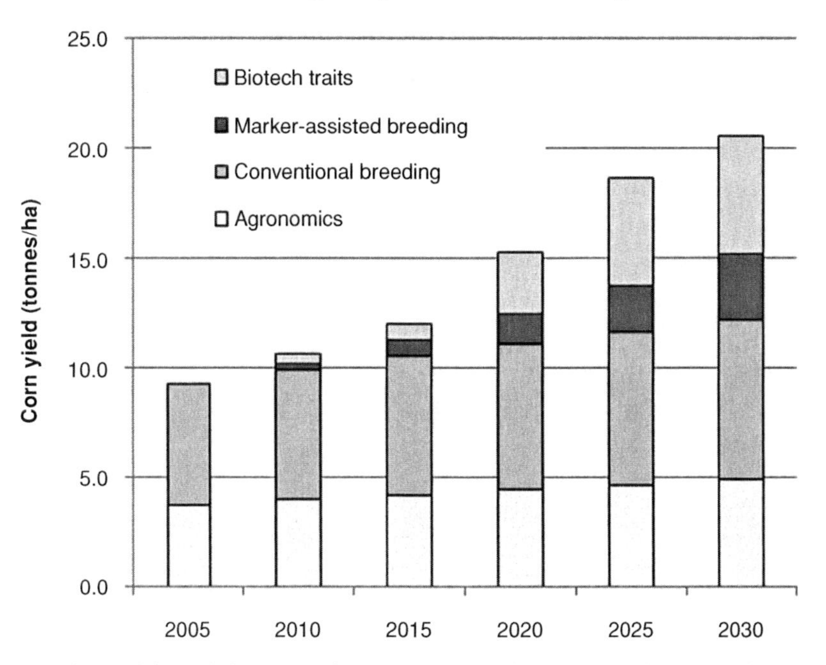

Figure 4.8. Anticipated Impact of Improvements in Agronomics, Breeding, and Biotechnology on Average Corn Yields in the United States. *Source:* Edgerton (2009).

existing cropland, thereby increasing grain supplies while conserving more diverse, native lands.

Continued advances in agronomic practices, marker-assisted breeding, and transgenic traits have the potential to double corn yields in the United States by 2030 (Figure 4.8). To achieve this goal, farmers will need access to significant improvements in stress tolerance, water-use efficiency, and broad dissemination of improved agronomic practices. To meet the global demand for feed, fuel, and food, while concurrently minimizing the need to bring large amounts of new land into crop production, farmers globally also will need to adopt new technologies in corn and other crops. Regardless of whether technology providers and farmers achieve a doubling of yields, public and private sector investment and progress in agricultural technology will aid in increasing productivity and sustainably supplying more food, meat, and energy to better meet global demand.

Biotechnology Adoption and Trait Stacking: The Global Adoption and Expansion of GM Crops

The first GM crop plants were produced in 1983, field-tested in 1987, and commercialized in 1996. Since that time, these crops have been rapidly adopted for cultivation in many countries around the world. According to the International Service for the Acquisition of Agribiotech Applications report, planting of GM crops in 2010 reached 148 million ha in 29 countries, representing a 10 percent increase in acreage from 2009 to 2010 (James 2010). In 2010, 19 developing countries and 10 industrial countries planted GM crops. Since then an additional 30 countries have approved the import of grain, food, and feed from these crops for domestic food and feed use. In the 15 years since commercialization of the first GM plants, farmers have planted more than 1 billion ha globally.

In 2010, 15.4 million farmers purchased and planted GM crops; the vast majority (14.4 million or 90 percent) of them were small, resource-poor farmers from developing countries, including 6.5 million in China, 6.3 million in India, and the balance of 1.6 million in Pakistan, Myanmar, Philippines, and the other 14 developing countries that have authorized GM crops for planting (James 2010). Globally, 11 countries have authorized stacked-trait GM crops: the use of both *Bt* and HT traits on their farms. Insect-resistance was the fastest growing trait group from 2009–10. The rapid adoption of GM crops by farmers in developing and industrial countries is strong evidence of the economic, environmental, and health benefits of growing GM crops.

Impacts of GM Crops

The cultivation of GM crops has created significant benefits for growers and society: increasing crop yield and farmer income, reducing chemical pesticide use and impacts of agricultural production on the environment, and producing food and feed with reduced levels of natural toxicants in specific products, making the resulting food and feed safer for human and animal consumption.

Transgenic cropping systems provide direct and indirect benefits compared to conventional cropping systems. Direct benefits to farmers include reduced pesticide, scouting, and application costs; lower pesticide residues in the environment; increased safety of nontarget species; reduced crop damage and losses to pests and weeds; higher yields; and increased net income. Indirect benefits to farmers and society include increased adoption of reduced/conservation tillage and soil conservation practices, reduced greenhouse gas emissions from agricultural practices, less farm worker exposure to pesticides, reduced levels of mycotoxins in corn used for feed and food, and shortening the time to harvest so that farmers in some parts of the world can plant a second crop in the same growing season (e.g., HT soybeans following wheat in Argentina or corn following *Bt* cotton in India). For many farmers, the convenience afforded by the ability to plant improved seeds that require less investment in time and money for the management of pests and weeds is a principal driver of biotech crop adoption; for many other farmers, the key driver is the added assurance of higher yields and profits.

Yield and Productivity

Since their introduction, GM crops have helped farmers harvest more yield than conventional crops employing traditional chemistry-based programs for managing pests and weeds (Brookes and Barfoot 2009a, 2009b, 2009d). Generally, *Bt* crops have provided higher yields by providing improved protection from pests. The yield increases have been higher in developing countries where various constraints on production, such as lower soil fertility and inadequate rainfall, typically increase losses due to both biotic and abiotic stressors. Industrial countries also benefit from positive impacts on yields, but the extent of this benefit is more dependent on the presence and level of the pest infestations, which can vary a great deal from field to field and year to year. In contrast, the yield impacts of HT crops have generally been neutral to positive, depending on the availability, investment in HT crops, number of herbicide applications, and presence of hard-to-control perennial weed species. Where traditional weed management practices have performed poorly, the use of HT crops has produced significant increases in yields, as a result of better weed control. For example, HT soybeans in Romania improved the average yield by 30 percent, HT corn in Argentina increased yield by 9 percent, and HT corn in the Philippines improved yield by 15 percent (Brookes and Barfoot 2009b).

Since 1996, the total production of soybeans, corn, cotton, and canola has increased by 141.5 Tg over what would have been produced using conventional crops (Brookes and Barfoot 2009a). Between 1996 and 2007, GM crops provided additional production equal to an extra 68 Tg of soybeans and 62 Tg of corn. In total, 51.1 percent ($22.1 million) of the cumulative farm income gain was derived by farmers in less developed countries (Brookes and Barfoot 2009b). The average yield increases reported for *Bt* cotton and *Bt* corn vary in each country depending on the magnitude of the pest problem, performance of alternative conventional pest management practices, local growing conditions, and germplasm (Brookes and Barfoot 2009a). Examples include a 54.8 percent yield increase for *Bt* cotton in India since its introduction in 2002, a 28.6 percent yield increase for *Bt* cotton in Argentina since its introduction in 1998, a 24.1 percent for *Bt* corn in the Philippines since its introduction in 2003, and 24.3 percent for *Bt* cotton and 15.3 percent for *Bt* corn in South Africa since their introduction in 1998 and 2000, respectively.

Environment

GM crops provide a number of significant positive environmental impacts. Generally, the benefits can be attributed to spraying less insecticide for pest management and using glyphosate- or gluphosinate-based herbicides for weed control, which provide greater flexibility and safety than many alternative herbicide products.

The environmental safety of *Bt* proteins has been well documented based on their high degree of selectivity for insect pests and no significant deleterious effects on other organisms, including mammals, fish, birds, or invertebrates (Betz et al. 2000), safety to nontarget insects (Head 2007; Saxena and Stotsky 2001), and rapid degradation in the environment (Head et al. 2002; Palm et al. 1996). Other benefits include reduced insecticide exposure to farm workers and the environment, improved habitat for agricultural wildlife, and lower environmental impact quotients.[1] Dozens of laboratory and field studies have confirmed via a weight of evidence that nontarget organisms are more abundant in *Bt* crops than conventional crops sprayed with insecticides (Marvier et al. 2007).

In India (Bennett et al. 2004) and in China (Huang et al. 2003), farmers planting *Bt* cotton have reduced pesticide applications by two-thirds, and in Australia, farmers have reduced pesticide application by half (Fitt 2003). Reducing pesticide use in agriculture conserves water and fuel because less energy and water are used in the manufacture, distribution, and application of these products. Fewer pesticide applications also result in less packaging entering the environment. In 2008, *Bt* corn with the corn rootworm trait in the United States eliminated the use of about 3.8 million plastic containers, 984 m^3 of aviation fuel, 72 km^3 of water used in insecticide formulations, 8.6 Tg of insecticides' active ingredient, and 72 km^3 of diesel fuel per year (Rice 2004).[2]

Similarly, there are important environmental benefits derived from the use of HT crops. The introduction of the Roundup Ready (RR) system – RR crops treated with the glyphosate family of herbicides – in soybean, corn, cotton, canola, alfalfa, and most recently sugar beet has provided farmers in the United States and in other countries with improved management options for controlling weeds. The RR system provides improved weed control and reduces the need for multiple herbicides to manage weeds. Better efficacy and fewer herbicide applications allow farmers to make fewer trips over their fields to apply herbicides, which reduces soil compaction and fuel use. The RR system also has facilitated the adoption of soil-saving reduced-tillage and no-tillage farming systems. As of 2004, RR soybean farmers had increased no-till acreage by 64 percent, and farmers growing RR corn and RR cotton increased no-till acres by 20 percent and 37 percent, respectively. Conservation tillage improves water quality and creates habitat for wildlife (CTIC 2009; Fawcett and Towry 2002).

The use of the RR system, coupled with conservation tillage practices, has significantly reduced the loss of topsoil caused by soil erosion (ASA 2001; Brookes and Barfoot 2009a; Keystone Center 2009), improved soil structure with a higher concentration of organic matter (CTIC 2009; Kay 1995), eliminated runoff of sediment and fertilizer, reduced on-farm fuel use (ASA 2001; Brookes and Barfoot 2007, 2008, 2009b, 2009c), reduced CO_2 emissions (Brookes and Barfoot 2007, 2008, 2009b, 2009c; CTIC 2009; Kern and Johnson 1993), and increased carbon sequestration in soil (Brookes and Barfoot 2007, 2008, 2009b, 2009c; Reicosky 1995; Reicosky and Lindstrom 1995). Reductions in fuel use and plowing/tillage combined to reduce

[1] More information about the benefits of *Bt* crops can be found in an extensive literature (Brookes and Barfoot 2007, 2008, 2009a, 2009b; OECD 2007; Qaim 2009; Romeis et al. 2008).

[2] These calculations were based on 38 million planted acres of *Bt* corn with the corn rootworm trait.

greenhouse gas emissions into the atmosphere by more than 14.2 Tg of carbon dioxide in 2007, equivalent to removing 6.3 million cars from the road for a year (Brookes and Barfoot 2009b, 2009c).

Overall, the cumulative global impact of *Bt* and HT crops between 1996 and 2007 can be measured in the reduction of pesticide usage by 8.8 percent (359 Gg; Gg = 1,000 tonnes) and of environmental degradation associated with insecticide and herbicide use by 17.2 percent (Brookes and Barfoot 2009b, 2009c). HT soybeans have been responsible for the largest environmental gain because they are planted on the largest area globally. In total, HT soybeans allowed farmers to reduce herbicide use by 73 Gg (a 4.6 percent reduction), and the overall environmental impact associated with herbicide use on these crops decreased by 20.9 percent, compared to the volume that would have probably been used if this cropping area had been planted to conventional soybeans. Similarly, since its introduction in 1996, *Bt* cotton has allowed farmers to reduce insecticide use by 147.6 Gg (a 23 percent reduction), which has reduced the associated environmental impact by 27.8 percent. There have been significant environmental benefits in corn and canola as well. In *Bt* corn, herbicide use decreased by 81.8 Gg and insecticide use decreased by 10.2 percent, and the associated environmental impact of herbicide use and insecticide use was reduced by 6.0 percent and 5.9 percent, respectively. For HT canola, farmers used 9.7 Gg (a 13.9 percent reduction) fewer herbicides, and the associated environmental impact of herbicide use fell by 25.8 percent, because of a switch to more environmentally benign herbicides.

Economy

GM crops improve farm income through enhanced productivity and improved efficiency. Between 1996 and 2007, farm incomes have increased by $44.6 billion globally (Brookes and Barfoot 2009a, 2009b, 2009d). The greatest income gains have been associated with HT soybeans ($21.8 billion), followed by *Bt* cotton ($12.7 billion), *Bt* corn ($5.7 billion), HT canola ($1.8 billion), HT corn ($1.5 billion), and HT cotton ($0.85 billion). In 2007, the direct global farm income benefit from growing all GM crops totaled $10.1 billion, including HT soybeans ($3.9 billion), *Bt* cotton ($3.2 billion), *Bt* corn ($2.1 billion), HT canola ($0.35 billion), HT corn ($0.44 billion), and HT cotton ($0.02 billion). Of the top four countries globally benefiting from the use of GM crops, two are developed countries with cumulative income benefits between 1996 and 2007 equal to $19.8 billion (United States) and $8.3 billion (Argentina), and two are developing countries with cumulative income benefits equal to $6.7 billion (China) and $3.2 billion (India).

Since 1996, farmers in developing countries gained a total of $22 billion, or about 50 percent of the total economic benefits provided by GM crops. At the same time, the total cost of the technology was 14 percent of the technology gains for farmers in developing countries, compared to a total cost of 34 percent of the technology gains in developed countries (Brookes and Barfoot 2009a). Naturally, circumstances vary a lot between countries, but these data demonstrate that GM crops provide substantial value to farmers in both developing and developed countries and that a much larger portion of the value offered by the technologies is delivered to the farmers in developing countries. However, important constraints remain on access to GM crops in many countries, including national research and regulatory capacity, effective support for intellectual property rights, and adequate infrastructure for agricultural inputs and seeds (Raney 2006). These challenges must be addressed to fully realize the benefits of GM crops in developing countries.

Brookes and Barfoot (2009b) also estimated the cumulative global value of non-pecuniary benefits associated with growing GM crops by taking into consideration less tangible reasons why farmers choose to plant GM crops, such as convenience, flexibility, and production risk management. The total cumulative nonpecuniary benefit between 1996 and 2007 was $5.1 billion, or 25 percent of the total cumulative direct farm income of GM crops. In another study in the United States, HT crops were associated with higher off-farm income (Fernandez-Cornejo et al. 2007). The researchers found that adoption of conservation tillage practices enabled by RR crops helps farmers economize on management time, which they can reinvest into other income generation.

The economic benefits of growing GM crops also extend into the social arena. *Bt* crops, *Bt* cotton in particular, have been studied to estimate the social benefits realized by farm families earning higher incomes from growing GM crops.[3] In 2007, a large study was conducted in India to better understand the impacts of *Bt* cotton on Indian farmers and communities. The Associated Chambers of Commerce and Industry of India (ASSOCHAM 2007) study included more than 9,000 farmers across 467 villages and 28 districts of the eight cotton growing states. It focused on the social benefits realized by farm families as a result of higher incomes from *Bt* cotton cultivation. The results showed that small farmers in villages growing more *Bt* cotton had an improved quality of life.

Generally, the ASSOCHAM (2007) study found that *Bt* cotton farming households tended to be better off across a range of socioeconomic indicators. An analysis of economic infrastructure and economic activity revealed that villages prominently growing *Bt* cotton were better off than those villages that did not adopt *Bt* cotton. Residents in *Bt* cotton villages had increased access to services, such as telephone systems, electricity, drinking water, banking services; better Internet connectivity; and better access to markets with a corresponding increase in shops and goods. In contrast to non-*Bt* cotton villages, *Bt* cotton villages had more permanent markets

[3] For more information on the social benefits to farm families producing biotech crops, see Qaim (2009); Raney (2006); Huesing and English (2004); Bennett et al. (2003); Huang et al. (2003); and Pray et al. (2002).

(44 percent versus 35 percent), a greater penetration of shops (24 percent versus 18 percent), and more banking (34 percent versus 28 percent). Women from *Bt* cotton households had greater access to maternal care services, and children from *Bt* cotton households were found to have a higher rate of immunization than children from non-*Bt* cotton households. Children belonging to *Bt* cotton households also showed significantly higher school enrollment in five of eight states than children in non-*Bt* cotton households.

Challenges to GM Crops

There continue to be concerns that GM crops and foods may be harmful to humans, animals, or the environment in spite of an abundance of scientific evidence and experience supporting the safety of GM crops and foods.

Potential for Health and Environmental Risks

The concerns generally relate to potential hazards that are monitored by the compulsory risk assessment conducted by regulatory authorities. Scientific and regulatory authorities have acknowledged the potential hazards associated with genetic modification of all kinds, including traditional cross-breeding, biotechnology, chemical mutagenesis, and seed radiation, and have established a safety assessment framework for GM crops designed to identify any food, feed, and environmental safety risks before commercial use. The intent of the safety assessment is to uncover the possibility of harm, to assess the likelihood that the harm will occur, and to draw a conclusion of whether a reasonable certainty of no harm exists. If regulatory authorities conclude that food and feed derived from the biotech crop are as safe as the traditional nonbiotech crop and determine that cultivation of the biotech crop in the agricultural landscape poses no significant risks to the environment, the regulatory authorities have a basis to authorize the GM crop for commercial use. Chapter 8 provides scientific evidence that many uses of biotechnology have been found to pose little risk to humans and the environment.

Risk Assessment, Regulation, and Policy

The current comparative safety assessment process has been developed to provide assurance of safety and nutritional quality by identifying similarities and differences between the new food or feed crop and the same conventional crop with a history of safe use (Chassy et al. 2004, 2008; Kok et al. 2008). Any differences are subjected to an extensive evaluation to determine whether there are any associated health or environmental risks and, if so, whether the identified risks can be mitigated through preventive management. GM crops undergo detailed phenotypic, agronomic, morphological, and compositional analyses to identify potential harmful effects that could affect product safety. This process is applicable to the next generation of GM crops that likely will include genetic changes that modulate the expression of one gene, several genes, or entire pathways. As part of the safety assessment, the nature of the inserted molecules, their function and effect within the plant, and the overall safety of the resulting crop are characterized. This process is designed to be adapted as needed for the next generation of GM crops, thereby enabling the development and commercial use of new products that are critical to meeting agriculture's challenges.

Commercial GM crops and foods have been assessed for food, feed, and environmental safety according to internationally accepted scientific standards and guidelines (Codex Alimentarius 2008). Over several years, a rigorous and comprehensive set of data is generated on every GM crop product based on extensive field and safety testing. Because GM crops are studied much more extensively than any other plant products, there is an equal or greater assurance of safety of these products compared to conventional plant varieties. In fact, the European Commission (2001) acknowledged that the greater regulatory scrutiny given to GM crops and foods probably makes them safer than conventional plants and foods.

Scientific and regulatory authorities worldwide have concluded that GM crops currently on the market are as safe as conventional crop varieties (NRC 2004; Paoletti et al. 2008; Royal Society 2009). The principal concerns about the safety of GM crops are the potential risks inherent in plant genetic modification and derive from an expectation of 100 percent certainty of safety that does not even apply to traditional foods.

Misleading Information and Fear

Advancements in science are proceeding at a rapid pace, and the public is often not equipped to discern fact from fiction (Sachs 2007). Scientists in the public and private sectors must accept more responsibility to serve public interests, to respond to concerns and allegations with facts and knowledge, and to communicate in a manner that is understandable and informative. The result will be a more informed public with improved critical thinking skills that will enable sound decision making – not based on ignorance and fear, but on knowledge and reason.

It is essential that individuals interested in learning more about GM crop safety and impacts seek information that is objective, based on widely accepted scientific methods, and supported by a weight of scientific evidence. The International Life Sciences Institute (ILSI) developed a *Resource Guide* of scientific publications in key topic areas on agricultural biotechnology, including general food safety, protein safety, allergy assessment, and substantial equivalence (ILSI 2004). This guide contains titles of more than 1,200 publications and international reports.

The Path Forward

GM crops (and, as presented in Chapter 6, animals) have provided important benefits to farmers, consumers, and the environment. More than 12 years of experience have established the safety of GM crops. Regulatory frameworks currently in place in countries worldwide are based on internationally established safety principles and guidelines that provide a proven foundation for assuring that future GM crops are as safe as conventionally bred crops for food, feed, and the environment.

Agriculture of the 21st century must pursue approaches to help solve urgent social, economic, and environmental challenges. Innovation in agriculture will be essential for increasing agricultural productivity and to meet the goals of sustainable development. Plant scientists are combining traditional and modern approaches to increase crop yields, make better use of agricultural inputs, and increase the economic returns for farmers. Breakthroughs in genetics are expanding the toolbox of genes available for reducing biotic and abiotic stressors that limit productivity. New traits are being tested that increase yield by reducing losses from pests, disease, and environmental stressors, while improving the efficiency of production. Further examples of the benefits of biotechnology to both crops and animal production are provided in the next two chapters. The future of agriculture is bright. Policies and actions should be focused on meeting the growing world's needs for feed, food and fuel.

References

ASA (American Soybean Association). 2001. *Conservation Tillage Study*. St. Louis: ASA. http://www.soygrowers.com/ctstudy/ctstudy_files/frame.htm.

ASSOCHAM (Associated Chambers of Commerce and Industry of India). 2007. *Bt Cotton Farming in India (Synopsis)*. New Delhi: ASSOCHAM.

Bennett, R., T. Buthelezi, Y. Ismael, and S. Morse. 2003. "*Bt* Cotton, Pesticides, Labour and Health – A Case Study of Smallholder Farmers in the Makhathini Flats, Republic of South Africa 2003." *Outlook on Agriculture* 32 (2): 123–8.

Bennett, R., Y. Ismael, U. Kambhampati, and S. Morse. 2004. "Economic Impact of Genetically Modified Cotton in India." *AgBioForum* 7 (3): 1–5.

Betz, F., B. Hammond, and R. Fuchs. 2000. "Safety and Advantages of *Bacillus thuringiensis*-Protected Plants to Control Insect Pests." *Regulatory Toxicology and Pharmacology* 32 (2): 156–73.

Borlaug, N. 2000. "The Green Revolution Revisited and the Road Ahead." Lecture presented at the Norwegian Nobel Institute, Oslo, Norway, September 8. http://nobelprize.org/nobel_prizes/peace/laureates/1970/borlaug-lecture.pdf.

______. 2001. "Feeding the World in the 21st Century: The Role of Agricultural Science and Technology." Lecture presented at Tuskegee University, Tuskegee, AL, April. http://www.highyieldconservation.org/articles/feeding_the_world.html.

______. 2004. "Foreword." In *The Frankenfood Myth: How Protest and Politics Threaten the Biotech Revolution*, edited by H. Miller and G. Conko, ix–xii. Westport, CT: Praeger Publishers.

Brookes, G., and P. Barfoot. 2007. "Global Impact of Biotech Crops: Socio-Economic and Environmental Effects in the First Ten Years of Commercial Use." *AgBioForum* 9 (3): 139–51.

______. 2008. "Global Impact of Biotech Crops: Socio-Economic and Environmental Effects 1996–2006." *AgBioForum* 11 (1): 21–38.

______. 2009a. "Global Impact of Biotech Crops: Income and Production Effects 1996–2007." *AgBioForum* 12 (2): 184–208.

______. 2009b. "Global Impact of Biotech Crops: Socio-Economic & Environmental Effects 1996–2007." *Outlooks on Pest Management* 20 (6): 1–7.

______. 2009c. *Focus on Environmental Impacts – Biotech Crops: Evidence, Outcomes and Impacts 1996–2007*. PG Economics. http://www.pgeconomics.co.uk/pdf/focusonenvimpacts2009.pdf.

______. 2009d. *Focus on Yield – Biotech Crops: Evidence, Outcomes and Impacts 1996–2007*. PG Economics. http://www.pgeconomics.co.uk/pdf/focusonyieldeffects2009.pdf.

Campos, H., M. Cooper, G. Edmeades, C. Loffler, J. Schussler, and M. Ibanez. 2006. "Changes in Drought Tolerance in Maize Associated with Fifty Years of Breeding for Yield in the U.S. Corn Belt." *Maydica* 51: 369–81.

Castiglioni, P., D. Warner, R. Bensen, D. Anstrom, J. Harrison, M. Stoecker, M. Abad, et al. 2008. "Bacterial RNA Chaperones Confer Abiotic Stress Tolerance in Plants and Improved Grain Yield in Maize under Water-Limited Conditions." *Plant Physiology* 147: 446–55.

Chassy, B., M. Egnin, Y. Gao, K. Glenn, G. Kleter, P. Nestel, M. Newell-McGloughlin, R. Phipps, and R. Shillito. 2008. "Nutritional and Safety Assessments of Foods and Feeds Nutritionally Improved through Biotechnology: Case Studies." *Comprehensive Reviews in Food Science and Food Safety* 7 (1): 50–113.

Chassy, B., J. Hlywka, G. Kleter, E. Kok, H. Kuiper, M. McGloughlin, I. Munro, R. Phipps, and J. Reid. 2004. "Nutritional and Safety Assessments of Foods and Feeds Nutritionally Improved through Biotechnology." *Comprehensive Reviews in Food Science and Food Safety* 3: 35–104.

Codex Alimentarius. 2008. *Guideline for the Conduct of Food Safety Assessment of Foods Derived from Recombinant-DNA Plants*. Reference CAC/GL 45-2003, adopted 2003, and amended 2008. http://www.codexalimentarius.net/download/standards/10021/CXG_045e.pdf.

Creelman, R., N. Gutterson, O. Ratcliffe, L. Reuber, E. Cerny, K. Duff, and S. Kjemtrup-Lovelace, et al. 2008. *Improved Yield and Stress Tolerance in Transgenic Plants*. WIPO Patent Publication WO/2008/005210, filed June 22, 2007, and published January 10, 2008.

Crosbie, T. M., S. R. Eathington, G. R. Johnson, M. Edwards, R. Reiter, S. Stark, R. G. Mohanty, et al. 2006. "Plant Breeding: Past, Present, and Future." In *Plant Breeding: The Arnel R. Hallauer International Symposium*, edited by K. R. Lamkey and M. Lee, 3–50. Ames, IA: Blackwell.

CTIC (Conservation Technology Information Center). 2009. *Top 10 Benefits of Conservation Tillage: Farm and Food Facts' 09*. Factsheet. West Lafayette, IN: Purdue University. http://www.ilfb.org/fff2009/37.pdf.

Duvick, D. N. 2005. "The Contribution of Breeding to Yield Advances in Maize (*Zea mays* L.)." *Advances in Agronomy* 86: 83–145.

Eathington, S., T. Crosbie, M. Edwards, R. Reiter, and J. Bull. 2007. "Molecular Markers in a Commercial Breeding Program." *Crop Science* 47 (S3): S154–63.

Edgerton, M. 2009. "Increasing Crop Productivity to Meet Global Needs for Feed, Food, and Fuel." *Plant Physiology* 149: 7–13.

EPA (Environmental Protection Agency). 2007. *EPA Finalizes Regulations for a Renewable Fuel Standard (RFS) Program for 2007 and Beyond*. Factsheet EPA420-F-07-019. http://www.epa.gov/OMS/renewablefuels/420f07019.pdf.

Evans, L. 1998. *Feeding the Ten Billion: Plants and Population Growth*. Cambridge: Cambridge University Press.

FAO (Food and Agriculture Organization). 2002. "What the New Figures on Hunger Mean." http://www.fao.org/english/newsroom/news/2002/9703-en.html.

FAPRI (Food and Agricultural Policy Research Institute). 2008. *U.S. and World Agricultural Outlook*. Staff Report 08-FSR 1. http://www.fapri.iastate.edu/outlook/2008/text/OutlookPub2008.pdf. Ames: FAPRI.

Fawcett, R., and D. Towry. 2002. *Conservation Tillage and Plant Biotechnology: How New Technologies Can Improve the Environment by Reducing the Need to Plow*. CTIC (Conservation Technology Information Center). West Lafayette, IN: Purdue University.

Fernandez-Cornejo, J., A. Mishra, R. Nehring, C. Hendricks, M. Southern, and A. Gregory. 2007. *Off-Farm Income, Technology Adoption, and Farm Economic Performance*. ERS (Economic Research Service) Report 36. http://www.ers.usda.gov/Publications/err36/err36_reportsummary.pdf.

Fitt, G. 2003. "Deployment and Impact of Transgenic Bt Cotton in Australia." In *The Economic and Environmental Impacts of Agbiotech: A Global Perspective*, edited by N. G. Kalaitzandonakes, 141–64. New York: Kluwer.

Frink, C., P. Waggoner, and J. Ausubel. 1999. "Nitrogen Fertilizer: Retrospect and Prospect." *Proceedings of the National Academy of Sciences of the United States of America* 96: 1175–80.

Head, G. 2007. "Soil Fate and Non-Target Impact of *Bt* Proteins in Microbial Sprays and Transgenic *Bt* Crops." In *Crop Protection Product for Organic Agriculture – Environmental, Health and Efficacy Assessment*, edited by A. Felsot and K. Racke, 212–21. ACS Symposium Series 947. Washington, DC: American Chemical Society. doi:10.1021/bk-2007-0947.ch015.

______, J. Surber, J. Watson, J. Martin, and J. Duan. 2002. "No Detection of Cry1Ac Protein in Soil after Multiple Years of Transgenic *Bt* Cotton (Bollgard) Use." *Environmental Entomology* 31 (1): 30–6.

Helentjaris, T., G. King, M. Slocum, C. Siedenstrang, and S. Wegman. 1985. "Restriction Fragment Length Polymorphisms as Probes for Plant Diversity and as Tools for Applied Plant Breeding." *Plant Molecular Biology* 5: 109–18.

Hu, H., M. Dai, J. Yao, B. Xiao, X. Li, Q. Zhang, and L. Xiong. 2006. "Overexpressing a NAM, ATAF, and CUC (NAC) Transcription Factor Enhances Drought Resistance and Salt Tolerance in Rice." *Proceedings of the National Academy of Sciences of the United States of America* 103: 12987–92.

Huang, J., R. Hu, C. Fan, C. Pray, S. Rozelle. 2003. *Bt Cotton Benefits, Costs and Impacts in China*. Working Paper 202. Brighton, UK: IDS (Institute of Development Studies).

Huesing, J., and L. English 2004. "The Impact of *Bt* Crops on the Developing World." *AgBioForum* 7 (1&2): 84–95.

ILSI (International Life Sciences Institute). 2004. *Resource Guide*. http://www.cropcomposition.org/query/file/ILSIBiotechResourceGuideThrough2004.pdf.

IPCC (Intergovernmental Panel on Climate Change). 2007. *Climate Change 2007: Synthesis Report: Summary for Policymakers*. IPCC. www.ipcc.ch/pdf/assessment-report/ar4/syr/ar4_syr_spm.pdf.

James, C. 2010. *Executive Summary, Global Status of Commercialized Biotech/GM Crops*. ISAAA Brief 42. International Service for the Acquisition of Agri-Biotech Applications (ISAAA). http://www.isaaa.org/resources/publications/briefs/42/executivesummary/default.asp.

Kay, B. D. 1995. "Soil Quality: Impact of Tillage on the Structure of Tilth of Soil." In *Farming for a Better Environment*, 7–8. SWCS (Soil and Water Conservation Society) White Paper. Ankeny, IA: SWCS.

Kern, J., and M. Johnson. 1993. "Conservation Tillage Impacts on National Soil and Atmospheric Carbon Levels." *Soil Science Society of America Journal* 57: 200–10.

Keystone Center. 2009. *Field to Market – The Keystone Alliance for Sustainable Agriculture.* http://keystone.org/files/file/SPP/environment/field-to-market/Field_to_Market_Background_December_2009.pdf.

Kok, E., J. Keijer, G. Kleter, and H. Kuiper. 2008. "Comparative Safety Assessment of Plant-Derived Foods." *Regulatory Toxicology and Pharmacology* 50 (1): 98–113.

Kucharik, C. 2008. "Contribution of Planting Date Trends to Increased Maize Yields in the Central United States." *Agronomy Journal* 100: 328–36.

Lardizabal, K., R. Effertz, C. Levering, J. Mai, M. Pedroso, T. Jury, E. Aasen, K. Gruys, and K. Bennett. 2008. "Expression of *Umbelopsis ramanniana* DGAT2A in Seed Increases Oil in Soybean." *Plant Physiology* 148: 89–96.

Lundry, D., W. Ridley, J. Meyer, S. Riordan, M. Nemeth, W. Trujillo, M. Breeze, and R. Sorbet. 2008. "Composition of Grain, Forage, and Processed Fractions from Second-Generation Glyphosate-Tolerant Soybean, MON 89788, Is Equivalent to that of Conventional Soybean (*Glycine max* L.)." *Journal of Agriculture and Food Chemistry* 56: 4611–22.

Marvier, M., C. McCreedy, J. Regetz, and P. Kareiva. 2007. "A Meta-Analysis of Effects of *Bt* Cotton and Maize on Nontarget Invertebrates." *Science* 316: 1475–7.

NASS (National Agricultural Statistics Service). 2007. *Data and Statistics.* http://www.nass.usda.gov/Data_and_Statistics/.

Nelson, D., P. Repetti, T. Adams, R. Creelman, J. Wu, D. Warner, D. Anstrom, et al. 2007. "Plant Nuclear Factor Y (NF-Y) B Subunits Confer Drought Tolerance and Lead to Improved Corn Yields on Water-Limited Acres." *Proceedings of the National Academy of Sciences of the United States of America* 104: 16450–5.

NOAA (National Oceanic and Atmospheric Administration) Climactic Data Center. 2010. *State of the Climate: Global Analysis for Annual 2010.* Asheville, NC: NOAA. http://www.ncdc.noaa.gov/sotc/global/2010/13.

NRC (National Research Council). 2004. *Safety of Genetically Engineered Foods: Approaches to Assessing Unintended Health Effects.* Washington, DC: National Academy Press. http://www.nap.edu/catalog/10977.html.

OECD (Organisation of Cooperation and Development). 2007. *Consensus Document on Safety Information on Transgenic Plants Expressing Bacillus thuringiensis-Derived Insect Control Protein.* OECD Environment, Health, and Safety Publications Series on Harmonization of Regulatory Oversight in Biotechnology, Number 42. Paris: OECD. http://www.oecd.org/officialdocuments/displaydocumentpdf?cote=env/jm/mono%282007%2914&doclanguage=en.

OECD–FAO (Organisation of Cooperation and Development–Food and Agriculture Organization). 2008. *Agricultural Outlook 2008–2017.* Paris: OECD/FAO. http://www.fao.org/es/esc/common/ecg/550/en/AgOut2017E.pdf.

Padgette, S. 2008. "Golden Opportunities: Working Jointly for Higher Yields." PowerPoint presentation. http://basf.com/group/corporate/de_DE/function/conversions:/publish/content/investor-relations/calendar/images/080916/080916_Presentation_Field_Trip_CropDesign.pdf.

———, K. Kolacz, X. Delannay, D. Re, B. LaValee, C. Tinius, W. Rhodes, et al. 1995. "Development, Identification, and Characterization of a Glyphosate-Tolerant Soybean Line." *Crop Science* 35: 1451–61.

Palm, C., R. Seidler, D. Schaller, and K. Donegan. 1996. "Persistence in Soil of Transgenic Plant Produced *Bacillus thuringiensis* var. *kurstaki* Delta-Endotoxin." *Canadian Journal of Microbiology* 42: 1258–62.

Paoletti, C., E. Flamm, W. Yan, S. Meek, S. Renckens, M. Fellous, and H. Kuiper. 2008. "GMO Risk Assessment around the World: Some Examples." *Trends in Food Science & Technology* 19: S66–74.

Paterson, A., E. Lander, J. Hewitt, S. Peterson, S. Lincoln, and S. Tanksley. 1988. "Resolution of Quantitative Traits into Mendelian Factors by Using a Complete Linkage Map of Restriction Fragment Length Polymorphisms." *Nature* 335: 721–6.

Perlak, F., R. Fuchs, D. Dean, S. McPherson, and D. Fischhoff. 1991. "Modification of the Coding Sequence Enhances Plant Expression of Insect Control Protein Genes." *Proceedings of the National Academy of Sciences of the United States of America* 88: 3324–8.

Pray, C., J. Huang, R. Hu, and S. Rozelle. 2002. "Five Years of *Bt* Cotton in China – The Benefits Continue." *Plant Journal* 31 (4): 423–30.

Qaim, M. 2009. "The Economics of Genetically Modified Crops." *Annual Review of Resource Economics* 1:665–93.

Raney, T. 2006. "Economic Impact of Transgenic Crops in Developing Countries." *Current Opinion in Biotechnology* 17: 174–8.

Raven, P. 2008. "The Environmental Challenge: The Role of GM Crops." In *Proceedings of the 53rd Brazilian Congress of Genetics*.

Reicosky, D. 1995. "Impact of Tillage on Soil as a Carbon Sink." In *Farming for a Better Environment*, 50–3. SWCS (Soil and Water Conservation Society) White Paper. Ankeny, IA: SWCS.

______, and M. J. Lindstrom. 1995. "Impact of Fall Tillage on Short-Term Carbon Dioxide Flux." In *Soils and Global Change*, edited by R. Lal, J. Kimbal, E. Levine, and B. A. Steard, 177–87. Chelsea, MI: Lewis Publishers.

Rice, M. 2004. "Transgenic Rootworm Corn: Assessing Potential Agronomic, Economic and Environmental Benefits." *Plant Health Progress*. Published electronically March 1. doi:10.1094/PHP-2004-0301-01-RV.

Riechmann, J., J. Heard, G. Martin, L. Reuber, C. Jiang, J. Keddie, L. Adam, et al. 2000. "Arabidopsis Transcription Factors: Genome-Wide Comparative Analysis among Eukaryotes." *Science* 290: 2105–10.

Romeis, J., D. Bartsch, F. Bigler, M. P. Candolfi, M. C. C. Gielkens, S. E. Hartley, R. L. Hellmich, et al. 2008. "Assessment of Risk of Insect-Resistant Transgenic Crops on Nontarget Arthropods." *Nature Biotechnology* 26 (2): 203–8.

Rosegrant, M., X. Cai, S. Clein. 2002. *World Water and Food to 2025: Dealing with Scarcity*. Washington, DC: International Food Policy Research Institute (IFPRI).

Roush, R. 1998. "Two Toxin Strategies for Management of Insecticidal Transgenic Crops: Can Pyramiding Succeed Where Pesticide Mixtures Have Not?" *Philosophical Transactions of the Royal Society B: Biological Sciences* 353: 1777–86.

Royal Society. 2009. *Reaping the Benefits: Science and the Sustainable Intensification of Global Agriculture*. Report 11/09 RS1608. London: The Royal Society. http://royalsociety.org/WorkArea/DownloadAsset.aspx?id=4294967942.

Sachs, E. 2007. "The Science Literacy Gap – Enabling Society to Critically Evaluate New Scientific Developments." In *Public Science in Liberal Democracy*, edited by J. Porter and P. Phillips, 205–14. Toronto: University of Toronto Press.

Saxena, D., and G. Stotzky. 2001. "*Bacillus thuringiensis* (*Bt*) Toxin Released from Root Exudates and Biomass of *Bt* Corn Has No Apparent Effect on Earthworms, Nematodes, Protozoa, Bacteria, and Fungi in Soil." *Soil Biology and Biochemistry* 33: 1225–30.

Scherr, S., and J. McNeely. 2007. "Biodiversity Conservation and Agricultural Sustainability: Towards a New Paradigm of 'Ecoagriculture' Landscapes." *Philosophic Transactions of the Royal Society B: Biological Sciences* 363: 477–94.

Secretariat of the Convention on Biological Diversity. 2000. *Cartagena Protocol on Biosafety to the Convention on Biological Diversity*. Montreal: Secretariat of the Convention on Biological Diversity.

Strange, A., J. Park, R. Bennett, and R. Phipps. 2008. "The Use of Life-Cycle Assessment to Evaluate the Environmental Impacts of Growing Genetically Modified Use-Efficient Canola." *Plant Biotechnology Journal* 6: 337–45.

Tacio, H., ed. 2009. "Feeding a World of 9 Billion." PeopleandPlanet.net. Accessed January 21. http://www.peopleandplanet.net/?lid=26107§ion=34&topic=44.

Thompson, W., ed. 2008. *The 30-Year Challenge – Agriculture's Strategic Role in Feeding and Fueling a Growing World*. Issue Report, Farm Foundation. http://www.farmfoundation.org/news/articlefiles/1694-Final%2030%20Year%20Challenge.pdf.

Tollenaar, M., and E. Lee. 2002. "Yield Potential, Yield Stability and Stress Tolerance in Maize." *Field Crops Research* 75: 161–9.

UN (United Nations). 1987. *Report of the World Commission on Environment and Development: Our Common Future*. General Assembly Resolution 42/187. New York: UN. http://www.un-documents.net/wced-ocf.htm.

UNFCC (United Nations Framework Convention on Climate Change). 2011. "Status of the Ratification of the Convention." Accessed October 3. http://unfccc.int/essential_background/convention/status_of_ratification/items/2631.php.

USCB (United States Census Bureau). 2009. *International Data Base*. (World population information; accessed January 21.) http://www.census.gov/ipc/www/idb/worldpopinfo.html.

USDA FCIC (United States Department of Agriculture Federal Crop Insurance Corporation). 2010. *Pilot Biotech Yield Endorsement*. Washington, DC: FCIC. http://www.rma.usda.gov/policies/2010/10-be.pdf.

Van Camp, W. 2005. "Yield Enhancement Genes: Seeds for Growth." *Current Opinion in Biotechnology* 16: 147–53.

5

Biotechnology and the Control of Viral Diseases of Crops

Jason R. Cavatorta, Stewart M. Gray, and Molly M. Jahn

The global impact of plant viruses on crop productivity is difficult to estimate because losses caused by viruses frequently go unnoticed. Viral infection may be inconspicuous – causing little or no obvious symptoms other than yield reduction (Waterworth and Hadidi 1998). The monetary impact of viral pathogens on food production systems is therefore chronically underreported and the calculations of impact underestimated. Yet that impact may be striking in magnitude. For instance, global annual reductions in yield due to virus are estimated at 7 percent in sugar beet and 8 percent in potato (Oerke and Dehne 2004). Of course, in specific cases of viral pathogens, the impact can be much more destructive than global averages. Tomato spotted wilt virus has been estimated to cause more than $1 billion in damage to cultivated plant species annually (Goldbach and Peters 1994). Another striking example is Cocoa swollen shoot virus, which has destroyed 200 million cocoa trees in Ghana alone (Lockhart and Sachey 2001).

Management options for viruses are relatively limited. They consist mainly of regulatory and cultural control methods such as exclusion of the viral pathogen from a geographic area and using disease-free seed (Agrios 2005). Unlike fungal pathogens, which may be controlled through the use of fungicides, there is no effective way to control viral pathogens directly. Chemical control of virus vectors has been widely adopted for controlling viral diseases despite limited success (Broadbent 1957). A literature search reveals that recommended rates of pesticide application can be very high, pesticides must be applied frequently, and often they do not effectively control virus spread (Table 5.1). Not only does this disease management strategy have potential negative environmental and health consequences but also viruses continue to be problematic in sprayed fields (Satapathy 1998). Indirect chemical control of nonpersistent viruses, which can be transmitted by a feeding insect in a matter of seconds, is particularly ineffective (Broadbent 1957).

One of the major goals of sustainable agriculture advocates is to reduce the amount of chemical inputs to farming systems. The U.S. Food, Agriculture, Conservation, and

Table 5.1. *Pesticide levels required for indirect control of virus diseases by targeting vector populations.*

Pesticide Class (specific chemical)	Virus	Vector	Application rate*	Reference
Carbamate (aldicarb)	*Potato leaf roll*	Aphid	3.36 kg/ha	Powell and Mondor (1973)
Carbamate (carbofuran)	*Maize chlorotic dwarf*	Leafhopper	2.24 kg/ha	Kuhn, Jellum, and All (1975)
Neonicotinoid (Imidacloprid)	*Tomato spotted wilt*	Thrips	Seedling drench	Coutts and Jones (2005)
Neonicotinoid (Imidacloprid)	*Barley yellow dwarf*	Aphid	Seed treatment	Makkouk and Kumari (2007)
Organophosphate (demephion)	*Potato leaf roll*	Aphid	0.25 kg/ha	Woodford et al. (1983)
Organophosphate (phosphamidon)	*Rice tungro*	Leafhopper	3.00 kg/ha	Pathak, Vea, and John (1967)
Phosphorodiamide (dimefox)	*Cocoa swollen shoot*	Mealybug	Trunk injections	Hanna (1954)
Pyrethroid (deltamethrin)	*Barley yellow dwarf*	Aphid	0.01–0.06 kg/ha**	McGrath and Bale (1990)
Pyrethroid (cypermethrin)	*Strawberry mild yellow edge*	Aphid	0.07 kg/ha	Converse and Aliniazee (1988)

* Application rate refers to the amount of active ingredient (not the commercial product) per acre.

** Application rate was not reported so the current recommended application rate by the producer is listed.

Trade Act of 1990 (1990, 3705–6) defines sustainable agriculture as "an integrated system of plant and animal production practices having a site-specific application that will... make the most efficient use of nonrenewable resources and on-farm resources and integrate, where appropriate, natural biological cycles and control." Chemical inputs have been shown to have a number of associated human health and environmental risks and can be quite expensive (Ahlborg et al. 1990; Ecobichon 2001; Pimentel et al. 1993). Thus, management of virus diseases by control of viral vectors using pesticide applications is not only unreliable but is also unsustainable, potentially dangerous, and inaccessible to resource-poor farmers.

Host Plant Resistance

Host Plant Resistance as a Sustainable Alternative

The most sustainable and effective method of virus disease control is the use of host plant resistance when available. In this approach, commercially acceptable cultivars are developed that have reduced disease severity or disease incidence. Development of resistant cultivars is time consuming and costly, but once completed and adopted by farmers, successful disease control can be achieved with little effort. The major limitations of host resistance are the identification of appropriate and effective genes that confer resistance to a virus or group of viruses, the expense of introgressing the resistance trait into commercial genetic backgrounds, and the risk of the viruses evolving to overcome genetic resistance (Kang, Yeam, and Jahn 2005). Some of these limitations may be mitigated by marker-assisted selection or genetic engineering (discussed later in this chapter). Resistance durability has proven difficult to predict, but may be maximized by the use of recessively inherited or horizontal disease resistance genes (Fraser 1990).

Host Plant Resistance through Conventional Breeding

Host plant resistance to viruses has traditionally been achieved by identifying economically important pathosystems, screening sexually compatible germplasm (often from wild sources) for appropriate virus resistance genes, studying the inheritance of resistance, and introgressing resistance into a commercially acceptable genetic background. In some cases, further study of the mode of action of resistance is performed to characterize the specificity of resistance alleles against multiple viral strains (Kang, Yeam, and Jahn 2005). This method has proven successful against a number of phytopathogenic viruses. To date, many resistance genes have been described with a diversity of inheritance types and modes of action (Diaz-Pendon et al. 2004). They have been used to develop virus-resistant cultivars in many crops (Zitter and McGrath 2006). For example, the I gene has been used effectively in bean breeding to develop bean varieties resistant to Bean common mosaic virus (Kang, Yeam, and Jahn 2005). Additionally, Tobacco mosaic virus has been successfully controlled in tobacco cultivars using the N gene (Holmes 1938).

Although conventional breeding is an effective method for managing many plant-virus pathosystems, it is limited by virus species or type specificity of resistance genes and by the substantial effort required to introgress each gene from wild sources (Ritzenthaler 2005). This process may require many years of backcrossing and selection to remove horticulturally undesirable traits. It is also possible that virus-resistant genes cannot be found within plant species that are sexually compatible with a given crop. Finally, breeding new virus-resistant cultivars, especially of clonally propagated crops (e.g., potato, grape, and fruit trees), may not be accepted by growers or consumers who are reluctant to adopt new cultivars, either because cultivar choice is important to the identity of their product or because other characteristics make existing cultivars preferable.

Host Plant Resistance through Genetic Engineering

Recently, advances in plant biotechnology have provided additional methods of creating virus-resistant plants. Genetic engineering of plants has allowed for the development of virus resistance that uses genes from many diverse organisms. One of the first uses of this technology was to express viral genes in plants to see what effect this had on virus infection. Tobacco plants expressing the coat protein from Tobacco mosaic virus were found to be resistant to that virus (Abel et al. 1986). Since then, other genes, from viruses or elsewhere, have been expressed in plants with similar results (reviewed in Sudarshana, Roy, and Falk 2007). Resistance has also been successfully achieved by transgenic expression of plant genes. Several resistance genes – for example N from tobacco and Tm-2^2 from tomato – have been shown to convey resistance when expressed in susceptible genotypes (Lanfermeijer et al. 2003; Whitham, McCormick, and Baker 1996). A plant protease inhibitor has been used successfully to confer transgenic resistance to multiple potyviruses (Gutierrez-Campos et al. 1999). Different ribosome-inactivating proteins are other examples of plant genes that have been used successfully to develop resistance in viral pathosystems: resistance to Potato virus Y (PVY) and Potato virus X (PVX) in tobacco and potato (Lodge, Kaniewski, and Tumer 1993), resistance to African cassava mosaic virus in *N. benthamiana* (Hong et al. 1996), and partial resistance to Cucumber mosaic virus and Tobacco mosaic virus in tobacco (Krishnan et al. 2002).

Despite the enthusiasm generated by these early results, the potential for commercialization of virus-resistant plant material has not been fully realized. Only several GM cultivars have been commercialized, which all rely on pathogen-derived resistance. Virus-resistant transgenic squash (*Cucurbita pepo*), developed by Asgrow Seed Co., has been approved for commercialization (APHIS 1996a). Cultivars developed from lines "ZW-20" and "CZW-3" contain coat-protein-mediated protection against combinations of Watermelon mosaic virus 2, Zucchini yellow mosaic virus, and Cucumber mosaic virus (APHIS 1996a; Tepfer 2002; Tricoll et al. 1995). The "Rainbow" papaya cultivar, developed by Cornell University and the University of Hawaii, has resistance to Papaya ringspot virus (PRSV; APHIS 1996b; Ferreira et al. 2002). Finally, the "Newleaf" potato cultivar contained the *Bt* gene and resistance to either Potato leafroll virus (cultivar "Newleaf Plus") or PVY (cultivar "Newleaf Y"), and was marketed briefly by the Monsanto subsidiary company, Naturemark (APHIS 1997; Kaniewski and Thomas 2004; Tepfer 2002). Although resistance to economically important viral diseases has been successfully obtained in many additional crops, regulatory approval and commercialization have not been achieved for several reasons (discussed later in this chapter).

Genomics of Virus Resistance

Dominantly Inherited Virus Resistance

Genomic research into virus resistance has led to the cloning and detailed study of an increasing number of virus resistance genes (Goldbach, Bucher, and Prins 2003). Several dominantly inherited plant virus resistance genes have been sequenced and studied in detail. Nearly all belong to the nucleotide binding site plus leucine rich repeat (NB-LRR) class of resistance genes with one of the following at the N terminus: a leucine zipper, a coiled coil, or a region similar to the Toll and Interleukin 1 receptor (Dangl and Jones 2001). Exceptions to this convention are the genes RTM1 and RTM2 (restricted TEV movement 1 and 2) against Tobacco etch virus (TEV) in *Arabidopsis*, which appear to be a lectin and a heat shock protein, respectively (Chisholm et al. 2000; Whitham et al. 2000). Dominant resistance is often associated with a hypersensitive response after specific recognition of a particular pathogen component (Goodman and Novacky 1994). Dominant resistance genes typically interact in some way with a specific effector molecule from an invading pathogen in a "gene-for-gene" manner, resulting in the activation of defense responses (Flor 1971; Jones and Dangl 2006). Detailed study of dominantly inherited resistance has facilitated the discovery of additional resistance genes that have offered an alternative to pathogen-derived transgenic resistance.

Recessively Inherited Virus Resistance

Study of recessively inherited resistance has lagged behind that of dominantly inherited resistance; thus less is understood about this class of resistance genes. Recessive resistance genes, in contrast to dominant ones, are thought to encode host proteins that are used by the pathogen to complete its life cycle and are often plant proteins involved in cellular processes (Truniger and Aranda 2009). If this host protein is mutated in such a way as to prevent the virus from performing an essential function (such as replication or movement) the host plant may become resistant (Diaz-Pendon et al. 2004; Fraser 1990). There is no associated programmed cell death and no activation of defense responses within the plants; therefore, recessive resistance may be referred to as "passive resistance" (Fraser 1990). The virus is simply unable to complete its life cycle in the host, so infection does not occur. Although recessive resistance is not unique to virus resistance, it does appear to be a more common defense mechanism against viruses than against other pathogen types (Diaz-Pendon et al. 2004; Provvidenti and Hampton 1992). Furthermore, recessive resistance, in general, seems to be more durable and less specific than dominant forms of resistance (Kang et al. 2005).

The virus life cycle may be disrupted at several different stages. Recessive resistance genes have been shown to disrupt viral replication (Murphy et al. 1998), cell-to-cell movement (Gao et al. 2004), and long-distance movement within plants (Schaad, Lellis, and Carrington 1997). A better understanding of the identity and function of recessive resistance genes will provide additional tools for developing virus-resistant crop varieties. By coupling knowledge of virus infection strategies with new advances in plant genomics, novel methods of combating virus infection will likely arise. In the following sections we discuss the most well-studied recessive resistance gene to illustrate this point. We describe the successful use of this gene to engineer virus-resistant pepper and potato and consider the prospects for developing transgenic virus resistance in other crops. Finally we discuss the potential impact of using plant genes on consumer acceptance of biotechnology.

Eukaryotic Translation Initiation Factor 4E

Detailed Understanding of a Recessive Virus Resistance Gene

Eukaryotic translation initiation factor 4E (eIF4E) is a plant host gene involved in binding to the 5' cap structure of a messenger RNA (Sonenberg et al. 1978). The eIF4E–cap interaction occurs during the initiation phase of translation and is required for recruitment of capped mRNA to the ribosomal complex (Browning 2004; Gingras, Raught, and Sonenberg 1999). Translation of the RNA genome of plant viruses in an infected host cell appears to share similarities to translation of host cellular mRNA, and a number of viruses have evolved to use eIF4E to complete their life cycle (Kang et al. 2005). Virus resistance alleles arise when natural eIF4E mutations disrupt the ability of the virus to interact with the modified eIF4E protein (Figure 5.1). The evolution of mutant eIF4E genes that function as virus resistance genes has occurred independently in multiple plant species. Because a single copy of the wildtype allele is sufficient to support viral infection in plants, this type of resistance is recessively inherited and has thus been difficult to study and define. In recent years, many recessive resistance alleles from several plant species have been cloned and found to encode eIF4E (Charron et al. 2008). A series of alleles conferring strain-specific resistance suggests that co-evolution between host and pathogen is focused around changes in potyviral genome-linked protein (VPg) and plant eIF4E. Detection of positive selection in both of these genes appears to confirm this hypothesis (Cavatorta et al. 2008; Moury et al. 2004). The interaction between host and pathogen in this system is therefore understood on an individual amino acid level, making this co-evolutionary pathosystem among the best understood in biology.

Transgenesis of eIF4E Controls Virus Infection

Many horticulturally important plant species are not known to have evolved eIF4E resistance alleles. Understanding the mechanisms of eIF4E-mediated resistance opens up opportunities to use biotechnology to express virus-resistant alleles from one species in another, as has been accomplished using dominant resistance genes (Whitham, McCormick, and Baker 1996). Tomato plants (*Solanum lycopersicum*) transformed with the pvr1 recessive resistance allele from *Capsicum chinense* were found to have broad-spectrum potyvirus resistance (Kang et al. 2007). A "dominant negative" model of protein interactions has been proposed to explain these observations and is continuing to be refined as eIF4E-mediated resistance is studied further.

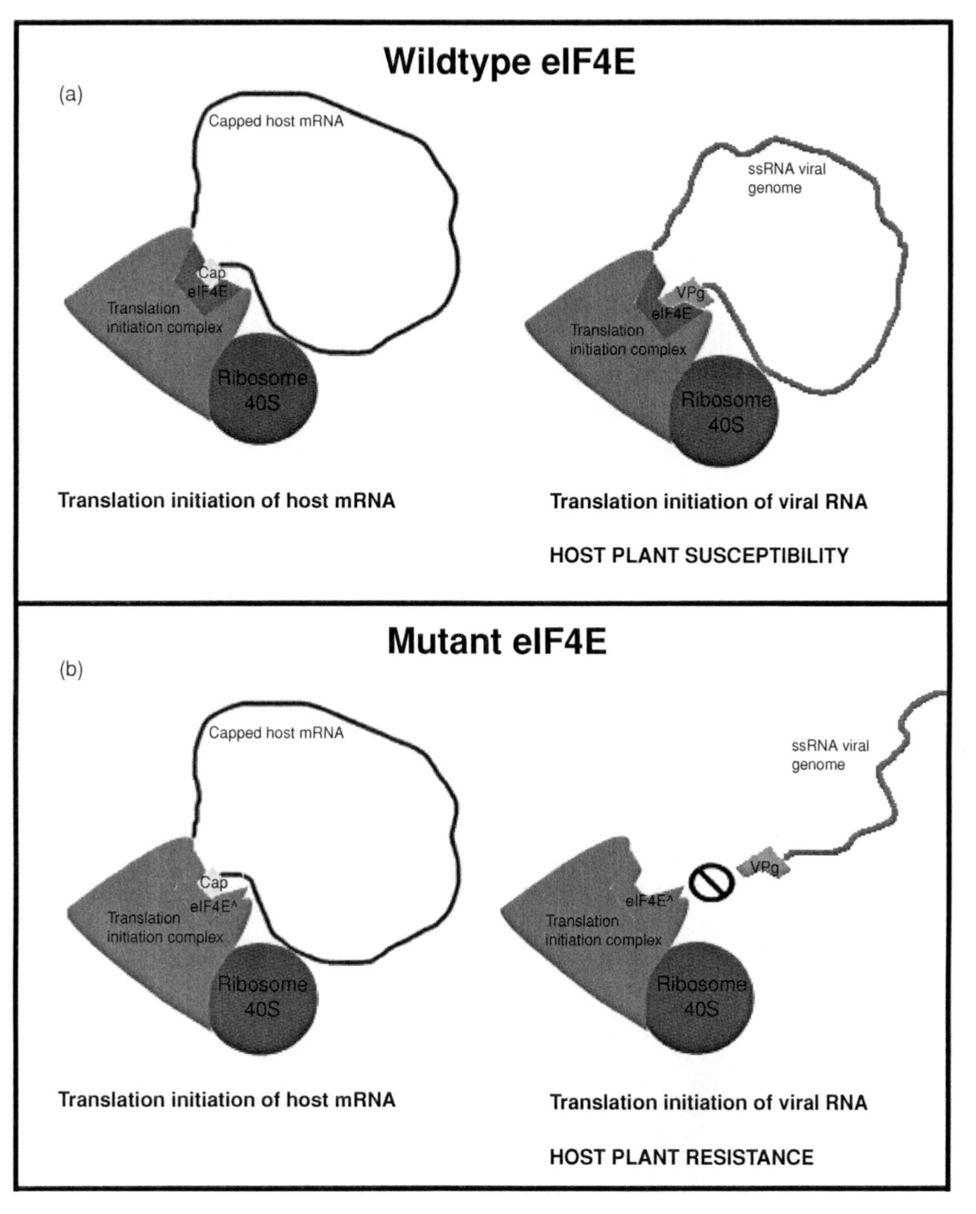

Figure 5.1. Mutations in Resistance Alleles Prevent an Invading Virus from Binding to eIF4E and Using It to Complete Its Own Life Cycle. A) Wildtype eIF4E is used by both capped mRNA and invading RNA viruses for efficient localization to the ribosome. B) Mutations in eIF4E resistance alleles appear to allow for the continued translation of capped host mRNA but do not bind to the viral protein VPg. This causes the virus to be unable to establish an infection in a resistant host. Mutant eIF4E is purple in color and labeled with an asterisk.

Intragenic Virus Resistance in Potato

PVY is the most economically important viral disease in potato (*Solanum tuberosum*) seed production and is a practical target for virus resistance. When the pepper eIF4E resistance allele $pvr1^2$ was expressed transgenically in potato, the resulting transgenic plants were found to be immune to mechanical inoculation of multiple strains of *PVY* (Cavatorta 2010). In addition, the endogenous potato eIF4E cDNA was isolated and mutated at sites homologous to mutations found in $pvr1^2$ relative to wildtype. When these mutant alleles were overexpressed in a susceptible potato cultivar it conferred virus resistance (Cavatorta et al. 2011). This technique is noteworthy because it mimics natural evolutionary mechanisms by working within the genome of a susceptible individual.

Consumer Acceptance Issues

A major advantage of transgenic expression of eIF4E or other plant genes that condition virus resistance is that it addresses some of the misgivings associated with using biotechnology to protect crops from virus infection. Because eIF4E from a susceptible host may be modified and used to generate virus resistance, the donor gene utilized is derived from the target crop. This so-called intragenic technology is predicted to be more acceptable to consumers (Rommens 2007). Empirical evidence suggests that consumers are less concerned about transgenic plants expressing plant genes compared to genes derived of viral origin. In one study, consumer willingness to eat a transgenic vegetable with a viral transgene was only 14.3 percent, but increased to 81.3 percent when the transgene was from within the same species (Lusk and Sullivan 2002). Nielsen (2003) and others have commented on these trends, and there is an increasing push to develop intragenic plants that express donor DNA from the same species (Rommens 2007, 2008). This area of research is particularly active in crops such as potato in which there has been opposition to previous attempts at commercialization of genetically engineered varieties (Rommens 2004).

As is true in other areas of biotechnology, consumer acceptance issues, more so than the lack of scientific progress, have limited commercialization of virus resistance in transgenic crops. Of the three GM species mentioned that were successfully approved by the U.S. Food and Drug Administration (FDA) (papaya, potato, and squash), only two remain commercially available. Lack of market acceptance prevented the "Newleaf" potato cultivar from reaching its commercial potential despite positive responses from consumers initially and successful adoption by a number of potato growers. Activists carried out a successful antibiotechnology campaign that ultimately convinced food service companies, potato vendors, and producers to adopt a GMO-free policy (Kaniewski and Thomas 2004). Their rejection resulted in the dissolution of Naturemark and the cessation of "Newleaf" potato sales. In addition, several promising areas of transgenic research in potato were abandoned because of the inhospitable marketing climate.

Transgenic papaya production has been crucial for the restoration of the Hawaiian papaya industry (Gonsalves 1998). Approximately three-quarters of Hawaiian papaya acres are planted with transgenic trees, but this accounts for less than 8.1 km^2 and only a fraction of a percent of global papaya production (NASS 2008, 2009). No transgenic trees are grown or their fruit sold outside of North America, and attempts to do so have met with considerable resistance (Davidson 2008; Gonsalves 2004).

A similar story exists for virus-resistant transgenic squash, as only 12 percent of U.S. summer squash acreage (about 28.3km^2) has been planted to genetically engineered squash (Fuchs and Gonsalves 2007; Shankula 2006). Resistant transgenic squash has not been adopted anywhere outside of North America and accounts for a marginal percentage of global summer squash production (James 2007).

The high adoption rate of transgenic virus-resistant crops by American farmers testifies to their importance in crop production. However, the failure of other countries to continue this trend underscores the concerns that global consumers continue to have about the use of pathogen-derived virus resistance. Development of transgenic virus resistance has proven to be an extremely effective and attainable method of developing host plant resistance in a large number of crops against a wide range of virus genera (Sudarshana et al. 2007). Adoption of virus resistance technology has not been limited by scientific progress, but rather by an unreceptive market environment. (Collinge, Lund, and Thordal-Christensen 2008). Although the scientific community largely agrees that most consumers' concerns are not scientifically valid (EU 2008; Fuchs and Gonsalves 2007), these concerns must nonetheless be addressed to enable the successful commercialization of transgenic products in food crops. History has proven that failure to account for consumer concerns translates into likely failure to transfer the technology from the researcher's bench to the farmer's field.

Conclusion

In this chapter, we argued that host plant resistance is the most effective and sustainable means of virus control. It can be developed by traditional breeding methods, using dominant or recessive resistance genes, or by multiple transgenic approaches. Expression of plant genes, such as eIF4E, has proven to be an effective and novel method of developing virus-resistant crops. However, unless consumers accept GM crops, the potential to use transgenic virus resistance to produce food more sustainably cannot be fully realized. We suggest that virus resistance using an intragenic strategy will produce an end product with an increased likelihood of market acceptance.

References

Abel, P. P., R. S. Nelson, B. De, N. Hoffmann, S. G. Rogers, R. T. Fraley, and R. N. Beachy. 1986. "Delay of Disease Development in Transgenic Plants that Express the Tobacco Mosaic Virus Coat Protein Gene." *Science* 232 (4751): 738–43.

Agrios, G. N. 2005. *Plant Pathology*. 5th ed. Burlington, MA: Elsevier Academic Press.

Ahlborg, U., M. Akerblom, G. Ekström, C. Hogstedt, M. Jaghabir, J. Jeyaratnam, F. Kaloyanova, et al. 1990. *Public Health Impact of Pesticides Used in Agriculture*. Geneva: WHO/UNEP (World Health Organization/United Nations Environment Programme). http://whqlibdoc.who.int/publications/1990/9241561394.pdf.

APHIS (Animal and Plant Health Inspection Service). 1996a. "Asgrow Seed Co.; Availability of Determination of Nonregulated Status for Squash Line Genetically Engineered for Virus Resistance." June 27. Docket No. 96-024-2. *Federal Register* 61 (125): 33484–5. http://www.aphis.usda.gov/brs/aphisdocs2/95_35201p_com.pdf.

______. 1996b. "Cornell University and University of Hawaii; Availability of Determination of Nonregulated Status for Papaya Lines Genetically Engineered for Virus Resistance." September 16. Docket No. 96-024-2. *Federal Register* 61 (180): 48663–4. http://www.aphis.usda.gov/brs/aphisdocs2/96_05101p_com.pdf.

______. 1997. "Monsanto Co.; Receipt of Petition for Determination of Nonregulated Status for Potato Lines Genetically Engineered for Insect and Virus Resistance." Docket No. 97-094-1, November 20. *Federal Register* 62 (224): 61961–92. http://www.gpo.gov/fdsys/pkg/FR-1997-11-20/pdf/97-30508.pdf.

Broadbent, L. 1957. "Insecticidal Control of the Spread of Plant Viruses." *Annual Review of Entomology* 2 (1): 339–54.

Browning, K. S. 2004. "Plant Translation Initiation Factors: It Is Not Easy to Be Green." *Biochemical Society Transactions* 32 (4): 589–91.

Cavatorta, J. R. 2010. "Intragenic Virus Resistance in Potato." PhD Dissertation, Cornell University.

______, K. W. Perez, S. M. Gray, J. Van Eck, I. Yeam, and M. Jahn, 2011. "Engineering Virus Resistance Using a Modified Potato Gene." *Plant Biotechnology Journal* doi: 10.1111/j.1467-7652.2011.00622.x

______, A. E. Savage, I. Yeam, S. M. Gray, and M. M. Jahn. 2008. "Positive Darwinian Selection at Single Amino Acid Sites Conferring Plant Virus Resistance." *Journal of Molecular Evolution* 67: 551–9.

Charron, C., M. Nicolai, J. L. Gallois, C. Robaglia, B. Moury, A. Palloix, and C. Caranta. 2008. "Natural Variation and Functional Analyses Provide Evidence for Co-Evolution between Plant eIF4E and Potyviral VPg." *Plant Journal* 54 (1): 56–68.

Chisholm, S. T., S. K. Mahajan, S. A. Whitham, M. L. Yamamoto, and J. C. Carrington. 2000. "Cloning of the *Arabidopsis* RTM1 Gene, Which Controls Restriction of Long-Distance Movement of Tobacco Etch Virus." *Proceedings of the National Academy of Sciences of the United States of America* 97 (1): 489–94.

Collinge, D., O. Lund, and H. Thordal-Christensen. 2008. "What are the Prospects for Genetically Engineered, Disease Resistant Plants?" *European Journal of Plant Pathology* 121 (3): 217–31.

Converse, R. H., and M. T. Aliniazee. 1988. "Effects of Insecticides Applied in the Field on Incidence of Aphid-Borne Viruses in Cultivated Strawberry." *Plant Disease* 72 (2): 127–9.

Coutts, B. A., and R. A. C. Jones. 2005. "Suppressing Spread of Tomato Spotted Wilt Virus by Drenching Infected Source or Healthy Recipient Plants with Neonicotinoid Insecticides to Control Thrips Vectors." *Annals of Applied Biology* 146 (1): 95–103.

Dangl, J. L., and J. D. G. Jones. 2001. "Plant Pathogens and Integrated Defence Responses to Infection." *Nature* 411 (6839): 826–33.

Davidson, S. N. 2008. "Forbidden Fruit: Transgenic Papaya in Thailand." *Plant Physiology* 147 (2): 487–93.

Diaz-Pendon, J. A., V. Truniger, C. Nieto, J. Garcia-Mas, A. Bendahmane, and M. A. Aranda. 2004. "Advances in Understanding Recessive Resistance to Plant Viruses." *Molecular Plant Pathology* 5 (3): 223–33.

Ecobichon, D. J. 2001. "Pesticide Use in Developing Countries." *Toxicology* 160: 27–33.

EU (European Union). 2008. "Scientific Opinion of the Panel on Genetically Modified Organisms." *EFSA Journal* 756: 1–18.

Ferreira, S. A., K. Y. Pitz, R. Manshardt, F. Zee, M. Fitch, and D. Gonsalves. 2002. "Virus Coat Protein Transgenic Papaya Provides Practical Control of Papaya Ringspot Virus in Hawaii." *Plant Disease* 86 (2): 101–5.

Flor, H. H. 1971. "Current Status of the Gene-for-Gene Concept." *Annual Review of Phytopathology* 9 (1): 275–96.

Fraser, R. S. S. 1990. "The Genetics of Resistance to Plant Viruses." *Annual Review of Phytopathology* 28 (1): 179–200.

Fuchs, M., and D. Gonsalves. 2007. "Safety of Virus-Resistant Transgenic Plants Two Decades after Their Introduction: Lessons from Realistic Field Risk Assessment Studies." *Annual Review of Phytopathology* 45 (1): 173–202.

Gao, Z., E. Johansen, S. Eyers, C. L. Thomas, T. H. N. Ellis, and A. J. Maule. 2004. "The Potyvirus Recessive Resistance Gene, sbm1, Identifies a Novel Role for Translation Initiation Factor eIF4E in Cell-to-Cell Trafficking." *Plant Journal* 40 (3): 376–85.

Gingras, A. C., B. Raught, and N. Sonenberg. 1999. "eIF4 Initiation Factors: Effectors of mRNA Recruitment to Ribosomes and Regulators of Translation." *Annual Review of Biochemistry* 68: 913–63.

Goldbach, R., E. Bucher, and M. Prins. 2003. "Resistance Mechanisms to Plant Viruses: An Overview." *Virus Research* 92 (2): 207–12.

———, and D. Peters. 1994. "Possible Causes of the Emergence of Tospovirus Diseases." *Seminars in Virology* 5 (2): 113–20.

Gonsalves, D. 1998. "Control of Papaya Ringspot Virus in Papaya: A Case Study." *Annual Review of Phytopathology* 36: 415–37.

———. 2004. "Transgenic Papaya in Hawaii and Beyond." *AgBioForum* 7 (1&2): 36–40. http://www.agbioforum.org/v7n12/v7n12a07-gonsalves.pdf.

Goodman, R. N., and A. J. Novacky. 1994. *The Hypersensitive Reaction in Plants to Pathogens: A Resistance Phenomenon*. St. Paul, MN: APS Press.

Gutierrez-Campos, R., J. A. Torres-Acosta, L. J. Saucedo-Arias, and M. A. Gomez-Lim. 1999. "The Use of Cysteine Proteinase Inhibitors to Engineer Resistance against Potyviruses in Transgenic Tobacco Plants." *Nature Biotechnology* 17 (12):1223–6.

Hanna, A. D. 1954. "Application of a Systemic Insecticide by Trunk Implantation to Control a Mealybug Vector of the Cacao Swollen Shoot Virus." *Nature* 173 (4407): 730–1.

Holmes, F. O. 1938. "Inheritance of Resistance to Tobacco-Mosaic Disease in Tobacco." *Phytopathology* 28: 553–61.

Hong, Y., K. Saunders, M. R. Hartley, and J. Stanley. 1996. "Resistance to Geminivirus Infection by Virus-Induced Expression of Dianthin in Transgenic Plants." *Virology* 220 (1): 119–27.

James, C. 2007. *Global Status of Biotech/GM Crops*. Brief 37. Ithaca, NY: ISAAA (International Service for the Acquisition of Agri-Biotech Applications). http://www.isaaa.org/resources/publications/briefs/37/download/isaaa-brief-37-2007.pdf.

Jones, J. D., and J. L. Dangl. 2006. "The Plant Immune System." *Nature* 444 (7117): 323–9.

Kang, B. C., I. Yeam, and M. M. Jahn. 2005. "Genetics of Plant Virus Resistance." *Annual Review of Phytopathology* 43: 581–621.

———, I. Yeam, H. Li, K. W. Perez, and M. M. Jahn. 2007. "Ectopic Expression of a Recessive Resistance Gene Generates Dominant Potyvirus Resistance in Plants." *Plant Biotechnology* 5 (4): 526–36.

Kaniewski, W. K., and P. E. Thomas. 2004. "The Potato Story." *AgBioForum* 7 (2&3): 41–6.

Krishnan, R., K. A. McDonald, A. M. Dandekar, A. P. Jackman, and B. Falk. 2002. "Expression of Recombinant Trichosanthin, a Ribosome-Inactivating Protein, in Transgenic Tobacco." *Journal of Biotechnology* 97 (1): 69–88.

Kuhn, C. W., M. D. Jellum, and J. N. All. 1975. "Effect of Carbofuran Treatment on Corn Yield, Maize Chlorotic Dwarf and Maize Dwarf Mosaic Virus Diseases, and Leafhopper Populations." *Phytopathology* 65: 1017–20.

Lanfermeijer, F. C., J. Dijkhuis, M. J. G. Sturre, P. de Haan, and J. Hille. 2003. "Cloning and Characterization of the Durable Tomato Mosaic Virus Resistance Gene Tm-22 from *Lycopersicon esculentum*." *Plant Molecular Biology* 52: 1039–51.

Lockhart, B. E., and S. T. Sachey. 2001. "Cacao Swollen Shoot." In *Encyclopedia of Plant Pathology*, edited by O. C. Maloy and T. D. Murray, 172–3. New York: John Wiley & Sons.

Lodge, J. K., W. K. Kaniewski, and N. E. Tumer. 1993. "Broad-Spectrum Virus Resistance in Transgenic Plants Expressing Pokeweed Antiviral Protein." *Proceedings of the National Academy of Sciences of the United States of America* 90 (15): 7089–93.

Lusk, J., and P. Sullivan. 2002. "Consumer Acceptance of Genetically Modified Foods." *Food Technology* 56: 32–7.

Makkouk, K. M., and S. G. Kumari. 2007. "Epidemiology and Integrated Management of Persistently Transmitted Aphid-Borne Viruses of Legume and Cereal Crops in West Asia and North Africa." *Virus Research* 141(2): 209–18.

McGrath, P. F., and J. S. Bale. 1990. "The Effects of Sowing Date and Choice of Insecticide on Cereal Aphids and Barley Yellow Dwarf Virus Epidemiology in Northern England." *Annals of Applied Biology* 117 (1): 31–43.

Moury, B., C. Morel, E. Johansen, L. Guilbaud, S. Souche, V. Ayme, C. Caranta, A. Palloix, and M. Jacquemond. 2004. "Mutations in Potato Virus Y Genome-Linked Protein Determine Virulence toward Recessive Resistances in *Capsicum annuum* and *Lycopersicon hirsutum*." *Molecular Plant-Microbe Interactions* 17 (3): 322–9.

Murphy, J. F., J. R. Blauth, K. D. Livingstone, V. K. Lackney, and M. K. Jahn. 1998. "Genetic Mapping of the pvr1 Locus in *Capsicum* spp. and Evidence that Distinct Potyvirus Resistance Loci Control Responses that Differ at the Whole Plant and Cellular Levels." *Molecular Plant-Microbe Interactions* 11 (10): 943–51.

NASS (National Agricultural Statistics Service). 2008. *Papaya Acreage Survey 2008 Results*. U.S. Department of Agriculture. http://www.nass.usda.gov/Statistics_by_State/Hawaii/Publications/Archive/xpap08.pdf.

———. 2009. *2009 Papaya Utilization Down*. U.S. Department of Agriculture. http://www.nass.usda.gov/Statistics_by_State/Hawaii/Publications/Archive/xpap09.pdf.

Nielsen, K. M. 2003. "Transgenic Organisms – Time for Conceptual Diversification?" *Nature Biotechnology* 21 (3): 227–8.

Oerke, E. C., and H. W. Dehne. 2004. "Safeguarding Production – Losses in Major Crops and the Role of Crop Protection." *Crop Protection* 23 (4): 275–85.

Pathak, M. D., E. Vea, and V. T. John. 1967. "Control of Insect Vectors to Prevent Virus Infection of Rice Plants." *Journal of Economic Entomology* 60 (1): 218–25.

Pimentel, D., H. Acquay, M. Biltonen, P. Rice, M. Silva, J. Nelson, V. Lipner, S. Giordano, A. Horowitz, and M. D'Amore. 1993. "Assessment of Environmental and Economic Impacts of Pesticide Use." In *The Pesticide Question*, edited by D. Pimentel and H. Lehman, 47-84. New York: Springer.

Powell, D. M., and T. W. Mondor. 1973. "Control of the Green Peach Aphid and Suppression of Leaf Roll on Potatoes by Systemic Soil Insecticides and Multiple Foliar Sprays." *Journal of Economic Entomology* 66 (1): 170–7.

Provvidenti, R., and R. O. Hampton. 1992. "Sources of Resistance to Viruses in the Potyviridae." *Archives of Virology Supplementum* 5: 189–211.

Ritzenthaler, Christophe. 2005. "Resistance to Plant Viruses: Old Issue, New Answers?" *Current Opinion in Biotechnology* 16 (2): 118–22.

Rommens, C. M. 2004. "All-Native DNA Transformation: A New Approach to Plant Genetic Engineering." *Trends in Plant Science* 9 (9):457–64.

______. 2007. "Intragenic Crop Improvement: Combining the Benefits of Traditional Breeding and Genetic Engineering." *Journal of Agricultural and Food Chemistry* 55 (11): 4281–8.

______. 2008. "The Need for Professional Guidelines in Plant Breeding." *Trends in Plant Science* 13 (6): 261–3.

Satapathy, M. K. 1998. "Chemical Control of Insect and Nematode Vectors of Plant Viruses." In *Plant Virus Disease Control*, edited by A. Hadidi, R. K. Khetarpal, and H. Koganezawa, 188–95. St. Paul, MN: APS Press.

Schaad, M. C., A. D. Lellis, and J. C. Carrington. 1997. "VPg of Tobacco Etch Potyvirus Is a Host Genotype-Specific Determinant for Long-Distance Movement." *Journal of Virology* 71 (11): 8624–31.

Shankula, S. 2006. *Quantification of the Impacts on U.S. Agriculture of Biotechnology-Derived Crops Planted in 2005*. Washington, DC: NCFAP (National Center for Food and Agricultural Policy). http://www.ncfap.org/documents/2005biotechimpacts-finalversion.pdf.

Sonenberg, N., M. A. Morgan, W. C. Merrick, and A. J. Shatkin. 1978. "A Polypeptide in Eukaryotic Initiation Factors that Crosslinks Specifically to the 5'-Terminal Cap in mRNA." *Proceedings of the National Academy of Science of the United States of America* 75 (10): 4843–7.

Sudarshana, M. R., G. Roy, and B. W. Falk. 2007. "Methods for Engineering Resistance to Plant Viruses." *Methods in Molecular Biology* 354: 183–95.

Tepfer, M. 2002. "Risk Assessment of Virus-Resistant Transgenic Plants." *Annual Review of Phytopathology* 40: 467–91.

Tricoll, D. M., K. J. Carney, P. F. Russell, J. R. McMaster, D. W. Groff, K. C. Hadden, P. T. Himmel, J. P. Hubbard, M. L. Boeshore, and H. D. Quemada. 1995. "Field Evaluation of Transgenic Squash Containing Single or Multiple Virus Coat Protein Gene Constructs for Resistance to Cucumber Mosaic Virus, Watermelon Mosaic Virus 2, and Zucchini Yellow Mosaic Virus." *Nature Biotechnology* 13 (12): 1458–65.

Truniger, V., and M. Aranda. 2009. "Recessive Resistance to Plant Viruses." *Advances in Virus Research* 75: 119–59.

Waterworth, H. E., and A. Hadidi. 1998. "Economic Losses due to Plant Viruses." In *Plant Virus Disease Control*, edited by A. Hadidi, R. K. Khetarpal, and H. Koganezawa, 1–13. St. Paul: APS Press.

Whitham, S. A., R. J. Anderberg, S. T. Chisholm, and J. C. Carrington. 2000. "*Arabidopsis* RTM2 Gene Is Necessary for Specific Restriction of Tobacco Etch Virus and Encodes an Unusual Small Heat Shock-Like Protein." *Plant Cell* 12 (4): 569–82.

Whitham, S. A., S. McCormick, and B. Baker. 1996. "The N Gene of Tobacco Confers Resistance to Tobacco Mosaic Virus in Transgenic Tomato." *Proceedings of the National Academy of Sciences of the United States of America* 93 (16): 8776–81.

Woodford, J. A., B. D. Harrison, C. S. Aveyard, and S. C. Gordon. 1983. "Insecticidal Control of Aphids and the Spread of Potato Leafroll Virus in Potato Crops in Eastern Scotland." *Annals of Applied Biology* 103 (1): 117–30.

Zitter, T., and M. McGrath. 2006. *Tables of Disease Resistant Varieties*. Ithaca, NY: Cornell University. http://vegetablemdonline.ppath.cornell.edu/Tables/TableList.htm

6

Animal Biotechnologies and Agricultural Sustainability

Alison Van Eenennaam and William Muir

Animal production systems can be broadly classified into three categories: grazing, mixed crop-livestock, and intensive. Originally, all livestock production was grassland based and fell into the grazing category. Grassland systems effectively convert human-inedible materials to high-quality human food. Where climatic, soil, and disease conditions permitted, grassland-based systems developed into mixed crop-livestock systems. On a global basis, mixed crop-livestock systems involve the largest number of animals, generate the most total production, and serve the largest number of people (Seré and Steinfeld 1996). Intensive animal agriculture systems developed more recently, mostly in the vicinity of urban centers when urbanization and income exceeded certain levels. Intensive livestock production is the fastest growing category as a result of several factors, including declining real prices for feed grains, and improved feed conversion ratios (unit feed per unit animal product), animal health, and reproductive rates (Naylor et al. 2005). That animals of most species produce more product per animal in less time when fed nutrient-dense grain diets is one of the factors favoring the growth of intensive systems.

Large-scale intensive operations, in which animals are raised in confinement, now account for three-quarters of the global poultry supply, 40 percent of the pork supply, and more than two-thirds of all eggs (Bruinsma 2003). Intensive livestock production systems have dramatically reduced the amount of land needed to produce a unit of animal product, such as a gallon of milk or a pound of meat. For example, over the last century, advances in the genetics, nutrition, and management of U.S. dairy cows have resulted in more than a fourfold increase in milk production per cow, and a threefold improvement in productive efficiency (milk output per feed resource input; VandeHaar and St-Pierre 2006). However, the environmental and ethical sustainability of these intensive production systems is coming increasingly under scrutiny. Large-scale animal operations concentrate environmental pollutants and result in ecological disturbances, and consumers in some countries are increasingly concerned about the health and well-being of animals raised in concentrated animal production systems.

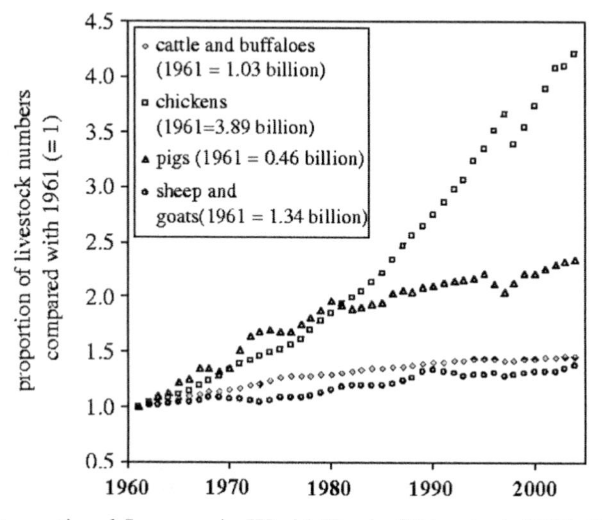

Figure 6.1. Proportional Increase in World Head of Livestock 1961–2004. *Source:* Pretty (2008).

These factors have led to opposition to the very existence of animal agriculture by some, whereas others question the production of meat in a world in which millions of people are starving. However, animal agriculture is an integral component of global food production systems. Animal products provide one-sixth of human food energy and more than one-third of the protein on a global basis (Bradford 1999). In less developed countries, much of this energy and protein has traditionally been derived from extensive, mixed-production systems in which livestock convert human-inedible materials (forages and byproducts) into high-quality human food. Animal agriculture also serves other functions, including the provision of draft power and transportation, nutrient recycling, wealth accumulation, and rangeland management functions, which are important to the efficiency and sustainability of food production systems. Evidence also exists to support the conclusion that the inclusion of foods of animal origin in the diets of young children with currently low levels of these foods leads to marked improvement in both physical and mental development (Allen et al. 1992; Grillenberger et al. 2007, 2006; Neumann et al. 2007).

Since the early 1960s, livestock production has grown rapidly, with a worldwide fourfold increase in the number of chickens, a twofold increase in the number of pigs, and a 40 to 50 percent increase in the numbers of cattle, sheep, and goats (Figure 6.1). Meat demand is expected to rise rapidly with continued economic growth, which will have important ramifications for world agricultural production systems. This rapidly growing demand for livestock products has been coined the "Livestock Revolution," after the better known "Green Revolution." To put this in perspective, from the beginning of the 1970s to the mid-1990s, consumption of meat and milk in developing countries increased by 70 and 105 Tg, respectively. The market value of that increase totaled approximately $155 billion (real 1990 dollars), which was more

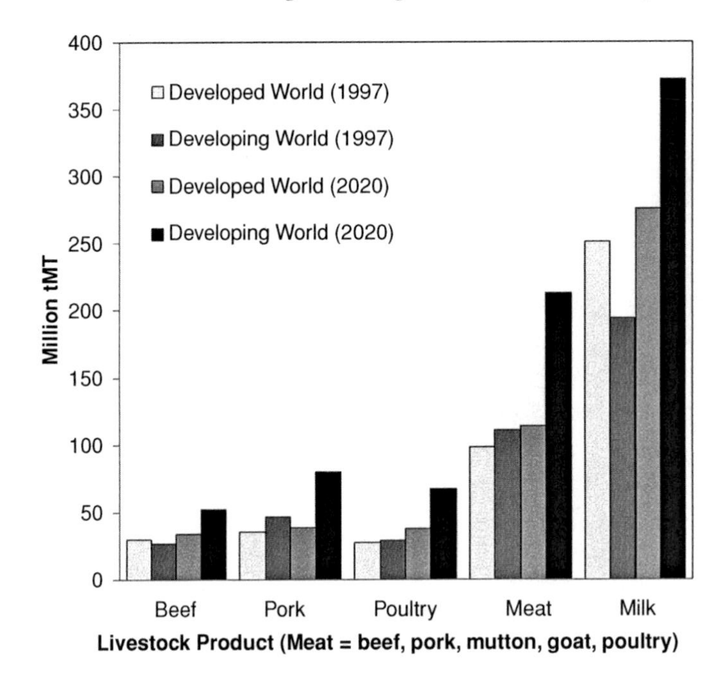

Figure 6.2. Projected Food Consumption Trends of Various Livestock Products to the Year 2020. *Source:* Based on data derived from Delgado (2003).

than *twice* the market value of increased wheat, rice, and maize consumption resulting from the Green Revolution (Delgado 2003). The Livestock Revolution is primarily being driven by demand: Lower income people everywhere are eating more animal products as their incomes rise. It is estimated that by 2020 developing countries will consume 107 Tg more meat and 177 Tg more milk than they did in 1996/1998, dwarfing developed-country increases of 19 Tg for meat and 32 Tg for milk (Figure 6.2). The growing demand for animal products in developing countries cannot be ignored and shows no evidence of diminishing.

Currently 40 percent of the land on earth is used for food production (Nonhebel 2005). It is estimated that more than 60 percent of the arable land is used for the production of animal feeds. In industrialized countries, 73 percent of cereals are fed to animals, whereas in developing countries some 37 percent are (Pretty 2008). Projections made by FAO show an approximate global doubling of the demand for animal food products in the period 2000–20 due to population growth and increases in consumption in developing countries, with poultry, meat, and egg consumption rising markedly more quickly than beef and pork (Bruinsma 2003). Table 6.1 shows the projected increase in the use of cereals as feed through 2020 needed to meet this increased demand for animal products.

On a global scale, doubling the land required for feed production is not possible simply because the quantity of good agricultural soils is insufficient. High-quality arable

Table 6.1. *Past and projected trends in use of cereal as animal feed to the year 2020*

	Annual growth rates (percent per year)			Per capita cereal use as feed (kg)	
	Cereal production	Cereal use as feed	Projected cereal use as feed		
Region	1982–93	1982–93	1993–2020	1993	2020
China	2.0	5.8	3.2	62	120
India	3.2	3.5	3.0	4	6
Other East Asia	−2.0	6.7	2.5	115	183
Other South Asia	2.1	1.5	2.9	7	8
Southeast Asia	2.4	8.6	2.9	32	49
Latin America	0.7	2.5	1.9	118	137
West Asia/North Africa	3.9	1.8	2.1	92	94
Sub-Saharan Africa	4.1	5.3	2.3	4	4
United States	0.0	1.0	0.9	603	622
Developing World	2.3	4.3	2.6	45	62
Developed World	0.2	−0.1	0.7	346	386
WORLD	1.3	0.9	1.4	115	120

Source: Bradford (1999).

land is becoming scarcer because of ongoing industrialization, urbanization, infrastructural development, and desertification. Given this fact, this chapter focuses on ways that biotechnologies may help animal agriculture meet the growing demand for animal products more sustainably – by balancing environmental, social, and economic goals. We recognize that there are often tradeoffs among these sustainability goals; therefore it is impossible to present a single animal production system or biotechnology that will satisfy all aspects of sustainability concurrently. We outline biotechnologies that may assist animal agriculture to become more efficient, decrease its impact on the environment, and improve animal well-being. Although some biotechnological approaches are prohibited by agricultural production systems that are purported to be sustainable, we consider that any biotechnology that works to improve efficiency or animal well-being and does not deleteriously affect the environment is likely to have some sustainability benefits.

What Is Animal Biotechnology?

Biotechnology can be defined as the application of science to living organisms. From this definition it is clear that a broad range of strategies for the genetic improvement of livestock, including widely used practices such as selective breeding, artificial insemination, and embryo transfer, qualify as animal biotechnologies. More recently, the term has become associated with the controversial technologies of GE and cloning. For the purposes of this chapter we discuss the subset of biotechnologies that focus on materials and methods related to the genetic manipulations of RNA and DNA and on genomic approaches to improve the sustainability of animal agriculture. However, we do recognize that traditional animal breeding and husbandry practices have resulted in major improvements in the efficiency of animal agriculture. For example, the average time to produce a broiler chicken in the United States was reduced from 72 days in 1960 to 48 days in 1995, and the slaughter weight rose from 1.8 to 2.2 kg. Concurrently, feed conversion ratios (kg feed/kg gain) were reduced by 15 percent. These remarkable improvements in production efficiency have dramatically decreased the inputs required to produce a unit of output, although it might be argued that the processes were employed without adequately considering environmental, social, and animal welfare goals. Broiler improvements highlight the point that environmental and animal well-being concerns associated with efficiency gains are not the sole purview of modern DNA-based biotechnologies. They are equally associated with the use of conventional breeding methods for the genetic improvement of livestock.

Animal breeders are always trying to maximize the response to selection. This is defined as the difference in the mean phenotypic value between the offspring of the selected parents as compared to that of the whole of the parental generation before selection. The genetic gain (ΔG) in animal breeding programs can be calculated as

$$\Delta G = \frac{i^* r^* \sigma_A}{L}$$

where i is the intensity of selection, r is the accuracy of selection, σ_A is the additive genetic standard deviation in the potential parent population, and L is the generation interval. If biotechnologies affect any of the variables in this equation, they can influence the genetic gain per generation. For example, increasing the intensity of selection (i.e., the proportion of animals in the parental population that are actually selected to produce offspring) can be achieved using a variety of approaches such as artificial insemination or cloning to maximize the use of superior breeding stock. The accuracy of selection can be increased through progeny testing programs or by using information from genetic markers. Breeders can increase the amount of genetic variability that exists in the prospective parental population (e.g., by increasing the number of breeds of potential selection candidates, or by using either GE to bring in new traits, or by increasing the additive genetic variability of existing traits). Likewise, they can also decrease the generation interval by selecting animals at a younger age or through the use of assisted reproductive technologies. Any biotechnology that can affect one of these four factors influencing genetic gain will be of potential value to animal breeders.

Genetic Engineering

The first GE livestock were generated more than 25 years ago using pronuclear microinjection techniques (Hammer et al. 1985). Since then a modest number of GE animals have been developed, and many more are envisioned for agricultural applications (Table 6.2). GE animals carry a segment of recombinant deoxyribonucleic acid (rDNA) – the transgene – that is stably transmitted to their offspring in a Mendelian fashion. The efficiency of pronuclear microinjection is low; usually only 3 to 5 percent of the animals born carry the rDNA, and not all lines express the transgene. An encouraging recent development is the use of viral vectors, particularly those based on lentiviruses, which results in much higher rates of germ-line positive GE animals (Golding et al. 2006). Additionally, nuclear transfer techniques (cloning) have been used to produce GE animals with precise genetic changes; for example, animals with targeted disruption of endogenous genes including cattle lacking the prion protein responsible for bovine spongiform encephalopathy (BSE) or "mad cow disease" (Kuroiwa et al. 2004; Richt et al. 2007). Despite these advances, the techniques for generating transgenic livestock and poultry remain somewhat inefficient and expensive, and at the current time no food products derived from GE livestock have reached the marketplace.

There are a number of GE livestock applications – both extant and proposed – that could be envisaged to align with agricultural sustainability goals (Table 6.2). These applications include GE animals with improved product quality, reduced environmental impact, and enhanced disease resistance. Additionally, there are GE examples where animals have improved productivity, including increased milk production and growth rate, improved feed utilization, enhanced reproductive performance, and/or increased prolificacy.

Potential Benefits

Breeding: Conventional breeding programs are limited to naturally occurring genetic variation in the parent population. GE offers a way to increase the genetic variability available for selection (i.e., the additive genetic variance term from Equation 1). It is likely to be most useful in developing novel genetic traits (e.g., providing mammals the ability to endogenously synthesize n-3 fatty acids) or genetic variation that did not exist in a specific population, breed, or even species. Transgenic laboratory animals have become increasingly important for biological and biomedical research, and the scientific literature associated with these applications is vast and growing. For example, there was a tenfold increase in the number of GE animals used in research in the United Kingdom between 1995–2005 (Lane 2005). Transgenic livestock are also increasingly being produced specifically as biomedical research models (Forsberg 2005; Petters et al. 1997). The public is mostly supportive of such applications, and

Table 6.2. *Extant and envisioned genetically engineered livestock applications for agriculture*

EXTANT APPLICATIONS	Species	Gene	Approach	Reference
PRODUCTIVITY				
Enhanced growth rate	Various fish species	Growth hormone	Transgene expression	Aerni (2004); Bessey et al. (2004); Cook et al. (2000); Martinez et al. (2000); Nam et al. (2001); Rahman et al. (1998)
Enhanced milk production	Swine	α-lactalbumin	Transgene expression	Marshall et al. (2006); Wheeler, Bleck, and Donovan (2001)
Enhanced growth rate	Swine	Growth hormone	Transgene expression	Pursel et al. (1989); Pursel et al. (1997)
Enhanced growth rate	Swine	Insulin-like growth factor (IGF1)	Transgene expression	Pursel et al. (2004)
DISEASE RESISTANCE				
BSE resistance	Cattle, goats, and sheep	Prion	Knockout	Denning et al. (2001); Richt et al. (2007); Yu et al. (2006)
Mastitis resistance	Cattle	Lysostaphin	Transgene expression	Wall et al. (2005)
Mastitis resistance	Cattle	Lactoferrin	Transgene expression	van Berkel et al. (2002)
BSE resistance	Goat	Prion	RNAi transgene	Golding et al. (2006)
Visna virus resistance	Sheep	Visna virus envelope gene	Transgene expression	Clements et al. (1994)
Mastitis resistance	Goats	Lysozyme	Transgene expression	Maga, Cullor, et al. (2006); Maga, Shoemaker, et al. (2006)
GCH virus resistance	Grass Carp	Lactoferrin	Transgene expression	Zhong et al. (2002)
Bacterial resistance	Channel Catfish	Cecropin B gene	Transgene expression	Dunham et al. (2002)
Prevent spread of avian influenza	Chicken	Decoy molecule	Transgene expression	Lyall et al. (2011)

ENVIRONMENTAL				
Decreased P in manure	Swine	Phytase	Transgene expression	Golovan et al. (2001)
PRODUCT QUALITY				
Increased ω-3 fatty acids in meat	Swine	n-3 fatty acid desaturase	Clone/transgene expression	Lai et al. (2006)
Increase cheese yield from milk	Cattle	β-casein, κ-casein	Clone/transgene expression	Brophy et al. (2003)
Increased mono-unsaturates in milk	Goat	Rat stearoyl-CoA desaturase	Transgene expression	Reh et al. (2004)

ENVISIONED APPLICATIONS	Species	Target	Proposed approach	Background information
Increased lean-muscle growth	Cattle	Myostatin	RNAi /knockout	McPherron and Lee (1997)
Increased postnatal growth	Various	Socs2	RNAi /knockout	Horvat and Medrano (2001)
Enhanced mammary gland development	Various	Socs1	RNAi /knockout	Lindeman et al. (2001)
Suppressing infectious pathogens	Various	RNA viruses (e.g., foot and mouth, fowl plague, swine fever)	RNAi	Clark and Whitelaw (2003); Whitelaw and Sang (2005)
Coronavirus resistance	Swine	Aminopeptidase N	RNAi /knockout	Schwegmann-Wessels et al. (2002)
Avian flu resistance	Poultry	Avian influenza	RNAi	Sang (1994); Tompkins et al. (2004)
Low-lactose milk	Cattle	Lactase	Transgene expression	Jost et al. (1999)
Low-lactose milk	Cattle	α-lactalbumin	RNAi /knockout	Stacey et al. (1995)
Increased ovulation rate	Sheep	GDF9, BMP15, ALK6/BMPR1B	RNAi /knockout	Melo et al. (2007)
High omega-3 fatty acid milk	Cattle	n-3 and n-6 fatty acid desaturase	Transgene expression	Morimoto et al. (2005)
Resistance to brucellosis	Cattle	NRAMP1	Transgene expression	Barthel et al. (2001)
Decreased P in manure	Poultry	MINPP	Transgene expression	Cho et al. (2005)
Decreased P in manure	Poultry	Phytase	Transgene expression	Guenther et al. (2005)

scientists pursuing such research are generally viewed as contributing toward societal good. Interestingly, the production and use of this vast number of transgenic animals for research purposes, estimated in Mak (2008) to be 10–50 million animals annually in the United States, have received little attention or comment from either the activist or the scientific community.

Sustainability: Whether GE livestock fit in with sustainability goals is greatly dependent on which goal and production system one is considering. However, some GE livestock applications (e.g., disease resistance) would seem to align with almost any definition of sustainability and clearly with the goal of improving animal well-being. Infectious diseases have major negative effects on poultry and livestock production, both in terms of economics and animal welfare. The costs of disease are estimated to be 35 to 50 percent of turnover in developing countries and 17 percent in the developed world. Improving animal health using GE has an added benefit in that it reduces the need for veterinary interventions and the use of antibiotics and other medicinal treatments. However, on ideological grounds, some may determine that disease-resistant GE animals have no place in sustainable production systems. For example, the standards of the U.S. Department of Agriculture (USDA) National Organic Program (NOP) specifically prohibit the use of GE. Such regulatory decisions should consider the fact that the combined employment of both disease-resistant livestock and improved animal management practices to minimize disease incidence is compatible with multiple sustainability goals. For example, the 2001 foot and mouth outbreak in the United Kingdom resulted in the slaughter and incineration of 4,078,000 animals (McConnell and Stark 2002). Clark and Whitelaw (2003) speculated that raising GE foot and mouth disease–resistant livestock might align more with sustainability goals than the mass slaughter and adverse environmental consequences that were associated with this disease outbreak in the United Kingdom.

Similarly, the use of more productive GE animals – animals that produce more units of output, such as gallons of milk or pounds of meat, with the same or fewer inputs – should be given due consideration in the context of sustainability. Consider the example of the AquaAdvantage™ salmon, the first GE food animal to go through the U.S. regulatory approval process. Since the mid-1980s, the yield of food fish from capture fisheries has been static at about 60 Tg per year. The growth of the fish supply since that time has largely come from aquaculture. Fletcher et al. (2004) calculated that an extra 52 Tg of aquaculture production will be needed by 2025 if the current rate of fish consumption is to be maintained. Atlantic salmon remains the most important farmed food fish in global trade. Salmon is a carnivorous fish, and aquaculturalists have been working to improve feed conversion rates and efficiencies through selective breeding and inclusion of plant-based protein (soy, rapeseed oil, and corn gluten) in feed formulations. As a consequence, feed input per fish has decreased to 44 percent of 1972 levels; likewise, current diets contain approximately half the content of fishmeal that they did at that time (Aerni 2004).

The AquaAdvantage salmon is an Atlantic salmon carrying a Chinook salmon growth hormone gene controlled by an antifreeze protein promoter from a third species, the ocean pout. The mature weight of these fish remains the same as other farmed salmon, but their growth rate is increased by 400 to 600 percent, with a concomitant 25 percent decrease in feed input, decreased waste per unit of product, and a shortened time to market (Cook et al. 2000; Du et al. 1992). Unlike other food animal species in which selective breeding programs have been ongoing for decades, many fish farmers are still reliant on brood fish collected from the wild. GE may offer one component of an approach to sustainably increase the efficiency of aquacultural production to meet the needs of the 21st century.

Concerns

Social acceptance and sustainability goals: After publication of a paper detailing the generation of a transgenic pig able to endogenously produce omega-3 fatty acids (Lai et al. 2006), a letter to the editor in that same journal criticized the research, stating that "the use of transgenic technology for this application represents the worst kind of research waste" and that "the animal biotech industry needs to confine its work to projects necessary for the achievement of important health, safety, or medical goals" (Fiester 2006). This criticism came despite the fact that an overwhelming number of studies document both the health benefits of increased omega-3 fatty acids in the diet (Connor 2000; Simopoulos 2004) and the fragility of the current supply of long-chain omega-3 fatty acids (Pauly et al. 2003). Additionally, traditional animal breeders are actively pursuing quantitative trait loci (QTL) associated with fatty acid composition in swine meat (Clop et al. 2003; Nii et al. 2006), and both academic and commercial plant metabolic engineering groups have been vigorously pursuing the land-based production of long-chain omega-3 fatty acids in plants (Domergue et al. 2005; Robert 2006; Robert et al. 2005; Wu et al. 2005). It is an interesting conundrum that agricultural research using GE animals to achieve a goal is considered a "research waste," a verboten mechanism to achieve the same goal that other researchers and companies consider to be an important one and that the scientific community is actively pursuing using a variety of non-GE approaches.

Other examples of transgenic livestock that have been developed for agricultural applications have likewise been subject to wide-ranging criticism from a variety of sources including activists, the popular media, and scientific colleagues. Transgenic animals with disease resistance attributes (Maga, Shoemaker, et al. 2006; Richt et al. 2007; Wall et al. 2005) and potential environmental benefit have been critiqued not for their phenotypes, but rather for the production systems that led to the problems that they attempt to address. Take the example of the GE "Enviropig." It produces dramatically lower manure phosphorus levels because it produces the enzyme phytase in its saliva and is therefore able to metabolize dietary phytate (Golovan et al. 2001).

Given the large increase that is expected in both pig and poultry production in the developing world over the next 20 years, decreasing the phosphorus levels in the manure of these monogastric species would likely have a huge worldwide environmental benefit (CAST 2006). However, using GE to reduce the levels of this important pollutant in swine manure has been subject to the criticism that this kind of approach encourages "non-sustainable, un-ecological approaches to livestock management," according to E. Ann Clark (Philipkoski 2001, para. 13). In that article, critics argue that if farmers really want to be environmentally friendly, they should let pigs graze on greens instead of feeding them grain. However, outdoor pig farming can exacerbate nutrient leaching into the soil and groundwater (Williams et al. 2000). This example highlights the fact that sustainability goals are sometimes in conflict, and managing a system for optimal environmental stewardship may clash with animal welfare objectives (Siegford, Powers, and Grimes-Casey 2008). Evaluation of the sustainability of animal agricultural systems therefore depends on the sustainability goal(s) under consideration.

To illustrate this point, an interesting case study from Sweden examined three scenarios for pig production based on different sustainability goals (Stern et al. 2005). The first focused on animal welfare and the natural behavior of the animals. The second focused on environmental goals and the efficient use of natural resources. The third focused on product quality and safety. The cost per pound of pork and land use was highest for the animal welfare scenario and similar for the other two scenarios. Not surprisingly, the environmental scenario had the lowest environmental impact using the life-cycle assessment (LCA) methodology. Stern et al. (2005, 402) summarized that "each scenario fulfilled different aspects of sustainability, but there were goal conflicts because no scenario fulfilled all sustainability goals." The authors also wrote that the evaluation and ranking of sustainability goals are mainly a political question. Leaving sustainability goal evaluation to the political process potentially exposes the process to subjective interpretation and political pressure from special interest groups. It would seem preferable to allow science to objectively evaluate the sustainability implications of different agricultural systems. This evaluation method would give the scientific community an opportunity to develop measures of product and system-level performance to assess and compare the ability of different systems to sustainably meet the needs of both animal and human populations.

Science-based concerns: The main science-based concerns associated with the use of GE food animals relate to food safety, the health and well-being of the animal, and the environment. A report by the National Academy of Sciences considered the ability of GE organisms, particularly fish and insects, to escape confinement and become feral to be the greatest concern facing the animal biotechnology industry (NRC 2002). Models have predicted that under certain circumstances, the interbreeding of GE fish with increased fitness attributes (e.g., younger age at sexual maturity or increased mating success) could have serious ecological consequences for native fish populations (Muir and Howard 1999, 2001, 2002).

The actual environmental risk posed by each species/transgene combination will depend on a number of factors, including the containment strategy(s), species mobility, ability to become feral, net fitness of the transgenic animal, genotype by environmental interactions, and the stability of the receiving community. Likewise, food safety concerns related to transgenic animals will be similarly case-specific depending on the attributes of the recombinant protein being expressed and whether it is intended to be a pharmaceutical, industrial, or food protein.

Regulatory concerns: There has been public discussion about whether the U.S. Coordinated Framework for the Regulation of Biotechnology, first published in the *Federal Register* on December 31, 1984, will be able to adequately address the safety and commercialization issues associated with the introduction of GE animals into the food supply. In January 2009, the U.S. Food and Drug Administration (FDA) issued a final guidance for industry on the regulation of GE animals (CVM of FDA 2009). The guidance explains the process by which FDA regulates GE animals and provides a set of recommendations to producers of GE animals to help them meet their obligations and responsibilities under the law. That document clarifies that the Center for Veterinary Medicine of the FDA plans to regulate GE animals under the new animal drug provisions of the Federal Food Drug and Cosmetics Act (FFDCA). The FFDCA requires that each new animal drug be approved through a new animal drug application (NADA) based on a demonstration that it is safe and effective for its intended use. The rationale behind regulating GE animals using the new animal drug approach is based on the fact that the rDNA construct in a GE animal is intended to affect the structure or function of the body of the GE animals. Under this interpretation, the rDNA construct meets the FFDCA definition of a drug. Use of a new animal drug is unsafe unless the FDA has approved a NADA based on a demonstration that it is safe and effective for its intended use. All transgenic animals are subject to these premarket approval requirements. The new animal drug regulatory approach focuses on three questions: (1) Is the new animal drug safe for the animal?; (2) is the new animal drug effective?; and (3) if the drug is for a food-producing animal, is the resulting food safe to eat? The FDA new animal drug approval process does not consider ethical and social concerns; regulatory approvals are based solely on safety and effectiveness.

Ethical concerns: These concerns include fundamental objections to the manipulation and use of animals, objections to specific modifications, and concerns about the consequences of genetic modifications, but many of these issues are not unique to GE livestock. Additionally, the current inefficiency of transgenic techniques results in the production of many more animals than would be necessary under higher success rates. To date, unknown consumer acceptance and uncertainties in the regulatory timeline have effectively halted commercial investment in the development of GE livestock for agricultural applications in the United States. There has been little public discussion of the consequences of not using this significant technology on animal well-being and especially for the development of disease-resistant animals (Murray and Maga 2009).

Cloning

A clone is an organism that is descended from and genetically identical to a single common ancestor. Cloning involves making genetically identical copies of an animal using asexual reproduction. Animals can be cloned by two different methods: mechanical embryo splitting or nuclear transfer. Embryo splitting involves bisecting the multicellular embryo at an early stage of development to generate clones or "twins." A 32-cell embryo, for example, might be bisected into two 16-cell twins. This type of cloning occurs naturally (human identical twins result from this process, but fraternal twins do not); it can also be performed in a laboratory, where it has been successfully used to produce clones from a number of different animal species. This technique was first used in agriculture to replicate valuable dairy breeding animals in the 1980s. The Holstein Association USA registered its first embryo split clone in 1982, and more than 2,300 had been registered by October 2002 (Norman and Walsh 2004). This method has a practical limitation in cattle (Johnson et al. 1995) and sheep (Willadsen 1981), in that a maximum of four clones can be produced from each embryo.

Cloning can also be accomplished by nuclear transfer, in which the genetic material from the nucleus of one cell is placed into a recipient egg. A recipient egg is an unfertilized egg that has had its own genetic material removed by enucleation. To begin the developmental process, the donor nucleus must be fused with the recipient egg through the administration of a brief electrical pulse or a chemical fusion process, after which the embryo starts to divide as if it had been fertilized. In the case of mammals, the embryo is then placed into a surrogate mother, where it will develop until birth and will be delivered as any newborn. Mammals were first cloned via nuclear transfer during the early 1980s, almost 30 years after the initial successful experiments with frogs (Briggs and King 1952). Numerous mammalian clones followed – including mice, rats, rabbits, pigs, goats, sheep (Willadsen 1986), cattle (Robl et al. 1987), and even two rhesus monkeys named Neti and Detto (Meng et al. 1997) – all as a result of nuclear transfer. The Holstein Association USA registered its first embryo nuclear transfer clone in 1989, and approximately 1,200–1,500 cows and bulls were produced by embryonic cell nuclear transfer in North America in the 1980s and 1990s (Yang et al. 2007). Because all of these clones were produced from the transfer of nuclei derived from early (8- to 32-cell) embryos, a theoretical maximum of only 32 clones could be produced from each individual embryo.

In 1996, the famous cloned sheep, Dolly, was born. Dolly was the first animal to be cloned via nuclear transfer from a cultured somatic cell derived from an adult (Wilmut et al. 1997). This process, known as somatic cell nuclear transfer (SCNT) cloning, opened the way for cloning to be performed on a potentially unlimited number of cells from adult animals. It allowed cloning technology to be extended to make copies of elite breeding animals with well-established breeding superiority based on their own performance records and those of their offspring. A diverse range of species have now been successfully cloned from adult tissues using SCNT, including cattle (Kato et al. 1998), mice (Wakayama et al. 1998), pigs (Polejaeva et al. 2000), cats (Shin et al. 2002), rabbits (Chesne et al. 2002), horses (Galli et al. 2003), goats (Keefer et al. 2001), dogs (B. C. Lee et al. 2005), rats (Zhou et al. 2003), and zebra fish (K. Y. Lee et al. 2002). In October 2007, there were approximately 500–600 SCNT livestock clones in the United States (B. Glenn, pers. comm.).[1] Very few of these clones of valuable breeding stock will enter the human food supply themselves; instead, food products like milk and meat will likely be derived from the sexually produced offspring of these SCNT clones.

Potential Benefits

Clones may provide a genetic insurance policy in cases of extremely valuable animals or by producing several identical genetically superior sires in production environments where artificial insemination (AI) is not a feasible option. This practice effectively provides a source of "proven" bulls with superior breeding values relative to the bulls that might otherwise have been used (i.e., increasing both the intensity and accuracy of selection terms from Equation 1). Clones could conceptually also be used to reproduce a genotype that is particularly well suited to a given environment. The advantage of this approach is that a genotype that is proven to do especially well in a particular location could be maintained indefinitely, without the genetic shuffle that normally occurs every generation with conventional reproduction. This genetic shuffle effectively decreases the accuracy term from Equation 1, until that young bull has been subsequently proven to carry superior genes though progeny testing.

Although cloning of elite breeding stock as a genetic insurance policy may provide a limited market for clones, the most significant impact of cloning will likely result from methods to make targeted genetic modifications to cells before using them for SCNT cloning (Forsberg 2005). Cloning enhances the efficiency of GE by offering the opportunity to produce 100 percent transgenic offspring from cell lines that are known to contain the transgene. This has already enabled the generation of animals

[1] Barbara Glenn is a former Managing Director, Animal Biotechnology at the Biotechnology Industry Organization.

with agricultural potential, including cattle without the prion protein responsible for bovine BSE (Richt et al. 2007), pigs able to endogenously produce meat with omega-3 fatty acids (Prather 2006), and dairy cows that express elevated levels of milk proteins in their milk (Brophy et al. 2003). The ability to make targeted changes in cell culture and its subsequent cloning opens the way for the previously impracticable targeted deletion of undesirable traits and for the more efficient addition of desirable traits using GE techniques.

Concerns

The proportion of adult somatic cell nuclei that successfully develop into live offspring after transfer into an enucleated egg is very low (Tsunoda and Kato 2002). High rates of pregnancy loss have been observed after transfer of the eggs containing the adult cell nuclei into recipient animals (Hill et al. 2000). On average, only 9 percent of transferred embryos result in calves, with efficiencies ranging from 0 to 45 percent depending on the type of somatic tissue from which the transferred nucleus was derived (Beyhan et al. 2007). The problems associated with the cloning process are not unique to SCNT cloning, and all have been observed in animals derived via other commonly used assisted reproductive technologies (e.g., embryo transfer, in vitro fertilization) and even natural mating (Rudenko, Matheson, and Sundlof 2007). However, the frequency of these problems tends to be higher in SCNT clones.

Various abnormalities, such as "large offspring syndrome" (in which cloned lambs and calves are often large at birth), placental abnormalities, edema, and perinatal deaths, have been observed in cloned animals with frequencies that are at least partially dependent on the type of somatic tissue from which the transferred nucleus was derived. On average, 42 percent of cloned calves die between delivery and 150 days of life (Panarace et al. 2007). Although cloning poses no risks that are unique or distinct from those encountered in modern agricultural practices, the frequency of the risks is increased in cattle during the early stages of the life cycle. However, some adult cloned cows have been observed to have normal breeding and calving rates, and cloned bulls produce high-quality semen and have normal fertility when used for artificial insemination and natural mating. To date, there has been no evidence of clone-associated abnormalities being passed on to their offspring following sexual reproduction. This suggests that abnormalities seen in clones are not heritable and appear to be corrected during gametogenesis (the formation of eggs and sperm).

Studies examining the composition of food products derived from clones have found that they have the same composition as milk or meat from conventionally produced animals (Heyman et al. 2007; Laible et al. 2007; Norman and Walsh 2004; Takahashi and Ito 2004; Tian et al. 2005; Tomé, Dubarry, and Fromentin 2004; Walker et al. 2007; Walsh et al. 2003; Yamaguchi, Ito, and Takahashi 2007; Yang et al. 2007). Milk and meat from clones produced by embryo splitting and nuclear transfer of embryonic cells have been entering the human food supply for more than 20 years with no evidence of problems. Nevertheless, in 2001 the Center for Veterinary Medicine at the FDA determined that it should complete a comprehensive risk assessment to identify hazards and characterize food consumption risks that may result from SCNT animal clones (Rudenko and Matheson 2007), and it asked companies not to introduce these cloned animals, their progeny, or their food products (e.g., milk or meat) into the human or animal food supply. Because there was no fundamental reason to suspect that clones would produce novel toxins or allergens, the main underlying food safety concern was whether the SCNT cloning process resulted in subtle changes in the composition of animal food products (Rudenko et al. 2004).

On January 15, 2008, the FDA published its final 968-page risk assessment on animal cloning, which examined all existing data relevant to the health of clones and their progeny or to food consumption risks resulting from their edible products. This report, which summarized all available data on clones and their progeny, concluded that meat and milk products from cloned cattle, swine, and goats, and the offspring of clones of any species traditionally consumed as food, were as safe to eat as those of conventionally bred animals (CVM of FDA 2008). This conclusion opened the door for animal products from SCNT clones and their offspring to enter the food supply.

Currently, the cost of obtaining cloned animals (~$15,000/head for cattle) is prohibitive for commercial cattle producers. However, if costs were to decrease so that cloning became more widely adopted, some have expressed concerns regarding its potential to decrease genetic diversity and render livestock populations vulnerable to a catastrophic disease outbreak or be singularly ill suited to changes that may occur in the environment. Although such criticisms could be equally relevant to many forms of assisted reproductive technology (e.g., artificial insemination, embryo transfer), for any single genotype to prove superior in all economically relevant traits across all production systems and environments is unlikely. Even if clones were to become widely used, producers in different regions would likely select different clonal lines from a range of breeds based on their decision as to which genotype best matched their region and production environment.

Genomic Selection

Traditional genetic improvement of livestock relies on developing accurate genetic merit predictions or "breeding values" for animals based on their performance and that of their ancestors and offspring. Selection of animals with the best breeding values for production traits has been very effective in improving the efficiency of livestock production. However, selection has not been as successful for traits that are difficult to measure such as disease resistance or traits that are not available until late in an animal's life, such as fertility or longevity.

The cow, chicken, and pig genomes have all recently been sequenced, and this has led to the discovery of many thousands of naturally occurring DNA sequence variations between individuals in the form of single nucleotide polymorphisms (SNPs). Researchers are now working to determine which variations are associated with desirable characteristics, such as disease resistance. It is hoped that using information on variation in DNA sequence between animals will help improve the accuracy of breeding values; that is, it will give breeders more confidence they are selecting the best animals. Additionally, because DNA is available from birth, it may be possible to predict the genetic potential of animals at a very young age and then keep only the very best animals for breeding purposes. This may pave the way for producers to select animals to become parents of the next generation based on breeding values calculated from DNA marker data alone, a process called "whole genome selection."

Whole genome selection involves the simultaneous use of a large number of markers (e.g., the 50,000 SNP bovine panel) to predict the genetic merit of genotyped animals for many different traits. This approach relies on a two-step analysis involving "training data" to estimate molecular breeding values (MBVs) of SNP haplotypes (Meuwissen, Hayes, and Goddard 2001) or alleles (Solberg et al. 2008). An overall measure of the genetic merit for genotyped individuals outside the training dataset can then be obtained by genotyping that animal and adding up the genetic merit of each of the chromosome fragments inherited. This process allows prediction of MBV at an early age, thereby removing many limitations of current phenotype-based breeding programs and providing a clear time advantage in developing genetic estimates for sex-limited traits or for traits that are not available until late in an animal's life, such as fertility or longevity. Genomic selection has the potential to affect the generation interval (age at sire selection), and both the intensity and accuracy of selection components of Equation 1.

Genomic technologies may also offer new opportunities to develop management systems to optimize the production environment based on an animal's DNA genotype. For example, the genotype of some beef and dairy cattle may be better suited to grass-based production systems. It may also be possible to select animals that are able to grow to a certain size using less feed or that are more resistant to certain diseases. These technologies have the potential to achieve sustainability goals, including the production of safer and more nutritious food with less environmental impact and improved animal welfare due to lower disease incidence.

Potential Benefits

If additive genetic merit can be precisely predicted from MBV (i.e., increase the accuracy of selection term in Equation 1), the design of breeding programs will rapidly evolve and rates of genetic improvement will increase. A major goal of the genomics programs in livestock and poultry is the identification of natural resistance genes or genes that enhance immune response (Müller and Brem 1998). Genomic selection offers a non-GE approach to improve selection for disease resistance.

It is also thought that the use of MBV may help decrease rates of inbreeding per generation because selection using this approach increases the emphasis on the Mendelian sampling of genes an individual receives from its parents, as distinct from emphasizing the parent average used by traditional selection methods (Daetwyler et al. 2007). Basically, genome-wide prediction increases the accuracy of breeding values through revealing which specific chromosome fragments an animal received from its parents, rather than estimating the value of an "average" offspring derived from those two parents. The latter approach favors the co-selection of siblings from elite parents in breeding programs, with a resultant emphasis on closely related individuals.

Concerns

Whole genome selection is an unproven technology. Although preliminary data coming from the dairy industry look promising (VanRaden et al. 2009), it is not known how well it will work in livestock industries with a wider diversity of breeds and less extensive phenotype and data collection resources. Additionally, simulations have shown that the process of selection itself based on MBVs rapidly reduces the accuracy of MBV, following selection on MBV (Muir 2007). Although genomic selection may initially encounter less public opposition because it uses naturally occurring genetic variation, some applications aimed at reducing generation interval, and hence increasing genetic gain per unit of time in Equation 1, may elicit public discomfort. For example, genomic selection enables an approach to decrease generation interval by harvesting immature oocytes from in utero calves (Georges and Massey 1991). Others have even proposed schemes in which breeding is essentially done in the laboratory, and genome scans allow for an estimation of the MBV of cells derived from in vitro meiosis events (Haley and Visscher 1998). Such animal breeding scenarios are largely hypothetical, but analogous manipulations in the world of plant breeding have certainly met with more success and both less regulation and public opposition than those engendered by the prospect of GE animals.

Functional Genomics

Genes usually influence the phenotype through regulatory networks. Functional genomics concentrates on mechanisms that regulate gene transcription and translation in these networks. Gene regulation can be modulated by environmental inputs as well as by DNA sequence variation. Whereas structural genomics is based on the discovery of alternative DNA sequences, functional genomics is based on the measurement of the abundance of mRNA associated with transcribed genes. The relative abundance of mRNA produced by each gene can be quantified in many ways, but for assessing large number of genes, the microarray is the current technology of choice. Gene expression profiles can be related to environmental perturbations and/or stage of development. Structural and functional genomics can be combined to find chromosomal locations or QTL associated with trait differences. Jansen and Nap (2001) defined the combination of these two fields as "genetical genomics." With this approach, DNA variation that controls gene expression, called expression quantitative trait loci (eQTL), can be mapped to a specific chromosomal location. One of the important outcomes of genetical genomics is the identification of eQTL that map to either the same (*cis*-acting loci) or different (*trans*-acting loci) genomic locations as the gene expressing the transcript being quantified (Pomp, Nehrenberg, and Estroda-Smith 2008). These data can then be combined to infer causal relationships among eQTL, *cis*- and *trans*- gene expression, and phenotypic traits (Sieberts and Schadt 2007).

Potential Benefits

Animal breeding programs that involve complex traits such as robustness, animal well-being, or disease resistance require a well-defined phenotype on which to base selection. For complex traits, identifying a selection criterion that has high repeatability and quantifies the breeding objective can be difficult, and ideal traits may be very expensive or impractical to measure. Functional genomics has the potential to provide biomarkers, which can be used to define complex traits such as behavior, stress, or disease in unique, quantifiable ways (Kadarmideen, von Rohr, and Janss 2006).

Most reported QTL in animals have large confidence intervals that possibly harbor hundreds of genes, making the determination of which is the causative gene difficult or unattainable (Kadarmideen et al. 2006). Genetical genomics may offer a potential solution to this problem. One of the first examples of the successful use of this approach was given by Liu et al. (2001), who mapped QTL in an F2 cross to find genes responsible for Marek's disease resistance in chickens. A concurrent microarray study was conducted on the founder lines for the cross to find genes that were differentially expressed after infection. Fifteen of these genes were subsequently mapped onto the chicken genome, and two of them mapped to a QTL region for Marek's disease resistance.

Genetical genomics can also generate substantial additional insight into the function and interrelation of gene products and gene action, which can then be used to unravel networks of gene regulation (de Koning, Carlborg, and Haley 2005; Jansen and

Nap 2001). These insights have importance to animal breeding because no selection program in commercial operations is based on a single trait (Emmerson 2003; Groen 2003). Pleiotropic effects due to common genes in pathways can result in either favorable or unfavorable correlations among traits. Understanding how genes interact in pathways on multiple traits through functional genomics can lead to the discovery of QTL that have either neutral or favorable pleiotropic effect, thus overcoming an unfavorable correlation.

One of the problems with applying molecular genetics to animal breeding stems from using linked markers, rather than tracking the actual DNA variant that causes the phenotypic difference. Linkage breaks down over time due to recombination, and markers may be in different phases in different families. Finding the causative genetic variants would greatly facilitate such breeding programs. *Cis*-eQTL are highly heritable and easier to identify because genetic control is generally highly robust. When a *cis*-eQTL localizes to the confidence interval of a phenotypic QTL, it becomes a relevant positional candidate (Pomp et al. 2008) and can be selected for without concern for linkage.

However, inclusion of molecular information in breeding programs requires an understanding of how genes interact, that is, do genes work independently or in certain combinations with others. It is increasingly apparent that genes interact extensively and that epistasis is common (Carlborg et al. 2006). Thus, using QTL in breeding programs requires understanding interacting networks of genes to determine the proper combinations for optimal response. The experimental basis for understanding heritable traits has largely involved studying biological systems one gene at a time. Yet the genome consists of tens of thousands of genes compounded by intricate interactions between genes (i.e., epistasis), and between genes and the environment (Pomp et al. 2008). All epistatic QTL would, by definition, be detected as a *trans*-eQTL in a microarray study (Kadarmideen et al. 2006). Aylor and Zeng (2008) proposed a framework for estimating and interpreting epistasis from expression data that combines quantitative genetics approaches with classical genetics associating genes with pathway regulation. With these approaches it may be possible to directly use epistatic variation in breeding programs.

Additionally, variation in gene expression between animals or different lines following disease challenge could unravel the genetics underlying immune response (de Koning et al. 2005), as well as characterize those parts of the molecular networks that help drive disease progression. Sieberts and Schadt (2007) concluded that the integration of gene expression and genotypic data will be critical to understanding how genetic and environmental perturbations lead to disease. A systematic approach is therefore needed to dissect the genetic basis for diseases and understand how genes interact with one another, and with environmental factors, to determine disease phenotypes (Zhu et al. 2007).

A final potential benefit of functional genomics was proposed by Haley and de Koning (2006) who suggested that certain genes and networks could be explored in tissue culture, thereby moving the focus of experimental studies from whole animals to in vitro systems. They concluded that functional genomics may offer "an animal-friendly means of tackling welfare and other problems and hence enhancing sustainable livestock agriculture in a world where the demand for animal protein is expected to increase substantially over the next decade" (Haley and de Koning 2006, 12).

Concerns

Concerns associated with functional genomics are generally related to cost, false positives, a lack of power, choice of tissue, and optimal time to collect samples. Compared to the experimental designs commonly encountered in QTL detection, eQTL experiments to date have been very small and therefore have had inadequate statistical power (de Koning et al. 2005). Multiple testing is a major problem in eQTL experiments. In eQTL mapping, this testing occurs on two levels. First, multiple correlated tests are carried out during the genome scan for eQTL. Second, eQTL analyses are performed for thousands of potentially highly correlated gene expression levels (de Koning et al. 2005). Another concern is the choice of tissue and the developmental stage at which to profile gene expression. This decision is critical, but proper choice requires a priori understanding of which tissue(s) is associated with regulation of the phenotype. A researcher's assumptions could be incorrect, multiple tissues may require examination, or several time points may be necessary (Doerge 2002).

Other Biotechnologies

RNAi

RNAi is a sequence-specific method to selectively knock down endogenous gene expression. It works by introducing transgenic homologous double-stranded gene constructs that enable the stable expression of small interfering RNA (siRNAs) that constitutively suppress target gene expression (Martin and Caplen 2007). Transgenic goats carrying lentivectors that express siRNAs against the prion protein have been reported (Golding et al. 2006). Similarly, knockdown of porcine endogenous retrovirus (PERV) expression was recently reported in transgenic pigs (Dieckhoff et al. 2008). In that study, pig fibroblasts were transfected using a lentiviral vector expressing a corresponding short hairpin RNA (shRNA), and transgenic pigs were produced by SCNT cloning. All seven of the piglets that were born had integrated the transgene. Expression of the shRNA was found in all tissues investigated, and PERV expression was significantly inhibited when compared with wildtype control animals. These recent developments suggest that this approach may be a highly efficient method to generate GE animals with targeted gene knockouts in the future, including GE animals that can knockdown infections caused by important contagious RNA viruses such as foot and mouth disease, classic swine fever, and fowl plague (Clark and Whitelaw 2003).

Modification of Rumen Microorganisms

Although not "animal" biotechnology per se, genetic manipulation of rumen microorganisms has enormous potential to reduce the environmental footprint of ruminant livestock agriculture, as well as enhance product quality (Edwards et al. 2008).

Recombinant Bovine Somatotropin

Perhaps no other animal biotechnology has stimulated more vigorous public debate than the use of recombinant bovine somatotropin (rBST) derived from GE bacteria. This protein, which results in increased milk production when administered to lactating cows, is widely used in the U.S. dairy industry. Administering the protein rBST does not modify the DNA of the cow, nor does the cow become GE. The use of rBST can increase milk production by as much as 30 percent in well-managed herds. Currently banned in Europe, the administration of rBST to dairy cows was approved by the U.S. FDA in 1993 after extensive testing by numerous medical associations and scientific societies revealed no health or safety concerns for consumers (Bauman 1999). Since then, there have been a number of negative campaigns targeting this product, which have resulted in the development of a value-added market for rBST-free milk. At least one paper has examined the environmental impact of rBST use in dairy production (Capper et al. 2008). Not surprisingly, the use of rBST not only markedly improved the efficiency of milk production but also mitigated the environmental impacts associated with the production of a gallon of milk (decreased eutrophication, acidification, greenhouse gas emissions, and fossil fuel use). This example emphasizes the need to weigh decisions to restrict producer access to high-yield technologies or genetic resources that improve productive efficiency against the potential negative impact such decisions may have on achieving environmental sustainability goals. Pretty (2008, 451) captured this idea succinctly when he wrote, "The idea of agricultural sustainability, though, does not mean ruling out any technologies or practices on ideological grounds. If a technology works to improve productivity for farmers and does not cause undue harm to the environment, then it is likely to have some sustainability benefits."

Conclusion

The demand for meat and dairy products is expected to rise rapidly with economic growth in the developing world. It is likely that this growth will increasingly occur in intensive systems where animals are fed with cereals and oils, rather than by using forage and byproducts that cannot be consumed by humans. Intensive systems have some sustainability benefits in that they minimize the resources required to produce a unit of animal product, but often have high external costs on the environment and may give rise to some animal health and well-being concerns. When the external costs of an agricultural system are high and can be reduced by the adoption of new practices and technologies, adoption of intensive systems is a move toward sustainability. A variety of animal biotechnologies offer sustainability benefits. Some are technologies that help animal breeders select the best genotypes to minimize the environmental footprint of animal agriculture, whereas others offer clear animal health and well-being benefits. These biotechnologies may allow intensive animal agricultural systems to proceed in a more sustainable direction. Given the projected demand for animal products in the future, serious consideration must be given to all technologies that can move animal agriculture toward production systems that integrate a sustainable balance of environmental, animal well-being, social, and economic goals.

References

Aerni, P. 2004. "Risk, Regulation and Innovation: The Case of Aquaculture and Transgenic Fish." *Aquatic Science* 66 (3): 327–41.

Allen, L. H., J. R. Backstrand, E. J. Stanek, G. H. Pelto, A. Chavez, E. Molina, J. B. Castillo, and A. Mata. 1992. "The Interactive Effects of Dietary Quality on the Growth and Attained Size of Young Mexican Children." *American Journal of Clinical Nutrition* 56 (2): 353–64.

Aylor, D. L., and Z. B. Zeng. 2008. "From Classical Genetics to Quantitative Genetics to Systems Biology: Modeling Epistasis." *PLoS Genetics* 4 (3). doi:10.1371/journal.pgen.1000029.

Barthel, R., J. Feng, J. A. Piedrahita, D. N. McMurray, J. W. Templeton, and L. G. Adams. 2001. "Stable Transfection of the Bovine NRAMP1 Gene into Murine RAW264.7 Cells: Effect on *Brucella Abortus* Survival." *Infection and Immunology* 69 (5): 3110–19. doi:10.1128/IAI.69.5.3110-3119.2001.

Bauman, D. E. 1999. "Bovine Somatotropin and Lactation: From Basic Science to Commercial Application." *Domestic Animal Endocrinology* 17: 101–16.

Bessey, C., R. H. Devlin, N. R. Liley, and C. A. Biagi. 2004. "Reproductive Performance of Growth-Enhanced Transgenic Coho Salmon." *Transactions of the American Fisheries Society* 133 (5): 1205–20.

Beyhan, Z., E. J. Forsberg, K. J. Eilertsen, M. Kent-First, and N. L. First. 2007. "Gene Expression in Bovine Nuclear Transfer Embryos in Relation to Donor Cell Efficiency in Producing Live Offspring." *Molecular Reproduction and Development* 74 (1): 18–27.

Bradford, G. E. 1999. "Contributions of Animal Agriculture to Meeting Global Human Food Demand." *Livestock Production Science* 59 (1–2): 95–112. doi:10.1016/S0301-6226(99)00019-6.

Briggs, R., and T. J. King. 1952. "Transplantation of Living Nuclei from Blastula Cells into Enucleated Frogs' Eggs." *Proceedings of the National Academy of Sciences of the United States of America* 38 (5): 455–63.

Brophy, B., G. Smolenski, T. Wheeler, D. Wells, P. L'Huillier, and G. Laible. 2003. "Cloned Transgenic Cattle Produce Milk with Higher Levels of Beta-Casein and Kappa-Casein." *Nature Biotechnology* 21 (2): 157–62.

Bruinsma, J., ed. 2003. *World Agriculture: Towards 2015/2030: An FAO Perspective.* London: Earthscan. ftp://ftp.fao.org/docrep/fao/005/y4252e/y4252e.pdf.

Capper, J. L., E. Castaneda-Gutierrez, R. A. Cady, and D. E. Bauman. 2008. "The Environmental Impact of Recombinant Bovine Somatotropin (rBST) Use in Dairy Production." *Proceedings of the National Academy of Sciences of the United States of America* 105 (28): 9668–73.

Carlborg, O., L. Jacobsson, P. Ahgren, P. Siegel, and L. Andersson. 2006. "Epistasis and the Release of Genetic Variation during Long-Term Selection." *Nature Genetics* 38 (4): 418–20. doi:10.1038/ng1761.

CAST (Council for Agricultural Science and Technology). 2006. *Biotechnological Approaches to Manure Nutrient Management.* Issue Paper 33. Ames, IA: CAST. http://www.cast-science.org/websiteUploads/publicationPDFs/manuremanagement_ip.pdf.

Chesne, P., P. G. Adenot, C. Viglietta, M. Baratte, L. Boulanger, and J. P. Renard. 2002. "Cloned Rabbits Produced by Nuclear Transfer from Adult Somatic Cells." *Nature Biotechnology* 20 (4): 366–9.

Cho, J., K. Choi, T. Darden, P. R. Reynolds, J. Petitte, and S. B. Shears. 2005. "The Avian MINPP Gene can Help to Alleviate the Planet's 'Phosphate Crisis.'" Abstract. In *Inositol Phosphates in the Soil-Plant-Animal System: Linking Agriculture and Environment*, edited by B. L. Turner, A. E. Richardson, and E. J. Mullaney, 27–8. Paper presented at the Bouyoucous Conference to Address the Biogeochemical Interaction of Inositol Phosphates in the Environment, Sun Valley, ID, August 22.

Clark, J., and B. Whitelaw. 2003. "A Future for Transgenic Livestock." *Nature Reviews Genetics* 4: 825–33.

Clements, J. E., R. J. Wall, O. Narayan, D. Hauer, R. Schoborg, D. Sheffer, A. Powell, et al. 1994. "Development of Transgenic Sheep That Express the Visna Virus Envelope Gene." *Virology* 200 (2): 370–80.

Clop, A., C. Ovilo, M. Perez-Enciso, A. Cercos, A. Tomas, A. Fernandez, A. Coll, et al. 2003. "Detection of QTL Affecting Fatty Acid Composition in the Pig." *Mammalian Genome* 14 (9): 650–6.

Connor, W. E. 2000. "Importance of N-3 Fatty Acids in Health and Disease." *American Journal of Clinical Nutrition* 71 (1): 171S–5S.

Cook, J. T., M. A. McNiven, G. F. Richardson, and A. M. Sutterlin. 2000. "Growth Rate, Body Composition and Feed Digestibility/Conversion of Growth-Enhanced Transgenic Atlantic Salmon (*Salmo salar*)." *Aquaculture* 188 (1–2): 15–32.

CVM of FDA (Center for Veterinary Medicine of the U.S. Food and Drug Administration). 2008. *Animal Cloning: A Risk Assessment.* Rockville, MD: U.S. Department of Health and Human Services. http://www.fda.gov/downloads/AnimalVeterinary/SafetyHealth/AnimalCloning/UCM124756.pdf.

———. 2009. *FDA Guidance for Industry: Regulation of Genetically Engineered Animals Containing Heritable Recombinant DNA Constructs.* January 15. Rockville, MD: U.S. Department of Health and Human Services. http://www.fda.gov/downloads/AnimalVeterinary/GuidanceComplianceEnforcement/GuidanceforIndustry/UCM113903.pdf.

Daetwyler, H. D., B. Villanueva, P. Bijma, and J. A. Woolliams. 2007. "Inbreeding in Genome-Wide Selection." *Journal of Animal Breeding and Genetics* 124: 369–76.

de Koning, D. J., O. Carlborg, and C. S. Haley. 2005. "The Genetic Dissection of Immune Response Using Gene-Expression Studies and Genome Mapping." *Veterinary Immunology and Immunopathology* 105: 343–52.

Delgado, C. L. 2003. "Rising Consumption of Meat and Milk in Developing Countries Has Created a New Food Revolution." *Journal of Nutrition* 133 (11): 3907S–10S.

Denning, C., S. Burl, A. Ainslie, J. Bracken, A. Dinnyes, J. Fletcher, T. King, et al. 2001. "Deletion of the Alpha(1,3) galactosyl transferase (GGTA1) Gene and the Prion Protein (PrP) Gene in Sheep." *Nature Biotechnology* 19 (6): 559–62.

Dieckhoff, B., B. Petersen, W. A. Kues, R. Kurth, H. Niemann, and J. Denner. 2008. "Knockdown of Porcine Endogenous Retrovirus (PERV) Expression by PERV-Specific shRNA in Transgenic Pigs." *Xenotransplantation* 15 (1): 36–45. doi:10.1111/j.1399-3089.2008.00442.x.

Doerge, R. W. 2002. "Mapping and Analysis of Quantitative Trait Loci in Experimental Populations." *Nature Reviews Genetics* 3: 43–52. doi:10.1038/nrg703.

Domergue, F., A. Abbadi, and E. Heinz. 2005. "Relief for Fish Stocks: Oceanic Fatty Acids in Transgenic Oilseeds." *Trends in Plant Science* 10 (3): 112–16.

Du, S. J., Z. Y. Gong, G. L. Fletcher, M. A. Shears, M. J. King, D. R. Idler, and C. L. Hew. 1992. "Growth Enhancement in Transgenic Atlantic Salmon by the Use of an All Fish Chimeric Growth-Hormone Gene Construct." *Biotechnology* 10: 176–81. doi:10.1038/nbt0292-176.

Dunham, R. A., G. W. Warr, A. Nichols, P. L. Duncan, B. Argue, D. Middleton, and H. Kucuktas. 2002. "Enhanced Bacterial Disease Resistance of Transgenic Channel Catfish *Ictalurus punctatus* Possessing Cecropin Genes." *Marine Biotechnology* 4 (3): 338–44.

Edwards, J. E., S. A. Huws, E. J. Kim, M. R. F. Lee, A. H. Kingston-Smith, and N. D. Scollan. 2008. "Advances in Microbial Ecosystem Concepts and Their Consequences for Ruminant Agriculture." *Animal* 2 (5): 653–60.

Emmerson, D. 2003. "Breeding Objectives and Selection Strategies for Broiler Production." In *Poultry Genetics, Breeding and Biotechnology*, edited by W. M. Muir and S. Aggrey, 113–26. Cambridge: CABI Press.

Fiester, A. 2006. "Why the Omega-3 Piggy Should Not Go to Market." *Nature Biotechnology* 24 (12): 1472–3.

Fletcher, G. L., M. A. Shears, E. S. Yaskowiak, M. J. King, and S. V. Goddard. 2004. "Gene Transfer: Potential to Enhance the Genome of Atlantic Salmon for Aquaculture." *Australian Journal of Experimental Agriculture* 44 (11): 1095–1100.

Forsberg, E. J. 2005. "Commercial Applications of Nuclear Transfer Cloning: Three Examples." *Reproductive Fertility and Development* 17 (1–2): 59–68.

Galli, C., I. Lagutina, G. Crotti, S. Colleoni, P. Turini, N. Ponderato, R. Duchi, and G. Lazzari. 2003. "A Cloned Horse Born to Its Dam Twin." *Nature* 425 (6949): 635. doi:10.1038/424635a.

Georges, M., and J. M. Massey. 1991. "Velogenetics or the Synergistic Use of Marker Assisted Selection and Germ-Line Manipulation." *Theriogenology* 35 (1): 151–6.

Golding, M. C., C. R. Long, M. A. Carmell, G. J. Hannon, and M. E. Westhusin. 2006. "Suppression of Prion Protein in Livestock by RNA Interference." *Proceedings of the National Academy of Sciences of the United States of America* 103 (14): 5285–90.

Golovan, S. P., R. G. Meidinger, A. Ajakaiye, M. Cottrill, M. Z. Wiederkehr, D. J. Barney, C. Plante, et al. 2001. "Pigs Expressing Salivary Phytase Produce Low-Phosphorus Manure." *Nature Biotechnology* 19: 741–5. doi:10.1038/90788.

Grillenberger, M., S. P. Murphy, C. G. Neumann, N. O. Bwibo, H. Verhoef, and J. G. Hautvast. 2007. "The Potential of Increased Meat Intake to Improve Iron Nutrition in Rural Kenyan Schoolchildren." *International Journal for Vitamin and Nutrition Research* 77 (3): 193–8.

———, C. G. Neumann, S. P. Murphy, N. O. Bwibo, R. E. Weiss, L. H. Jiang, J. G. Hautvast, and C. E. West. 2006. "Intake of Micronutrients High in Animal-Source Foods Is Associated with Better Growth in Rural Kenyan School Children." *British Journal of Nutrition* 95 (2): 379–90.

Groen, A. F. 2003. "Breeding Objectives and Selection Strategies for Layer Production." In *Poultry Genetics, Breeding and Biotechnology*, edited by W. M. Muir and S. Aggrey, 101. Cambridge: CABI Press.

Guenther, G. G., L. M. Hylle, C. H. Stahl, E. A. Koutsos, and D. G. Peterson. 2005. "Development of Methods for the Production of Transgenic Quail Expressing an *E. coli* Phytase Gene." Abstract. In *Meeting Report and Abstracts of the 2005 UC Davis Transgenic Animal Research Conference V*, edited by J. D. Murray, 128. Paper presented at the Transgenic Animal Research Conference V, Tahoe City, CA, August 14–18. doi:10.1007/s11248-005-3259-3.

Haley, C. S., and D. J. de Koning. 2006. "Genetical Genomics in Livestock: Potentials and Pitfalls." *Animal Genetics* 37: 10–12. doi:10.1111/j.1365-2052.2006.01470.x.

———, and P. M. Visscher. 1998. "Strategies to Utilize Marker-Quantitative Trait Loci Associations." *Journal of Dairy Science* 81: 85–97. doi:10.3168/jds.S0022-0302(98)70157-2.

Hammer, R. E., V. G. Pursel, C. E. Rexroad, R. J. Wall, D. J. Bolt, K. M. Ebert, R. D. Palmiter, and R. L. Brinster. 1985. "Production of Transgenic Rabbits, Sheep and Pigs by Microinjection." *Nature* 315: 680–3. doi:10.1038/315680a0.

Heyman, Y., R. Chavatte-Palmer, G. Fromentin, V. Berthelot, C. Jurie, P. Bas, M. Dubarry, et al. 2007. "Quality and Safety of Bovine Clones and Their Products." *Animal* 1: 963–72. doi:10.1017/S1751731107000171.

Hill, J. R., R. C. Burghardt, K. Jones, C. R. Long, C. R. Looney, T. Shin, T. E. Spencer, et al. 2000. "Evidence for Placental Abnormality as the Major Cause of Mortality in First-Trimester Somatic Cell Cloned Bovine Fetuses." *Biology of Reproduction* 63 (6): 1787–94.

Horvat, S., and J. F. Medrano. 2001. "Lack of Socs2 Expression Causes the High-Growth Phenotype in Mice." *Genomics* 72 (2): 209–12.

Jansen, R. C., and J. P. Nap. 2001. "Genetical Genomics: The Added Value from Segregation." *Trends in Genetics* 17 (7): 388–91.

Johnson, W. H., N. M. Loskutoff, Y. Plante, and K. J. Betteridge. 1995. "Production of 4 Identical Calves by the Separation of Blastomeres from an In-Vitro Derived 4-Cell Embryo." *Veterinary Record* 137 (1): 15–16.

Jost, B., J. L. Vilotte, I. Duluc, J. L. Rodeau, and J. N. Freund. 1999. "Production of Low-Lactose Milk by Ectopic Expression of Intestinal Lactase in the Mouse Mammary Gland." *Nature Biotechnology* 17: 160–4. doi:10.1038/6158.

Kadarmideen, H. N., P. von Rohr, and L. L. G. Janss. 2006. "From Genetical Genomics to Systems Genetics: Potential Applications in Quantitative Genomics and Animal Breeding." *Mammalian Genome* 17 (6): 548–64. doi:10.1007/s00335-005-0169-x.

Kato, Y., T. Tani, Y. Sotomaru, K. Kurokawa, J. Y. Kato, H. Doguchi, H. Yasue, and Y. Tsunoda. 1998. "Eight Calves Cloned from Somatic Cells of a Single Adult." *Science* 282: 2095–8.

Keefer, C. L., H. Baldassarre, R. Keyston, B. Wang, B. Bhatia, A. S. Bilodeau, J. F. Zhou, et al. 2001. "Generation of Dwarf Goat (*Capra hircus*) Clones Following Nuclear Transfer with Transfected and Nontransfected Fetal Fibroblasts and In Vitro-Matured Oocytes." *Biology of Reproduction* 64 (3): 849–56. doi:10.1095/ biolreprod64.3.849.

Kuroiwa, Y., P. Kasinathan, H. Matsushita, J. Sathiyaselan, E. J. Sullivan, M. Kakitani, K. Tomizuka, I. Ishida, and J. M. Robl. 2004. "Sequential Targeting of the Genes Encoding

Immunoglobulin-μ and Prion Protein in Cattle." *Nature Genetics* 36 (7): 775–80. doi:10.1038/ng1373.

Lai, L. X., J. X. Kang, R. F. Li, J. D. Wang, W. T. Witt, H. Y. Yong, Y. H. Hao, et al. 2006. "Generation of Cloned Transgenic Pigs Rich in Omega-3 Fatty Acids." *Nature Biotechnology* 24: 435–6.

Laible, G., B. Brophy, D. Knighton, and D. N. Wells. 2007. "Compositional Analysis of Dairy Products Derived from Clones and Cloned Transgenic Cattle." *Theriogenology* 67: 166–77.

Lane, N. 2005. "Welfare Issues during the Production and Maintenance of GM Mouse Strains and Recommendations for Current Best Practice." *AgBiotechNet*. 7. http://www.cabi.org/cabreviews/FullTextPDF/Alphabetical/ABN135.pdf.

Lee, B. C., M. K. Kim, G. Jang, H. J. Oh, F. Yuda, H. J. Kim, M. S. Hossein, et al. 2005. "Dogs Cloned from Adult Somatic Cells." *Nature* 436: 641. doi:10.1038/436641a.

Lee, K. Y., H. G. Huang, B. S. Ju, Z. G. Yang, and S. Lin. 2002. "Cloned Zebrafish by Nuclear Transfer from Long-Term-Cultured Cells." *Nature Biotechnology* 20: 795–9. doi:10.1038/nbt721.

Lindeman, G. J., S. Wittlin, H. Lada, M. J. Naylor, M. Santamaria, J. G. Zhang, R. Starr, et al. 2001. "SOCS1 Deficiency Results in Accelerated Mammary Gland Development and Rescues Lactation in Prolactin Receptor-Deficient Mice." *Genes & Development* 15: 1631–6. doi:10.1101/gad.880801.

Liu, H. C., H. H. Cheng, L. Sofer, and J. Burnside. 2001. "Identification of Genes and Pathways for Resistance to Marek's Disease through DNA Microarrays." Abstract. In *Proceedings of the 6th International Symposium on Marek's Disease*, edited by K. A. Schat, R. W. Morgan, M. S. Parcells, and J. L. Spencer, 157–62. Kennett Square, PA: American Association of Avian Pathologists.

Lyall, J., R. M. Irvine, A. Sherman, T. J. McKinley, A. Núñez, A. Purdie, L. Outtrim, I. H. Brown, G. Rolleston-Smith, H. Sang, and L. Tiley. 2011. "Suppression of Avian Influenza Transmission in Genetically Modified Chickens." *Science* 331 (6014): 223-226. doi: 10.1126/science.1198020.

Maga, E. A., J. S. Cullor, W. Smith, G. B. Anderson, and J. D. Murray. 2006. "Human Lysozyme Expressed in the Mammary Gland of Transgenic Dairy Goats Can Inhibit the Growth of Bacteria That Cause Mastitis and the Cold-Spoilage of Milk." *Foodborne Pathogens and Disease* 3 (4): 384–92. doi:10.1089/fpd.2006.3.384.

______, C. F. Shoemaker, J. D. Rowe, R. H. BonDurant, G. B. Anderson, and J. D. Murray. 2006. "Production and Processing of Milk from Transgenic Goats Expressing Human Lysozyme in the Mammary Gland." *Journal of Dairy Science* 89 (2): 518–24.

Mak, N. 2008. *Genetically Modified Animals: Animal Welfare, Ethics, and Regulation*. Report of the American Anti-Vivisection Society. Presented at the USDA Advisory Committee on Biotechnology & 21st Century Agriculture (AC21) Meeting. Washington, DC, August 26.

Marshall, K. M., W. L. Hurley, R. D. Shanks, and M. B. Wheeler. 2006. "Effects of Suckling Intensity on Milk Yield and Piglet Growth from Lactation-Enhanced Gilts." *Journal of Animal Science* 84: 2346–51. doi:10.2527/jas.2005-764.

Martin, S. E., and N. J. Caplen. 2007. "Applications of RNA Interference in Mammalian Systems." *Annual Review of Genomics and Human Genetics* 8: 81–108. doi:10.1146/annurev.genom.8.080706.092424.

Martinez, R., J. Juncal, C. Zaldivar, A. Arenal, I. Guillen, V. Morera, O. Carrillo, M. Estrada, A. Morales, and M. P. Estrada. 2000. "Growth Efficiency in Transgenic Tilapia (*Oreochromis* sp.) Carrying a Single Copy of a Homologous cDNA Growth Hormone." *Biochemical and Biophysical Research Communications* 267 (1): 466–72.

McConnell, A., and A. Stark. 2002. "Foot-and-Mouth 2001: The Politics of Crisis Management." *Parliamentary Affairs* 55: 664–81.

McPherron, A. C., and S. J. Lee. 1997. "Double Muscling in Cattle due to Mutations in the Myostatin Gene." *Proceedings of the National Academy of Sciences of the United States of America* 94 (23): 12457–61.

Melo, E. O., A. M. O. Canavessi, M. M. Franco, and R. Rumpf. 2007. "Animal Transgenesis: State of the Art and Applications." *Journal of Applied Genetics* 48 (1): 47–61.

Meng, L., J. J. Ely, R. L. Stouffer, and D. P. Wolf. 1997. "Rhesus Monkeys Produced by Nuclear Transfer." *Biology of Reproduction* 57 (2): 454–9. doi:10.1095/biolreprod57.2.454.

Meuwissen, T. H., B. J. Hayes, and M. E. Goddard. 2001. "Prediction of Total Genetic Value Using Genome-Wide Dense Marker Maps." *Genetics* 157 (2): 1819–29.

Morimoto, K. C., A. L. Van Eenennaam, E. J. DePeters, and J. F. Medrano. 2005. "Hot Topic: Endogenous Production of n-3 and n-6 Fatty Acids in Mammalian Cells." *Journal of Dairy Science* 88: 1142–6.

Muir, W. M. 2007. "Comparison of Genomic and Traditional BLUP-Estimated Breeding Value Accuracy and Selection Response under Alternative Trait and Genomic Parameters." *Journal of Animal Breeding and Genetics* 124 (6): 342–55.

______, and R. D. Howard. 1999. "Possible Ecological Risks of Transgenic Organism Release When Transgenes Affect Mating Success: Sexual Selection and the Trojan Gene Hypothesis." *Proceedings of the National Academy of Sciences of the United States of America* 96 (24): 13853–6.

______. 2001. "Fitness Components and Ecological Risk of Transgenic Release: A Model Using Japanese Medaka (*Oryzias latipes*)." *American Naturalist* 158 (1): 1–16.

______. 2002. "Assessment of Possible Ecological Risks and Hazards of Transgenic Fish with Implications for Other Sexually Reproducing Organisms." *Transgenic Research* 11 (2): 101–14. doi:10.1023/A:1015203812200.

Müller, M., and G. Brem. 1998. "Transgenic Approaches to the Increase of Disease Resistance in Farm Animals." *Revue Scientifique et Technique* 17 (1): 365–78.

Murray, J. D., and E. A. Maga. 2009. "Is There a Risk from Not Using GE Animals?" *Transgenic Research* 19 (3): 357–61. doi:10.1007/s11248-009-9341-5.

Nam, Y. K., J. K. Noh, Y. S. Cho, H. J. Cho, K. N. Cho, C. G. Kim, and D. S. Kim. 2001. "Dramatically Accelerated Growth and Extraordinary Gigantism of Transgenic Mud Loach *Misgurnus mizolepis*." *Transgenic Research* 10 (4): 353–62. doi:10.1023/A:1016696104185.

National Research Council (NRC). 2002. *Animal Biotechnology: Science-Based Concerns*. Washington, DC: National Academy Press.

Naylor, R., H. Steinfeid, W. Falcon, J. Galloways, V. Smil, E. Bradford, J. Alder, and H. Mooney. 2005. "Losing the Links between Livestock and Land." *Science* 310 (5754): 1621–2.

Neumann, C. G., S. P. Murphy, C. Gewa, M. Grillenberger, and N. O. Bwibo. 2007. "Meat Supplementation Improves Growth, Cognitive, and Behavioral Outcomes in Kenyan Children." *Journal of Nutrition* 137: 1119–23.

Nii, M., T. Hayashi, F. Tani, A. Niki, N. Mori, N. Fujishima-Kanaya, M. Komatsu, et al. 2006. "Quantitative Trait Loci Mapping for Fatty Acid Composition Traits in Perirenal and Back Fat Using a Japanese Wild Boar x Large White Intercross." *Animal Genetics* 37 (4): 342–7.

Nonhebel, S. 2005. "Renewable Energy and Food Supply: Will There Be Enough Land?" *Renewable and Sustainable Energy Reviews* 9 (2): 191–201. doi:10.1016/j.rser.2004.02.003.

Norman, H. D., and M. K. Walsh. 2004. "Performance of Dairy Cattle Clones and Evaluation of Their Milk Composition." *Cloning and Stem Cells* 6 (2): 157–64.

Panarace, M., J. I. Aguero, M. Garrote, G. Jauregui, A. Segovia, L. Cane, J. Gutierrez, et al. 2007. "How Healthy Are Clones and Their Progeny: 5 Years of Field Experience." *Theriogenology* 67 (1): 142–51.

Pauly, D., J. Alder, E. Bennett, V. Christensen, P. Tyedmers, and R. Watson. 2003. "The Future for Fisheries." *Science* 302 (5649):1359–61. doi:10.1126/science.1088667.

Petters, R. M., C. A. Alexander, K. D. Wells, E. B. Collins, J. R. Sommer, M. R. Blanton, G. Rojas, et al. 1997. "Genetically Engineered Large Animal Model for Studying Cone Photoreceptor Survival and Degeneration in Retinitis Pigmentosa." *Nature Biotechnology* 15: 965–70. doi:10.1038/nbt1097-965.

Philipkoski, K. 2001. "Toxic Manure: Endangered Feces?" *Wired*, May 1. http://www.wired.com/medtech/health/news/2001/05/43447.

Polejaeva, I. A., S. H. Chen, T. D. Vaught, R. L. Page, J. Mullins, S. Ball, Y. Dai, et al. 2000. "Cloned Pigs Produced by Nuclear Transfer from Adult Somatic Cells." *Nature* 407: 86–90. doi:10.1038/35024082.

Pomp, D., D. Nehrenberg, and D. Estroda-Smith. 2008. "Complex Genetics of Obesity in Mouse Models." *Annual Review of Nutrition* 28: 331–45. doi:10.1146/annurev.nutr.27.061406.093552.

Prather, R. S. 2006. "Cloned Transgenic Heart-Healthy Pork?" *Transgenic Research* 15 (4): 405–7. doi:10.1007/s11248-006-0022-3.

Pretty, J. 2008. "Agricultural Sustainability: Concepts, Principles and Evidence." *Philosophical Transactions of the Royal Society B: Biological Sciences* 363 (1491): 447–65. doi:10.1098/rstb.2007.2163.

Pursel, V. G., A. D. Mitchell, G. Bee, T. H. Elsasser, J. P. McMurtry, R. J. Wall, M. E. Coleman, and R. J. Schwartz. 2004. "Growth and Tissue Accretion Rates of Swine Expressing an Insulin-Like Growth Factor I Transgene." *Animal Biotechnology* 15 (1): 33–45. doi:10.1081/ABIO-120029812.

______, C. A. Pinkert, K. F. Miller, D. J. Bolt, R. G. Campbell, R. D. Palmiter, R. L. Brinster, and R. E. Hammer. 1989. "Genetic Engineering of Livestock." *Science* 244 (4910): 1281–8. doi:10.1126/science.2499927.

______, R. J. Wall, M. B. Solomon, D. J. Bolt, J. E. Murray, and K. A. Ward. 1997. "Transfer of an Ovine Metallothionein-Ovine Growth Hormone Fusion Gene into Swine." *Journal of Animal Science* 75 (8): 2208–14.

Rahman, M. A., R. Mak, H. Ayad, A. Smith, and N. Maclean. 1998. "Expression of a Novel Piscine Growth Hormone Gene Results in Growth Enhancement in Transgenic Tilapia (*Oreochromis niloticus*)." *Transgenic Research* 7 (5): 357–69. doi:10.1023/A:1008837105299.

Reh, W. A., E. A. Maga, N. M. B. Collette, A. Moyer, J. S. Conrad-Brink, S. J. Taylor, E. J. DePeters, et al. 2004. "Hot Topic: Using a Stearoyl-CoA Desaturase Transgene to Alter Milk Fatty Acid Composition." *Journal of Dairy Science* 87 (10): 3510–14.

Richt, J. A., P. Kasinathan, A. N. Hamir, J. Castilla, T. Sathiyaseelan, F. Vargas, J. Sathiyaseelan, et al. 2007. "Production of Cattle Lacking Prion Protein." *Nature Biotechnology* 25 (1): 132–8. doi:10.1038/nbt1271.

Robert, S. S. 2006. "Production of Eicosapentaenoic and Docosahexaenoic Acid-Containing Oils in Transgenic Land Plants for Human and Aquaculture Nutrition." *Marine Biotechnology* 8 (2): 103–9.

______, S. P. Singh, X. R. Zhou, J. R. Petrie, S. I. Blackburn, P. M. Mansour, P. D. Nichols, Q. Liu, and A. G. Green. 2005. "Metabolic Engineering of *Arabidopsis* to Produce Nutritionally Important DHA in Seed Oil." *Functional Plant Biology* 32 (6): 473–9. doi:10.1071/FP05084.

Robl, J. M., R. Prather, F. Barnes, W. Eyestone, D. Northey, B. Gilligan, and N. L. First. 1987. "Nuclear Transplantation in Bovine Embryos." *Journal of Animal Science* 64: 642–7.

Rudenko, L., and J. C. Matheson. 2007. "The US FDA and Animal Cloning: Risk and Regulatory Approach." *Theriogenology* 67 (1): 198–206. doi:10.1016/j.theriogenology.2006.09.033.

———, J. C. Matheson, A. L. Adams, E. S. Dubbin, and K. J. Greenlees. 2004. "Food Consumption Risks Associated with Animal Clones: What Should Be Investigated?" *Cloning Stem Cells* 6 (2): 79–93.

———, J. C. Matheson, and S. F. Sundlof. 2007. "Animal Cloning and the FDA – The Risk Assessment Paradigm under Public Scrutiny." *Nature Biotechnology* 25 (1): 39–43.

Sang, H. 1994. "Transgenic Chickens – Methods and Potential Applications." *Trends in Biotechnology* 12 (10): 415–20.

Schwegmann-Wessels, C., G. Zimmer, H. Laude, L. Enjuanes, and G. Herrler. 2002. "Binding of Transmissible Gastroenteritis Coronavirus to Cell Surface Sialoglycoproteins." *Journal of Virology* 76 (12): 6037–43. doi:10.1128/JVI.76.12.6037-6043.2002.

Seré, C., and H. Steinfeld. 1996. *World Livestock Production Systems*. In collaboration with J. Groenewold. FAO Animal Production and Health Paper 127. Rome: Food and Agriculture Organization (FAO). ftp://ftp.fao.org/docrep/fao/005/w0027e/w0027e00.pdf.

Shin, T., D. Kraemer, J. Pryor, L. Liu, J. Rugila, L. Howe, S. Buck, et al. 2002. "A Cat Cloned by Nuclear Transplantation." *Nature* 415: 859. doi:10.1038/nature723.

Sieberts, S. K., and E. E. Schadt. 2007. "Moving toward a System Genetics View of Disease." *Mammalian Genome* 18 (6–7): 389–401. doi:10.1007/s00335-007-9040-6.

Siegford, J. M., W. Powers, and H. G. Grimes-Casey. 2008. "Environmental Aspects of Ethical Animal Production." *Poultry Science* 87: 380–6. doi:10.3382/ps.2007-00351.

Simopoulos, A. P. 2004. "Omega-6/Omega-3 Essential Fatty Acid Ratio and Chronic Diseases." *Food Reviews International* 20 (1): 77–90. doi:10.1081/FRI-120028831.

Solberg, T. R., A. K. Sonesson, J. A. Woolliams, and T. H. Meuwissen. 2008. "Genomic Selection Using Different Marker Types and Densities." *Journal of Animal Science* 86: 2447–54. doi:10.2527/jas.2007-0010.

Stacey, A., A. Schnieke, H. Kerr, A. Scott, C. Mckee, I. Cottingham, B. Binas, C. Wilde, and A. Colman. 1995. "Lactation Is Disrupted by Alpha-Lactalbumin Deficiency and Can Be Restored by Human Alpha-Lactalbumin Gene Replacement in Mice." *Proceedings of the National Academy of Sciences of the United States of America* 92 (7): 2835–9.

Stern, S., U. Sonesson, S. Gunnarsson, I. Oborn, K. I. Kumm, and T. Nybrant. 2005. "Sustainable Development of Food Production: A Case Study on Scenarios for Pig Production." *Ambio* 34 (4–5): 402–7.

Takahashi, S., and Y. Ito. 2004. "Evaluation of Meat Products from Cloned Cattle: Biological and Biochemical Properties." *Cloning and Stem Cells* 6 (2): 165–71.

Tian, X. C., C. Kubota, K. Sakashita, Y. Izaike, R. Okano, N. Tabara, C. Curchoe, et al. 2005. "Meat and Milk Compositions of Bovine Clones." *Proceedings of the National Academy of Sciences of the United States of America* 102 (18): 6261–6. doi:10.1073/pnas.0500140102.

Tomé, D., M. Dubarry, and G. Fromentin. 2004. "Nutritional Value of Milk and Meat Products Derived from Cloning." *Cloning Stem Cells* 6 (2): 172–7.

Tompkins, S. M., C. Y. Lo, T. M. Tumpey, and S. L. Epstein. 2004. "Protection against Lethal Influenza Virus Challenge by RNA Interference in Vivo." *Proceedings of the National Academy of Sciences of the United States of America* 101 (23): 8682–6. doi:10.1073/pnas.0402630101.

Tsunoda, Y., and Y. Kato. 2002. "Recent Progress and Problems in Animal Cloning." *Differentiation* 69 (4–5): 158–61.

van Berkel, P. H., M. M. Welling, M. Geerts, H. A. van Veen, B. Ravensbergen, M. Salaheddine, E. K. Pauwels, et al. 2002. "Large Scale Production of Recombinant Human Lactoferrin in the Milk of Transgenic Cows." *Nature Biotechnology* 20:484–7. doi:10.1038/nbt0502-484.

VandeHaar, M. J., and N. St-Pierre. 2006. "Major Advances in Nutrition: Relevance to the Sustainability of the Dairy Industry." *Journal of Dairy Science* 89 (4):1280–91.

VanRaden, P. M., C. P. Van Tassell, G. R. Wiggans, T. S. Sonstegard, R. D. Schnabel, J. F. Taylor, and F. S. Schenkel. 2009. "Reliability of Genomic Predictions for North American Holstein Bulls." *Journal of Dairy Science* 92 (1):16–24. doi:10.3168/jds.2008-1514.

Wakayama, T., A. C. Perry, M. Zuccotti, K. R. Johnson, and R. Yanagimachi. 1998. "Full-Term Development of Mice from Enucleated Oocytes Injected with Cumulus Cell Nuclei." *Nature* 394: 369–74. doi:10.1038/28615.

Walker, S. C., R. K. Christenson, R. P. Ruiz, D. E. Reeves, S. L. Pratt, F. Arenivas, N. E. Williams, B. L. Bruner, and I. A. Polejaeva. 2007. "Comparison of Meat Composition from Offspring of Cloned and Conventionally Produced Boars." *Theriogenology* 67 (1):178–84.

Wall, R. J., A. M. Powell, M. J. Paape, D. E. Kerr, D. D. Bannerman, V. G. Pursel, K. D. Wells, et al. 2005. "Genetically Enhanced Cows Resist Intramammary *Staphylococcus aureus* Infection." *Nature Biotechnology* 23: 445–51. doi:10.1038/nbt1078.

Walsh, M. K., J. A. Lucey, S. Govindasamy-Lucey, M. M. Pace, and M. D. Bishop. 2003. "Comparison of Milk Produced by Cows Cloned by Nuclear Transfer with Milk from Non-Cloned Cows." *Cloning and Stem Cells* 5 (3): 213–19. doi:10.1089/153623003769645875.

Wheeler, M. B., G. T. Bleck, and S. M. Donovan. 2001. "Transgenic Alteration of Sow Milk to Improve Piglet Growth and Health." *Reproduction (Cambridge, England) Supplement* 58: 313–24.

Whitelaw, C. B., and H. M. Sang. 2005. "Disease-Resistant Genetically Modified Animals." *Revue Scientifique et Technique* 24 (1): 275–83.

Willadsen, S. M. 1981. "The Developmental Capacity of Blastomeres from 4-Cell and 8-Cell Sheep Embryos." *Journal of Embryology and Experimental Morphology* 65: 165–72.

———. 1986. "Nuclear Transplantation in Sheep Embryos." *Nature* 320: 63–5. doi:10.1038/320063a0.

Williams, J. R., B. J. Chambers, A. R. Hartley, S. Ellis, and H. J. Guise. 2000. "Nitrogen Losses from Outdoor Pig Farming Systems." *Soil Use and Management* 16 (4): 237–43. doi:10.1111/j.1475-2743.2000.tb00202.x.

Wilmut, I., A. E. Schnieke, J. McWhir, A. J. Kind, and K. H. Campbell. 1997. "Viable Offspring Derived from Fetal and Adult Mammalian Cells." *Nature* 385 (27): 810–13. doi:10.1038/385810a0.

Wu, G. H., M. Truksa, N. Datla, P. Vrinten, J. Bauer, T. Zank, P. Cirpus, E. Heinz, and X. Qiu. 2005. "Stepwise Engineering to Produce High Yields of Very Long-Chain Polyunsaturated Fatty Acids in Plants." *Nature Biotechnology* 23: 1013–17. doi:10.1038/nbt1107.

Yamaguchi, M., Y. Ito, and S. Takahashi. 2007. "Fourteen-Week Feeding Test of Meat and Milk Derived from Cloned Cattle in the Rat." *Theriogenology* 67 (1): 152–65.

Yang, X. Z., X. C. Tian, C. Kubota, R. Page, J. Xu, J. Cibelli, and G. Seidel. 2007. "Risk Assessment of Meat and Milk from Cloned Animals." *Nature Biotechnology* 25: 77–83. doi:10.1038/nbt1276.

Yu, G. H., J. Q. Chen, H. Q. Yu, S. G. Liu, J. Chen, X. J. Xu, H. Y. Sha, et al. 2006. "Functional Disruption of the Prion Protein Gene in Cloned Goats." *Journal of General Virology* 87: 1019–27. doi:10.1099/vir.0.81384-0.

Zhong, J. Y., Y. P. Wang, and Z. Y. Zhu. 2002. "Introduction of the Human Lactoferrin Gene into Grass Carp (*Ctenopharyngodon idellus*) to Increase Resistance against GCH Virus." *Aquaculture* 214 (1): 93–101. doi:10.1016/S0044-8486(02)00395-2.

Zhou, Q., J. P. Renard, G. Le Friec, V. Brochard, N. Beaujean, Y. Cherifi, A. Fraichard, and J. Cozzi. 2003. "Generation of Fertile Cloned Rats by Regulating Oocyte Activation." *Science* 302: 1179.

Zhu, J., M. C. Wiener, C. Zhang, A. Fridman, E. Minch, P. Y. Lum, J. R. Sachs, and E. E. Schadt. 2007. "Increasing the Power to Detect Causal Associations by Combining Genotypic and Expression Data in Segregating Populations." *PLoS Computational Biology* 3 (4): 692–703. doi:10.1371/journal.pcbi.0030069.

7

Genetically Engineered Crops Can Be Part of a Sustainable Food Supply

Food and Food Safety Issues

Peggy G. Lemaux, Ph.D.

Although the concept of sustainability has long been discussed and people today frequently read information on the topic, do they really know what the practice of sustainable food production is? A 2008 consumer poll revealed that 41 percent of individuals feel they know some or a lot about sustainable food production, compared to only 30 percent in 2007 (Food Insight 2008). Another question that could be raised is whether consumers understand how GE crops could play a role in sustainable food production. And would their concern about possible food safety risks of GE foods influence their acceptance of this approach? Their apprehension could lead to reluctance to use products of this technology as a part of the solution to sustainability.

Genetic engineering allows introduction of genes from one organism to those in another kingdom, a situation that does not occur using classical genetic technologies (Lemaux 2008). This technology is unsettling to some, who consider it a dangerous and "unnatural" process. From a scientific standpoint, many of the concerns raised about GE crops could be applied in much the same way to crops created through classical methods, like induced mutation and cross-hybridization with wild species (NRC 2004).

The intent of this chapter is to present as accurate a scientific picture as possible of the issues related to the safety of GE foods in the hope that scientific information will be used as part of consumers' decision-making process about the desirability of consuming foods made from GE crops; however, issues beyond the technical, science-based facts are also important. Topics in this chapter represent major issues raised regarding GE foods and their safety. However, not all issues raised are discussed, and not all aspects of the issues raised are addressed.

Will Introducing Fish Genes into Strawberries Result in Health Risks?

Engineering using rDNA methods enables the introduction of genes into plants from all living organisms. Such exchanges are possible because the genetic code in all living organisms is principally the same; however, issues with such foods go beyond scientific risk and concern whether exchanges between certain organisms should be performed. In fact, cross-kingdom transfer of animal genes to plants is not popular with consumers worldwide (Macer 2003); comfort levels are much higher for gene transfer among plants or between plants and bacteria (James and Burton 2003).

Given that exchanges between plants and animals are possible, the question is whether such exchanges have been done in any commercialized GE crop. Early in the 1990s, an antifreeze gene from Arctic flounder was introduced into tobacco and tomato (Hightower et al. 1991) and field-tested in tobacco (Lee et al. 1990) and tomato to inhibit ice recrystallization. Although often depicted otherwise, the gene was not introduced into strawberries, and it was not commercialized in any crop. Although humans have already been consuming flounder genes and proteins in their food, substantial equivalence and lack of allergenicity and toxicity would still have to be proven for a food engineered with the fish gene before such a GE product could be commercialized.

Can GE Foods Have Nutritional Differences That Cause Health Risks?

Preventing adverse health effects of foods due to nutritional differences requires use of appropriate scientific methods to predict and identify unintended compositional changes caused by genetic modification – whether by conventional or rDNA methods. It is the final product, rather than the modification process itself, that is more likely to result in unintended effects (NRC 2004).

Levels of individual components in GE foods, except for those purposely engineered, must be equal to levels in conventional counterparts: so-called substantial equivalence. Thus to determine substantial equivalence it is first necessary to determine the nutritional composition of GE foods – its protein, carbohydrate, fat, vitamin, mineral, fiber, moisture, and phytochemical content. When considering substantial equivalence, it is important to note that a range of natural variation is already observed in conventionally bred cultivars even when grown under similar conditions (Shewry et al. 2006), with even more marked variation when environmental conditions vary. Therefore, substantial equivalence of nutritional content of GE foods must be measured against naturally occurring variation in conventional foods grown under comparable conditions.

Extensive nutritional equivalence studies have been conducted on RR soybeans and include analyses of protein, oil, fiber, ash, carbohydrates, moisture content, and the amino acid and fatty acid composition in both seeds and toasted soybean meal, in comparison to conventional soybeans. These analyses were published in a peer-reviewed journal in the same year that RR soybeans were commercially introduced (Padgette et al. 1996). Special attention was given to anti-nutrients and phytonutrients levels (e.g., trypsin inhibitors, lectins, and isoflavones; http://ucbiotech.org:8080/qilan/

ReturnRef.jsp?01RefID=416). Of the dozens of comparisons, only one significant increase was noted: trypsin inhibitor levels were 11 to 26 percent higher in defatted, nontoasted soybean meal. However, levels in seeds and in defatted, toasted soybean meal, the form used in foods, were similar for all lines. Equivalence as a feed was demonstrated with rat, chicken, catfish, and dairy cattle feeding (Hammond et al. 1996). Another broader study looking at the feeding value of *Bt* corn and RR corn and soybean to sheep, chickens, and beef and dairy cattle concluded that GE grain was substantially equivalent to grain from non-GE varieties (Clark and Ipharraguerre 2001).

In a 1999 study (Lappé et al.1999), often cited by those concerned about GE crops, data were presented that RR soybeans had reduced levels of isoflavones – claiming significant implications for human health because of potential positive health benefits of isoflavones. However, a response to this study found that the variation in isoflavone levels was within variability limits for conventional soybean varieties (Ag BioTech InfoNet 1999).

It is important to note that genetic engineering can purposefully be used to introduce positive changes in nutritional profiles. Examples of such foods include increased β-carotene in rice, flavonoids in alfalfa and tomato, calcium in potato, folate in tomato, and iron availability in corn (Lemaux 2008). FDA policy requires that GE foods with altered nutrition be labeled indicating those nutritional differences.

Does a Lack of GE Food Labeling Raise Risks for Human Consumption?

The FDA's labeling policy for GE foods, which is the same as for conventional foods, was created to assure that consumers are informed about changes in nutritional content, health safety, or food quality. Legally, labels are not required to provide information about the production process. However, if a GE food has a significantly different nutritional profile from its conventional counterpart, the food must be labeled to indicate the difference; for example, when nutritional profile changes are created with genetic information from a previously recognized allergenic source, such as peanut, soy, or wheat, or the new protein has characteristics of known allergens. Examples where labeling would be required would be food from a GE crop with a gene inserted from peanut and wheat, or oils derived from GE varieties engineered with changes in fatty acid composition. One such product, a low-linoleic acid soybean oil, has already been deregulated (Monsanto 2010); foods containing those oils must be labeled and a new name used – in this case Vistive™.

Did People Die after Consuming the Nutritional Supplement Tryptophan?

In 1989 a nutritional supplement, L-tryptophan, was blamed for eosinophilia-myalgia syndrome (EMS) by those using the supplement. The number reported to be affected was "between 5000 and 10 000 people and the number of deaths near 40" (Smith 2003, 111) – all had consumed tryptophan made by one Japanese company (Roufs 1992) that was produced using GE bacteria that had been used without incident before that date. Before the deaths occurred, the company had changed GE bacterial strains and its manufacturing practices by eliminating some filtration and purification steps. Although 99.6 percent pure, the final product still contained 60 impurities (Mayeno and Gleich 1994), any one of which could have caused the illness. The illness was never conclusively linked to the organism or the manufacturing process; however, reconstruction experiments (Mayeno and Gleich 1994) found it was most likely that the illness was caused by the changes in processing, and legal documents so stated (Hawes 2010). Perhaps the lesson to be learned is that nutritional supplements should undergo the same rigorous premarket testing that GE foods do.

Were Potatoes Engineered to Produce a Lectin Unsafe to Eat?

In the late 1990s studies on rats fed potatoes engineered with a lectin gene from snowdrop (Ewen and Pusztai 1999) described the development of stomach lesions. Authors concluded that "damage to the rats did not come from the lectin, but apparently from the same process of genetic engineering that is used to create the GM foods everyone was already eating" (Smith 2003, 17). The study was strongly criticized by the scientific community as being conducted poorly with too few animals and inadequate controls. After being released to the popular press, the findings were published to allow others to view the data, but researchers were unable to confirm or deny the results (Lachmann 1999). Dutch scientists suggested the toxic effects might have been due to nutritional differences between control and GE potatoes, and not the GE process (Kuiper, Noteborn, and Peijnenburg 1999). These experiments should have been repeated on larger numbers of animals with proper controls before public release of their results, an approach that should be the standard methodology to avoid reaching potentially errant conclusions.

Have Any Food Safety Studies Been Done on GE Foods?

At present all GE crops used commercially in foods undergo safety testing by the companies or institutions developing them, and the data are then reviewed by federal regulatory agencies. In many cases the products are also tested by outside groups and published in peer-reviewed journals. Although consultation with and submission to regulatory agencies of certain safety data for GE foods are voluntary, to date all commercial products have undergone full review (for searchable data on specific events, see the crop database at CERA [2010]). This compliance is due at least in part to legal liabilities that would result should a safety problem arise after market introduction.

Health safety assessments of GE foods are based in part on substantial equivalence (Kuiper et al. 2001), such that if the food is substantially equivalent to existing foods, it is treated like conventional foods with respect to certain aspects of its safety (Kessler et al. 1992). Food or food ingredients used safely for long periods do not require additional extensive safety testing; however, substances that raise scientifically based safety issues require extensive laboratory or animal testing. GE foods can be designated substantially equivalent to existing counterparts, substantially equivalent except for certain defined differences (on which safety assessments are separately needed), or not substantially equivalent, for which additional safety testing and review are necessary.

Large numbers of animal tests on GE foods and GE ingredients have been conducted; for reviews, see Chassy et al. (2004), König et al. (2004), Preston (2005), Van Eenennaam (2006), and Flachowsky et al. (2007). In these reviews, both chemical analyses and studies in a variety of animals – dairy cows, beef cattle, pigs, laying hens, broilers, fish, and rabbits – revealed no significant, unintended differences between GE and conventional varieties in composition, digestibility, or animal health and performance. Food-safety testing in animals is used to determine toxicity and allergenicity; however, such testing of whole GE foods is difficult because animals must consume large amounts of food to obtain sufficient quantities of the GE ingredient. Compositional analyses and toxicity testing of individual components of foods are actually more sensitive and accurate in assessing safety (Chassy et al. 2004), and thus safety tests are conducted on the products encoded in both target and selectable marker genes. According to the food additive provision, Section 409, of the 1992 Federal Food Drug and Cosmetics Act (FFDCA; FDA 2001), substances intentionally added to food are food additives, unless they are *g*enerally *r*ecognized *a*s *s*afe (GRAS) or are exempt. GRAS status is established either by a long history of food use or when the nature of the substance does not raise significant, scientifically based safety issues (FDA 1992).

Does the Transgene DNA in Foods Cause Safety Problems?

Humans eat foods and their DNA every day. What happens to this DNA? Mostly all DNA is degraded, no matter its source, both during industrial processing of foods and in the digestive tract, because all DNA is chemically identical. Yet small DNA fragments of orally administered phage M13 and plant DNA were found in phagocytes, because they normally function to clear the body of foreign agents (Schubbert et al. 1997, 1998). In these studies, fragments, but not intact genes, passed in rare instances to other organs, including the fetus, although others challenged these observations as resulting from sample contamination (Beever and Kemp 2000; Goldstein et al. 2005; Jonas et al. 2001). In July 2007, the European Food Safety Authority (EFSA) stated the following about the fate of genes and proteins in food and feed:

After ingestion, a rapid degradation into short DNA or peptide fragments is observed in the gastrointestinal tract of animals and humans. To date a large number of experimental studies with livestock have shown that recombinant DNA fragments or proteins derived from GM plants have not been detected in tissues, fluids or edible products of farm animals like broilers, cattle, pigs or quail.

EFSA (2007b, 5–6)

Although no data exist that transgene DNA has unique behavior relative to native DNA, in 2006 Dr. Irina Ermakova of the Russian Academy of Sciences publicly announced the results of her study in which she described stunted development and higher infant mortality in rats fed diets containing RR soybeans (55.6 percent mortality), versus conventional soybeans (9 percent mortality). She claimed that animals died of mutations induced solely by the transgene DNA. Although not published in a peer-reviewed scientific journal, her results were presented at international symposia (Pravda 2005) and during a parliamentary debate in New South Wales where they were used to push for a ban on GE crop cultivation (Parliament of New South Wales 2005). In fact, numerous other studies, some of which were multigenerational, on rats and mice fed RR soybean found no adverse effects on litter size, histological appearance of tissues, or numbers of deaths of progeny (Brake and Evenson 2004; Teshima et al. 2000; Zhu et al. 2004; for a review, see Soybean Tissue Culture and Genetic Engineering Center 2007). Differences between these studies and Ermakova's likely relate to shortcomings in her experimental procedures. As suggested for the lectin potato, studies like those of Ermakova's should be repeated with proper controls and be subjected to peer review before the results are released.

Can Eating *Bt* Protein Cause Food Safety Issues for Consumers?

Bt toxins, naturally occurring insecticides produced by the soil bacterium, *Bacillus thuringiensis*, have been used to control crop pests since the 1920s (Glazer and Nikaido 1995). Many strains of *B. thuringiensis* exist, producing different *Bt* toxins that vary in their target insects (e.g., butterfly larvae, beetles, and mosquitoes). *Bt* toxins form crystals inside the bacterium; hence the name Cry proteins. As full-sized proteins they are inactive against their insect targets. When inside the insect's midgut they are cleaved and become active, binding to specialized receptors and creating holes in the gut and killing the larvae. Specificity of different proteins for their targets and the lack of effect on nontarget hosts, like mammals, reside in their specific, tight binding to receptors (Federici 2002).

Numerous safety studies have been conducted on native *Bt* proteins showing that they do not have characteristics of toxins or food allergens (for a review, see Federici 2002; Mendelsohn et al. 2003). *Bt* proteins are termed plant-incorporated protectants (PIPs), so that safety considerations come under the Environmental Protection Agency (EPA), which stated that the data must show that *Bt* proteins "behave as would be expected of a dietary protein, are not structurally related to any known food allergen or protein toxin, and do not display any oral toxicity when administered at high doses" (EPA 2001, 12). To provide the needed data, in vitro digestion assays were used to confirm normal degradation characteristics; murine feeding studies confirmed the lack of acute oral toxicity (Betz, Hammond, and Fuchs 2000; Nawaz et al. 2001). The EPA did not require long-term studies because the protein's instability in digestive fluids makes such studies meaningless in terms of human health impacts (Shelton, Zhoa, and Roush 2002).

Possible allergenic effects of four maize *Bt* varieties were investigated in potentially sensitive populations (Batista et al. 2005). Skin prick tests were performed in two sensitive groups: children with food and inhalant allergies and individuals with asthma-rhinitis. IgE immunoblot reactivity of sera from patients with food allergies was tested versus *Bt* maize and pure CryIA[b] protein. No individual reacted differently to transgenic and nontransgenic samples; none had detectable IgE antibodies against pure *Bt* protein.

Some *Bt* crops produce a truncated version of the full-length *Bt* containing the toxic fragment. Assessment of the truncated protein's mammalian toxicity was done by feeding truncated Cry protein to groups of mice at $\leq$4 g/kg body weight – doses representing a 200- to 1,000-fold excess over exposures predicted based on human consumption (AGBIOS 2005). No treatment-related effects on body weight, food consumption, survival, or gross pathology on necropsy were observed for these mice.

A positive safety aspect of *Bt* corn is its lower levels of mycotoxins: toxic and carcinogenic chemicals produced by fungal pathogens that colonize corn after insect damage (Wu 2006, 2007, 2008). In some cases, reduction of mycotoxins in *Bt* corn has resulted in positive economic impacts on domestic and international markets for U.S. corn. More importantly, in less developed countries certain mycotoxins are significant food contaminants and result in adverse health effects; their reduction could improve human and animal health.

In 2002, another *Bt* corn variety with increased rootworm (*Diabrotica* spp.) resistance was deregulated. This variety contains a Cry protein distinguished from the native protein by the addition of an alanine residue and seven amino acid differences that enhance expression and insecticidal activity (AGBIOS 2006). Company food safety assessments used 90-day mouse feeding trials (Monsanto 2005); independent safety assessments were also conducted, showing no adverse health effects (Grant et al. 2003; Hyun et al. 2005; Taylor et al. 2003). In 2007, a statistical reanalysis of the safety data was published that differed from the original analysis, leading to the conclusion that "with the present data it cannot be concluded that GM corn MON 863 is a safe product" (Séralini, Cellier, and Vendomois 2007, 596). Later in 2007, the European Community (EC) requested EFSA to determine what impact the reanalysis had on the EC's earlier decision regarding the safety of MON 863. EFSA concluded that the reanalysis did not raise new safety concerns (EFSA 2007a).

Are Allergens Being Introduced into GE Foods?

Using genetic engineering to create new varieties has the potential to introduce allergens, just as do classical breeding and mutagenized populations. Under FDA's biotechnology food policy, GE foods must be labeled if the source of the gene is from one of the common allergy-causing foods, such as cow's milk, eggs, fish and shellfish, tree nuts, wheat, and especially peanuts and soybeans (Clydessdale 1996), unless the new food is proven not to be allergenic through additional safety testing. Although not mandatory, to date all new marketed GE foods have been analyzed to determine if introduced proteins have properties indicating they might be allergenic (i.e., similarities to known allergens, small size, slow digestibility, and high heat stability; Taylor and Hefle 2002). Although each category has exceptions, these characteristics indicate a protein might be allergenic and thus merits further study.

This type of analysis was performed on Starlink™ corn, but only after the corn ended up in human food. Another example of testing for allergenicity that occurred before the product was grown commercially involved soybeans engineered with a Brazil nut protein, used to improve soy's nutritional quality. Because allergies to nuts are among the most common allergies and Brazil nut allergies had been previously documented, testing of the GE soybeans for allergenicity was conducted before commercial release (Arshad et al. 1991). Serum from people allergic to Brazil nuts reacted with the GE soybean; the soy variety was never marketed (Nordlee et al. 1996).

It is important to point out that GE technology can also be used to engineer foods with lower allergenicity (i.e., hypoallergenicity). A number of foods cause human allergies, and in some cases the nature of the proteins causing these reactions is known, making it possible to engineer lower levels of the allergenic proteins or change their conformation to reduce allergenic responses (Buchanan 2001). Examples include reducing allergenicity in grass pollen and in foods like wheat, rice, and peanuts (Lemaux 2008).

Were Foods with *Bt* Corn Removed from the Market Due to Safety Concerns?

Starlink™ corn contained a Cry toxin with no previous history of human dietary exposure, was not readily digested, and was stable at 90°C (EPA 1998). The latter two characteristics are potential hallmarks of some allergens, and Cry9C also had biochemical characteristics different from other previously reviewed Cry proteins (EPA 1998). Because of these characteristics, the corn variety with the modified *Bt* was approved initially only for animal feed, because farm animals do not have food allergies. While additional allergenicity testing was being conducted, Starlink™ was found in human food. The FDA then issued a recall of food products containing Starlink™.

In October 2000 the FDA asked the Centers for Disease Control and Prevention (CDC) to investigate cases of human illness claimed to be related to the consumption of Starlink™ (CDC 2001). Some of these individuals described symptoms consistent with possible allergic reactions to corn products; blood serum samples of some of the patients were tested to detect antibodies to *Bt*. The CDC study indicated that Starlink™-specific antibodies were not found in those human sera; however, there were weaknesses in the study (i.e., food allergies can occur in individuals in the absence of detectable allergy-specific antibodies [Ogura et al. 1993], and the protein used to make antibodies was of bacterial, not plant, origin). Researchers also analyzed the corn-containing foods consumed by some of the test subjects and did not find *Bt* protein (FIFRA SAP 2001). Taken together, these results suggest that Starlink™ was not involved in the allergic reactions of some of the individuals, but uncertainty still remains because blood and food samples were not received from all individuals experiencing a true allergic reaction. In separate studies, an EPA scientific advisory panel concluded that the *Bt* in Starlink™ had a moderate chance to cause allergies, based on its biochemical nature (FIFRA SAP 2000). However, Starlink™ corn was removed from the market in 2000, and on the basis of USDA monitoring, the food supply in 2007 was 99.99 percent Starlink™-free (EPA 2008). That this situation occurred points to the necessity for complete regulatory review of both human and animal safety before allowing commercial production of crops to avoid safety and financial impacts.

Can GE Crops Engineered to Make Pharmaceuticals Contaminate Foods?

The first examples of plants engineered to deliver vaccines involved modifying tobacco to express a protein to prevent dental caries and a hepatitis B surface antigen (Curtiss and Cardineau 1990; Mason, Lam, and Arntzen 1992). Since then, maize, potato, rice, soybean and tomato have been used to produce both human and animal vaccines (Pascual 2007). Examples include vaccines against pneumonic and bubonic plague, a potato-based vaccine for hepatitis B, a pollen vaccine that reduces symptoms in allergy sufferers, and an edible rice-based vaccine targeted to allergic diseases, such as asthma, seasonal allergies, and atopic dermatitis (Lemaux 2008). Plant vaccines have the advantage of being consumed by humans or animals and requiring limited or no processing to vaccinate the consumer and do not require cold storage – both advantages for vaccine delivery in developing countries.

With these advantages, however, comes the possibility that such products could enter the food supply, despite U.S. regulations prohibiting GE plants containing pharmaceutical or industrial products from doing so (FDA 2004). The Animal and Plant Health Inspection Service (APHIS), which regulates the movement and field-testing of GE plants, requires special steps to prevent plants that produce drugs or industrial enzymes from contaminating food crops. These steps include (a) labeling, packaging, and segregating regulated plant materials; (b) reproductive isolation to prevent GE pollen from fertilizing conventional plants; (c) postharvest monitoring to remove volunteer plants; and (d) proper disposal of transgenic material. In 2005 these rules were tightened (a) to exclude field growth without a permit; (b) to include crop inspections seven times per year, two after harvest; (c) to increase field isolation distances; and (d) to use dedicated farm equipment (APHIS 2005). This tightening of the rules resulted from early violations of field-testing permits (APHIS 2002).

Such violations demonstrate that "pharming" in food plants could result in mixing with food. The Grocery Manufacturers of America urged the USDA to restrict plant-made pharmaceutical production to nonfood crops (Childs 2002). The National Corn Growers Association, however, countered by proposing safeguards such as using (a) plants that are male-sterile or that produce non-GE pollen, (b) dedicated production systems that isolate pharma crops, (c) third-party verification, and (d) grower training programs (Cassidy and Powell 2002). In September 2002, the FDA released a guidance document recommending strategies to prevent pharma plants from contaminating human or animal feed (FDA 2002). For example, those growing drug-producing plants that cross-pollinate, such as corn and canola, must strengthen containment procedures by growing the crop in geographical regions where little or none is grown for food. Despite these precautions, it is not 100 percent guaranteed that GE crops would not mix with food. In deciding whether the crop should be grown in the field the focus should be on possible consequences of such mixing.

Are GE Foods 100 Percent Safe?

Almost everything in our technologically complex world comes with risks; for example, the automobile, pasteurization, vaccination. Only after individuals gained experience with these new technologies did many realize that their benefits outweighed risks and they were accepted for common use. Despite this history, there are still individuals who do not want to use these technologies.

Acceptance of GE foods depends on perceptions of their benefit and risk. The first commercialized GE crops benefited farmers, companies, and often the environment, but consumers saw little benefit from them. In the development pipeline are GE crops and foods that might benefit consumers, but how much they do so depends on which crops, how they are deployed, and how they are valued. Another important acceptance factor is assurance of safety. Commercialized GE crops go through extensive testing, which is reviewed by the USDA, FDA, and/or EPA. Despite this testing, there is no guarantee of 100 percent safety; even conventionally bred foods and those grown organically cannot be declared completely safe (Avery 2006). Given the extensive safety testing and determination of substantial equivalence of GE foods, it is unlikely that safety issues will arise that are greater than with non-GE foods produced through conventional or organic farming practices (Lemaux 2008; NRC

2004). A final acceptance factor has to do with individual values – which cannot be addressed with scientific data.

Are Organic Foods Safer and Healthier Than Those Grown Conventionally?

The U.S. food supply is among the safest in the world, regardless of how agricultural production is practiced (FDA 2008). Organic farming methods differ from conventional farming in that, although pesticides can be used, they cannot be synthetic; synthetic fertilizers and growth enhancers are also not allowed. Organic certification guarantees the manner in which foods are grown, handled, and processed, but the label "organic" does not guarantee the nature of the product and does not imply the foods are safer or more nutritious (Ronald and Fouche 2006).

The reasons people consume organic foods vary, but many consumers perceive them to be healthier, taste better, be better for the environment and for animal welfare, and have lower pesticide levels and fewer food additives (Soil Association 2007). In general, however, there are not a large number of peer-reviewed studies that analyze nutritional differences between foods produced conventionally and organically. Of the limited studies, some show differences, and others do not. For example, in one study profiles of 44 metabolites in wheat grown under comparable organic and conventional conditions in Switzerland showed no difference (Zörb et al. 2006). Another study showed increases in vitamin C levels and antioxidant activities in kiwifruit grown organically relative to conventionally grown kiwifruit on the same farm and harvested at the same maturity stage (Amodio et al. 2007). Levels of the flavonoids, quercitin and kaempferol aglycones, in archived samples of organically produced tomatoes in a decade-long comparison were shown to be statistically higher than in those tomatoes grown using conventional practices (Mitchell et al. 2007). Increases in total antioxidant activity were also found in red oranges grown organically relative to nonorganic oranges (Tarozzi et al. 2005). There is "moderately strong and consistent data showing that organic potatoes are richer sources of vitamin C than their conventionally grown counterparts" (Williamson 2007, 105).

Although differences have been reported in nutrient composition between organically and conventionally produced foods, there are many important nutrients for which no significant differences have been found. Much more research is needed to determine whether nutritional differences between organic and conventional foods are reproducible across environments and growing practices and whether they have significant impacts on human health. Much more importantly, there *is* convincing epidemiological evidence that diets rich in fresh fruits and vegetables improve health and reduce frequency and severity of a number of health conditions (Riboli and Norat 2003). Thus, if the goal is to promote healthy eating, it is more critical for consumers to eat a diet rich in fruits and vegetables than to focus on whether foods are organic or conventional.

What Complexities Do Future GE Trait Introductions Present?

The first GE crops involved introduction of single, simple traits with defined biochemical outcomes, such as *Bt* insect resistance and herbicide tolerance traits. Future efforts will use genes for traits with more complicated outcomes that engage regulatory cascades – like those for abiotic and biotic stress (Korban 2005). Plants engineered in such cascades will require more complicated safety assessments because they could introduce "unforeseen metabolic perturbations" (Querci et al. 2008, 8). Such new varieties will require more sophisticated analyses to identify unforeseen effects that might occur in addition to those expected with the gene introduction. It is important to note, however, that such perturbations also occur during classical breeding and mutational strategies focused on crop improvement.

In addition to traits that engage more complex strategies, a number of recent GE varieties have multiple, stacked traits (i.e., combinations of two or more single traits in a single crop plant). For example, recent varieties have multiple *Bt* genes or one or more *Bt* genes with an herbicide tolerance gene. In 2008 stacked varieties of corn represented 40 percent of U.S. acreage and 45 percent of cotton acreage, but no stacked varieties exist for soybean (ERS 2010a, 2010b). In terms of regulatory oversight, varieties with multiple traits could require more complex regulatory scenarios than single-trait varieties. This could be particularly problematic in the EU where approval of GE crops is slow, and therefore testing for both EU-authorized and EU-nonauthorized traits is necessary. Because of this situation, crops with individual traits could be approved before crops with the same traits as a stacked variety. In this situation the precise nature of the GE food must be known to determine whether the imported food has EU approval. More robust sampling methods will be required to distinguish stacked events from simple mixtures of crops containing the corresponding single events. Recently new methods have been developed that can be used for multiplex detection of GE traits in complex raw materials and derived products (Prins et al. 2008).

The potential use of a wider range of organisms as sources of genes to introduce new traits and the creation of GE crops and foods by countries with less rigorous regulatory structures present new identification and safety assessment challenges for foods imported into the United States. Although an internationally harmonized approach to premarket safety assessment of GE foods could be created and adhered to, this is not yet in place, and developers of GE crops and products, risk assessors, and risk managers must stay alert to the nature of the new products and develop appropriate assessment tools to assure food safety.

Conclusion

Researchers using engineering methods now have the technology to transfer genes not only within a species but also from one kingdom to another. This can lead to changes in agricultural crops that were not previously possible and result in plants that are better able to survive pest attack and abiotic stresses, are enhanced nutritionally, or can immunize humans and animals. However, with this capacity comes the responsibility to proceed with caution, investigating possible outcomes carefully. Conversely, there is also a responsibility to use the technology where it can provide positive change for human health, the environment, and farmers' productivity.

Based on the intensive study of the data and peer-reviewed research needed to write this chapter, it seems to me that GE crop development to date has been conducted responsibly and regulatory agencies have, in general, proceeded with caution in releasing GE varieties. Although no human activity can be guaranteed 100 percent safe, commercial GE crops and products available today are at least as safe as those produced by conventional methods. This does not mean we should relax our vigilance in investigating such products or those from more time-honored methods. However, we should not hold GE products to standards not required for food and feed products produced by other methods. With the proper balance of caution and scrutiny, we can take advantage of the power of this technology without compromising the health of humans, animals, or the environment.

Acknowledgments

The author thanks the *Annual Review of Plant Biology* for allowing me to use many of the ideas contained in an article written for that series (Lemaux 2008). The author is also indebted to Ms. Barbara Alonso, who helped in the preparation of this article in many ways, but most importantly by constantly updating a scientific database that was instrumental in being able to cite the myriad of articles from the scientific literature.

References

Ag BioTech InfoNet. 1999. "ASA Response to: Alterations in Clinically Important Phytoestrogens in Genetically Modified, Herbicide-Tolerant Soybeans." http://www.biotech-info.net/ASA_response.html.

AGBIOS. 2005. GM Crop Database. ("Mon 810"; accessed December 10). http://cera-gmc.org/index.php?action=gm_crop_database&mode=Submit&evidcode=MON810.

______. 2006. GM Crop Database. ("Mon863"; accessed December 10). http://cera-gmc.org/index.php?action=gm_crop_database&mode=Submit&evidcode=MON863.

Amodio, M. L., G. Colelli, J. K. Hasey, and A. A. Kader. 2007. "A Comparative Study of Composition and Postharvest Performance of Organically and Conventionally Grown Kiwifruits." *Journal of the Science of Food and Agriculture* 87 (7): 1228–36.

APHIS (Animal and Plant Health Inspection Service). 2002. "USDA Investigates Biotech Company for Possible Permit Violations." Press release, November 13. http://www.aphis.usda.gov/lpa/news/2002/11/prodigene.html.

———. 2005. "Introductions of Plants Genetically Engineered to Produce Industrial Compounds." 7 CFR Part 340 [Docket No. 03-038-2] RIN 0579-AB89. *Federal Register* 70 (85): 23009–11. http://edocket.access.gpo.gov/2005/pdf/05-8860 .pdf.

Arshad, S. H., E. Malmberg, K. Krapf, and D.W. Hide. 1991. "Clinical and Immunological Characteristics of Brazil Nut Allergy." *Clinical and Experimental Allergy* 21 (3): 373–6.

Avery, A. A. 2006. *The Truth about Organic Foods*. Chesterfield, MO: Henderson Communications.

Batista, R., B. Nunes, M. Carmo, C. Cardoso, H. S. Jose, A. B. de Almeida, A. Manique, L. Bento, C. P. Ricardo, and M. M. Oliveira. 2005. "Lack of Detectable Allergenicity of Transgenic Maize and Soya Samples." *Journal of Allergy and Clinical Immunology* 116 (2): 403–10.

Beever, D., and C. Kemp. 2000. "Safety Issues Associated with the DNA in Animal Feed Derived from Genetically Modified Crops: A Review of Scientific and Regulatory Procedures." *Nutrition Abstracts and Reviews Series B: Livestock Feeds and Feeding* 70 (3): 175–82. http://www.biotech-info.net/BeeverKemp1.pdf.

Betz, F. S., B. F. Hammond, and R. L. Fuchs. 2000. "Safety and Advantages of *Bacillus thuringiensis*-Protected Plants to Control Insect Pests." *Regulatory Toxicology and Pharmacology* 32 (2): 156–73.

Brake, D. G., and D.P. Evenson. 2004. "A Generational Study of Glyphosate-Tolerant Soybeans on Mouse Fetal, Postnatal, Pubertal and Adult Testicular Development." *Food and Chemical Toxicology* 42 (1): 29–36.

Buchanan, B. B. 2001. "Genetic Engineering and the Allergy Issue." *Plant Physiology* 126: 5–7.

Cassidy, B., and D. Powell. 2002. "Pharmaceuticals from Plants: The Prodigene Affair." *Food Safety Network* Agnet. Ontario: University of Guelph. http://archives.foodsafety.ksu.edu/agnet/2002/12-2002/agnet_december_3.htm.

CDC (Centers for Disease Control and Prevention). 2001. *Investigation of Human Health Effects Associated with Potential Exposure to Genetically Modified Corn*. Atlanta: CDC. http://www.cdc.gov/nceh/ehhe/Cry9cReport/pdfs/cry9creport.pdf.

CERA (Center for Environmental Risk Assessment). 2010. GM Crop Database. (Accessed December 10). Washington, DC: ILSI (International Life Sciences Institute). http://cera-gmc.org/index.php?action=gm_crop_database.

Chassy, B., J. J. Hlywka, G. A. Kleter, E. J. Kok, H. A. Kuiper, M. McGloughlin, I. C. Munro, R. H. Phipps, and J. E. Reid. 2004. *Nutritional and Safety Assessments of Foods and Feeds Nutritionally Improved through Biotechnology: An Executive Summary*. With contributions by K. Glenn, B. Henry, and R. Shillito. Chicago: Institute of Food Technologists.

Childs, S. 2002. "GMA Urges the Use of Non-Food Crops for Biotech Drugs; ProdiGene's Errors Raise Serious Concerns, Says GMA." *PR Newswire*, Nov 14. http://www.thefreelibrary.com/GMA+Urges+the+Use+of+Non-Food+Crops+for+Biotech+Drugspercent3b+ProdiGenepercent27s...-a094228770.

Clark, J. H., and I. R. Ipharraguerre. 2001. "Livestock Performance: Feeding Biotech Crops." *Journal of Dairy Science* 84 (E. Suppl.): E9–E18. http://www.agbioworld.org/pdf/livestockperformance.pdf.

Clydessdale, F. M. 1996. "Allergenicity of Foods Produced by Genetic Modification." *Food Science and Nutrition* 36 (Supplement): 1–186.

Curtiss, R. I., III, and C. A. Cardineau. 1990. "Genetically Modified Plants for Use as Oral Immunogens." World Patent Application WO 90/02484.

EFSA (European Food Safety Authority). 2007a. "Statement of the Scientific Panel on Genetically Modified Organisms on the Analysis of Data from a 90-Day Rat Feeding

Study with MON 863 Maize." Adopted June 25. Parma, Italy: EFSA. http://www.efsa.europa.eu/EFSA/Statement/GMO_statement_MON863.pdf.

______. 2007b. "EFSA Statement on the Fate of Recombinant DNA or Proteins in Meat, Milk and Eggs from Animals Fed with GM Feed." Parma, Italy: EFSA. http://www.fve.org/veterinary/pdf/food/efsa_statement_recombinant_dna_food_safety.pdf.

EPA (Environmental Protection Agency). 1998. "*Bacillus thuringiensis* Subspecies *tolworthi* Cry9C Protein and Genetic Material Necessary for Production in Corn." Federal Register Document 98-13604. May 22. *Federal Register* 63 (99): 28258–61. http://frwebgate.access.gpo.gov/cgi-bin/getdoc.cgi?IPaddress=frwais.access.gpo.gov&dbname=1998_register&docid=98-13604-filed.

______. 2001. "Biopesticides Registration Action Document. Preliminary Risks and Benefits Section: *Bacillus thurigiensis* Plant-Pesticides." *Biotechnology Law Report* 20 (1): 57–69. doi:10.1089/073003101300053013.

______. 2008. "EPA White Paper Regarding StarLink™ Corn Dietary Exposure and Risk; Availability." April 25. *Federal Register* 73 (81): 22716.

ERS (Economic Research Service). 2010a. "Adoption of Genetically Engineered Crops in the U.S.: Corn Varieties." http://www.ers.usda.gov/data/biotechcrops/extentofadoptiontable1.htm.

______. 2010b. "Adoption of Genetically Engineered Crops in the U.S.: Upland Cotton Varieties." http://www.ers.usda.gov/data/biotechcrops/ExtentofAdoptionTable2.htm.

Ewen, S. W. B., and A. Pusztai. 1999. "Effect of Diets Containing Genetically Modified Potatoes Expressing Galanthus Nivalis Lectin on Rat Small Intestine." *Lancet* 354 (9187): 1353–4.

FDA (Food and Drug Administration). 1992. "Statement of Policy: Foods Derived from New Plant Varieties." Docket No. 92N-0139. May 29. *Federal Register* 54 (104): 22984. http://www.fda.gov/Food/GuidanceComplianceRegulatoryInformation/GuidanceDocuments/Biotechnology/ucm096095.htm.

______. 2001. "Premarket Notice Concerning Bioengineered Foods." http://www.reginfo.gov/public/do/eAgendaViewRule?pubId=200104&RIN=0910-AC15.

______. 2002. *Guidance for Industry Drugs, Biologics, and Medical Devices Derived from Bioengineered Plants for Use in Humans and Animals*. Washington, DC: FDA. http://www.fda.gov/downloads/AnimalVeterinary/GuidanceComplianceEnforcement/GuidanceforIndustry/UCM055424.pdf.

______. 2004. "Chapter IV: Food." In *Federal Food, Drug, and Cosmetic Act*. Washington, DC: FDA. http://www.fda.gov/RegulatoryInformation/Legislation/FederalFoodDrugandCosmeticActFDCAct/FDCActChapterIVFood/default.htm.

______. 2008. *Food Protection Plan*. Washington, DC: FDA. http://www.fda.gov/oc/initiatives/advance/food.html.

Federici, B. 2002. "Case Study: *Bt* Crops A Novel Mode of Insect Control." In *Genetically Modified Crops: Assessing Safety*, edited by K. T. Atherton, 164–200. London: Taylor and Francis.

FIFRA SAP (Federal Insecticide, Fungicide, and Rodenticide Act Scientific Advisory Panel). 2000. "A Set of Scientific Issues Being Considered by the Environmental Protection Agency Regarding: Assessment of Scientific Information Concerning StarLink™ Corn." SAP Report No. 2000-06. Nov. 28. http://www.epa.gov/scipoly/sap/meetings/2000/november/one.pdf.

______. 2001. "A Set of Scientific Issues Being Considered by the Environmental Protection Agency Regarding: Assessment of Scientific Information Concerning StarLink™ Corn." SAP Report No. 2001-09. July 17–18. http://www.epa.gov/scipoly/sap/meetings/2001/july/julyfinal.pdf.

Flachowsky, G., K. Aulrich, H. Böhme, and I. Halle. 2007. "Studies on Feeds from Genetically Modified Plants (GMP) – Contributions to Nutritional and Safety Assessment." *Animal Feed Science and Technology* 133 (2007): 2–30.

Food Insight. 2008. "Food Biotechnology: A Study of U.S. Consumer Attitudinal Trends, 2008 Report." *Food Insight*, October 21. http://www.foodinsight.org/Resources/Detail.aspx?topic=Food_Biotechnology_A_Study_of_U_S_Consumer_Attitudinal_Trends_2008_REPORT.

Glazer, A. N., and H. Nikaido. 1995. *Microbial Biotechnology: Fundamentals of Applied Microbiology*. New York: W. H. Freeman.

Goldstein, D. A., B. Tinland, L. A. Gilbertson, J. M. Staub, G. A. Bannon, R. E. Goodman, R. L. McCoy, and A. Silvanovich. 2005. "Human Safety and Genetically Modified Plants: A Review of Antibiotic Resistance Markers and Future Transformation Selection Technologies." *Journal of Applied Microbiology* 99 (1): 7–23.

Grant, R. J., K. C. Fanning, D. Kleinschmit, E. P. Stanisiewski, and G. F. Hartnell. 2003. "Influence of Glyphosate-Tolerant (event nk603) and Corn Rootworm Protected (event MON863) Corn Silage and Grain on Feed Consumption and Milk Production in Holstein Cattle." *Journal of Dairy Science* 86 (5): 1707–15.

Hammond, B. G., J. L. Vicini, G. F. Hartnell, M. W. Naylor, C. D. Knight, E. H. Robinson, R. L. Fuchs, and S. R. Padgette. 1996. "The Feeding Value of Soybeans Fed to Rats, Chickens, Catfish and Dairy Cattle is Not Altered by Genetic Incorporation of Glyphosate Tolerance." *Journal of Nutrition* 126 (3): 717–27.

Hawes, A. 2010. "Contaminated L-Tryptophan and 5-Hydroxy-L-Tryptophan, Eosinophilia Myalgia Syndrome [EMS], the 1989 Epidemic and the 1998 Warning." Accessed December 10. http://www.alexanderinjury.com/library/library_toxic_10.shtml.

Hightower, R., C. Baden, E. Penzes, P. Lund, and P. Dunsmuir. 1991. "Expression of Antifreeze Proteins in Transgenic Plants." *Plant & Molecular Biology* 17 (5): 1013–21.

Hyun, Y., G. E. Bressner, R. L. Fischer, P. S. Miller, M. Ellis, B. A. Peterson, E. P. Stanisiewski, and G. F. Hartnell. 2005. "Performance of Growing-Finishing Pigs Fed Diets Containing YieldGard Rootworm Corn (MON 863), a Nontransgenic Genetically Similar Corn, or Conventional Corn Hybrids." *Journal of Animal Science* 83 (7): 1581–90.

James, S., and M. Burton. 2003. "Consumer Preferences for GM Food and Other Attributes of the Food System." *Australian Journal of Agricultural Research and Economics* 47 (4): 501–18.

Jonas, D. A., I. Elmadfa, K. H. Engle, K. J. Heller, G. Kozianowski, A. Konig, D. Muller, J. F. Narbonne, W. Wackernagel, and J. Kleiner. 2001. "Safety Considerations of DNA in Food." *Annals of Nutrition and Metabolism* 45 (6): 235–54.

Kessler, D. A., M. R. Taylor, J. H. Maryanski, E. L. Flamm, L. S. Kahl. 1992. "The Safety of Foods Developed by Biotechnology." *Science* 256 (5065): 1747–9.

König, A., A. Cockburn, R. W. R. Crevel, E. Debruyne, R. Grafstroem, U. Hammerling, I. Kimber, et al. 2004. "Assessment of the Safety of Foods Derived from Genetically Modified (GM) Crops." *Food and Chemical Toxicology* 42 (7): 1047–88.

Korban, S. S. 2005. "Genetic and Metabolic Engineering for Value-Added Traits." *Crop Science* 45 (2): 435–6. doi: 10.2135/cropsci2005.0435.

Kuiper, H. A., G. A. Kleter, H. P. Noteborn, and E. J. Kok. 2001. "Assessment of the Food Safety Issues Related to Genetically Modified Foods." *Plant Journal* 27 (6): 503–28.

———, H. P. Noteborn, and A. A. Peijnenburg. 1999. "Adequacy of Methods for Testing the Safety of Genetically Modified Foods." *Lancet* 354 (9187): 1315–16.

Lachmann, A. 1999. "GM Food Debate." *Lancet* 354 (9191): 1726.

Lappé, M. A., E. B. Bailey, C. Childress, and K. D. R. Setchell. 1999. "Alterations in Clinically Important Phytoestrogens in Genetically Modified, Herbicide Tolerant Soybeans." *Journal of Medicinal Food* 1 (4): 241–5.

Lee, J., M. S. Cetiner, W. J. Blackmon, and J. M. Jaynes. 1990. "The Reduction of the Freezing Point of Tobacco Plants Transformed with the Gene Encoding for the

Antifreeze Protein from Winter Flounder." *Journal of Cellular Biochemistry* 44 (S14E): 303.

Lemaux, P. G. 2008. "Genetically Engineered Plants and Foods: A Scientist's Analysis of the Issues (Part I)." *Annual Review of Plant Biology* 59: 771–812.

Macer, D. R. J. 2003. "Genetic Engineering: Cross Species and Cross Cultural Perspectives." In *Dialog der Kulturen*, edited by S. Fritsch-Oppermann, 159–80. Tsukuba: Evangelische Akademie Loccum.

Mason, H. S., D. M. Lam, and C. J. Arntzen. 1992. "Expression of Hepatitis B Surface Antigen in Transgenic Plants." *Proceedings of the National Academy of Sciences of the United States of America* 89 (24): 11745–9.

Mayeno, A. N., and G. J. Gleich. 1994. "Eosinophilia-Myalgia Syndrome and Tryptophan Production: A Cautionary Tale." *Trends in Biotechnology* 12 (9): 346–52.

Mendelsohn, M., J. Kough, Z. Vaituzis, and K. Matthews. 2003. "Are *Bt* Crops Safe?" *Nature Biotechnology* 21 (9): 1003–9.

Mitchell, A. E., Y. J. Hong, E. Koh, D. M. Barrett, D. E. Bryant, R. F. Denison, and S. Kaffka. 2007. "Ten-Year Comparison of the Influence of Organic and Conventional Crop Management Practices on the Content of Flavonoids in Tomatoes." *Journal of Agricultural and Food Chemistry* 55 (15): 6154–9.

Monsanto. 2005. "Backgrounder on the Safety Assessment and Regulatory Status of MON 863 YieldGard® Rootworm Maize." April 17. www.monsanto.co.uk/news/ukshowlib.phtml?uid=8846.

______. 2010. "Vistive Low Linolenic Soybeans." Accessed December 10. http://www.monsanto.com/products/Pages/vistive-soybeans.aspx.

Nawaz, M. S., B. D. Erickson, A. A. Khan, S. A. Khan, J. V. Pothuluri, F. Rafii, J. B. Sutherland, R. D. Wagner, and C. E. Cerniglia. 2001. "Human Health Impact and Regulatory Issues Involving Antimicrobial Resistance in the Food Animal Production Environment." *Regulatory Research Perspectives* 1 (1): 1–10.

Nordlee, J. A., S. L. Taylor, J. A. Townsend, L. A. Thomas, and R. K. Bush. 1996. "Identification of a Brazil-Nut Allergen in Transgenic Soybeans." *New England Journal of Medicine* 334 (11): 688–92.

NRC (National Research Council). 2004. *Safety of Genetically Engineered Foods: Approaches to Assessing Unintended Health Effects*. Washington, DC: National Academy Press.

Ogura, Y., H. Ogura, N. Zushi, H. Morita, and T. Kurashige. 1993. "The Usefulness and the Limitations of the Radioallergosorbent Test in Diagnosing Food Allergy in Atopic Dermatitis." *Arerugi - Japanese Journal of Allergology* 46 (6): 748–56.

Padgette, S. R., N. B. Taylor, D. L. Nida, M. R. Bailey, J. MacDonald, L. R. Holden, and R. L. Fuchs. 1996. "The Composition of Glyphosate-Tolerant Soybean Seeds Is Equivalent to That of Conventional Soybeans." *Journal of Nutrition* 126 (3): 702–16.

Parliament of New South Wales. 2005. *Legislation Review Digest*. No. 12. http://www.parliament.nsw.gov.au/Prod/parlment/committee.nsf/0/5688893E71F8AAB7CA25709E00236589.

Pascual, D. W. 2007. "Vaccines Are for Dinner." *Proceedings of the National Academy of Science of the United States of America* 104 (26): 10757–8.

Pravda. 2005. "People Eating Genetically Modified Food May Have Rat-Short Lifespan." Pravda, October 27. http://english.pravda.ru/science/tech/27-10-2005/9136-gmf-0/.

Preston, C. 2005. "Peer Reviewed Publications on Safety of GM Foods." *AgBioWorld*. http://www.agbioworld.org/biotech-info/articles/biotech-art/peer-reviewed-pubs.html.

Prins, T. W., J. P. van Dijk, H. G. Beenen, A. M. A. Van Hoef, M. M. Voorhuijzen, C. D. Schoen, H. J. M. Aarts, and E. J. Kok. 2008. "Optimised Padlock Probe Ligation and Microarray Detection of Multiple (Non-Authorised) GMOs in a Single Reaction." *BMC Genomics* 9 (1): 584. doi:10.1186/1471-2164-9-584.

Querci, M., G. Kleter, J.-P. Malingreau, H. Broll, and G. Van den Eede. 2008. *Executive Summary of Scientific and Technical Contribution to the Development of an Overall Health Strategy in the Area of GMOs*, 41. Report EUR 23542 EN. Luxembourg: Office for Official Publications of the European Communities. doi:10.2788/16411.

Riboli, E., and T. Norat. 2003. "Epidemiologic Evidence of the Protective Effect of Fruit and Vegetables on Cancer Risk." *American Journal of Clinical Nutrition* 78 (3): 559S–69S.

Ronald, P., and B. Fouche. 2006. *Genetic Engineering and Organic Production Systems*. Agricultural Biotechnology in California Series Publication 8188. Oakland: University of California. http://ucanr.org/freepubs/docs/8188.pdf.

Roufs, J. B. 1992. "Review of L-Tryptophan and Eosinophilia-Myalgia Syndrome." *Journal of the American Dietetic Association* 92 (7): 844–50.

Schubbert, R., U. Hohlweg, D. Renz, and W. Doerfler. 1998. "On the Fate of Orally Ingested Foreign DNA in Mice: Chromosomal Association and Placental Transmission in the Fetus." *Molecular and General Genetics* 259 (6): 569–76.

———, B. Renz, B. Schmitz, and W. Doerfler. 1997. "Foreign (M13) DNA Ingested by Mice Reaches Peripheral Leukocytes, Spleen, and Liver via the Intestinal Wall Mucosa and Can Be Covalently Linked to Mouse DNA." *Proceedings of the National Academy of Sciences of the United States of America* 94 (3): 961–6.

Séralini, G. E., D. Cellier, and J. S. de Vendomois. 2007. "New Analysis of a Rat Feeding Study with a Genetically Modified Maize Reveals Signs of Hepatorenal Toxicity." *Archives of Environmental Contamination Toxicology* 52 (4): 596–602.

Shelton, A. M., J.-Z. Zhoa, and R. T. Roush. 2002. "Economic, Ecological, Food Safety and Social Consequences of the Deployment of *Bt* Transgenic Plants." *Annual Review of Entomology* 47 (1): 845–81.

Shewry, P. R., M. Baudo, A. Lovegrove, S. Powers, J. A. Napiera, J. L. Ward, J. M. Baker, and M. H. Beale. 2006. "Are GM and Conventionally Bred Cereals Really Different?" *Trends Food Science & Technology* 18 (4): 201–9.

Smith, J. M. 2003. *Seeds of Deception*. Fairfield, IA: Chelsea Green Publishing.

Soil Association. 2007. "10 Reasons to Eat Organic Food." http://www.whyorganic.org/healthy_tenReasons.asp.

Soybean Tissue Culture and Genetic Engineering Center. 2007. "Roundup Ready Soybean Selected References." http://mulch.cropsoil.uga.edu/soy-engineering/RoundupReady.html.

Tarozzi, A., S. Hrelia, C. Angeloni, F. Morroni, P. Biagi, M. Guardigli, G. Cantelli-Forti, and P. Hrelia. 2005. "Antioxidant Effectiveness of Organically and Non-Organically Grown Red Oranges in Cell Culture Systems." *European Journal of Nutrition* 45 (3): 152–8.

Taylor, M. L., Y. Hyun, G. F. Hartnell, S. G. Riordan, M. A. Nemeth, K. Karunanandaa, B. George, and J. D. Astwood. 2003. "Comparison of Broiler Performance When Fed Diets Containing Grain from YieldGard Rootworm (MON863), YieldGard Plus (MON810 x MON863), Nontransgenic Control, or Commercial Reference Corn Hybrids." *Poultry Science* 82 (12): 1948–56.

Taylor, S. L., and S. L. Hefle. 2002. "Genetically Engineered Foods: Implications for Food Allergy." *Current Opinion in Allergy and Clinical Immunology* 2 (3): 249–52.

Teshima, R., H. Akiyama, H. Okunuki, J. Sakushima, Y. Goda, H. Onodera, J. Sawada, and M. Toyoda. 2000. "Effect of GM and Non-GM Soybeans on the Immune System of BN Rats and B10A Mice." *Journal of the Food Hygienic Society of Japan* 41 (3): 188–93.

Van Eenennaam, A. L. 2006. *Genetic Engineering and Animal Agriculture*. Genetic Engineering Fact Sheet 7 Publication 8184. Oakland: University of California Agriculture and Natural Resources Communication Services. http://ucanr.org/freepubs/docs/8184.pdf.

Williamson, C. 2007. "Is Organic Food Better for Our Health?" *Nutrition Bulletin* 32 (2): 104–8.

Wu, F. 2006. "Mycotoxin Reduction in *Bt* Corn: Potential Economic, Health, and Regulatory Impacts." *Transgenic Research* 15 (3): 277–89.

______. 2007. "*Bt* Corn and Impact on Mycotoxins." *CAB Reviews: Perspectives in Agriculture, Veterinary Science, Nutrition and Natural Resources* 2 (060): 1–8.

______. 2008. "Field Evidence: *Bt* Corn and Mycotoxin Reduction." *Information Systems for Biotechnology* News Report, February. Blacksburg: Virginia Tech. http://croplife.intraspin.com/Biotech/papers/ID279.pdf.

Zhu, Y. Z., D. F. Li, F. L. Wang, J. D. Yin, and H. Jin. 2004. "Nutritional Assessment and Fate of DNA of Soybean Meal from Roundup Ready or Conventional Soybeans Using Rats." *Archives of Animal Nutrition* 58 (4): 295–310.

Zörb, C., G. Langenkämper, T. Betxche, K. Niehaus, and A. Barsch. 2006. "Metabolite Profiling of Wheat Grains (*Triticum aestivum* L.) from Organic and Conventional Agriculture." *Journal of Agricultural and Food Chemistry* 54 (21): 8301–6.

8

Ecological Considerations in Biotechnology

Ecological Concerns and Environmental Risks of Transgenic Crops

Mark K. Sears and Jeffrey D. Wolt

Genetically Engineered Products Now and in the Future

To date, the dominant products of GE are primarily herbicide-tolerant and insecticide-expressing plants from toxins produced by the bacterium, *Bacillus thuringiensis* either as crystalline (Cry) proteins or vegetative insecticidal proteins (Vip). The impact that this technology has had on food production of major crops has been significant, whereas the impact these products have had on the environment has been minimal, if at all (Brookes and Barfoot 2006). Most effort has been focused on how biotechnology can best be incorporated into integrated pest management (IPM) systems and on the impact of these crops in reducing pesticide use or conserving natural enemy activity in multiple pest crops systems (Romeis, Shelton, and Kennedy 2008). Future traits will diverge from a pest management focus to that of value-added traits such as nutraceuticals and pharmaceuticals, tolerance to growing conditions such as drought and salt tolerance, and nutrient enhancement.

Glyphosate- and Gluphosinate-Tolerant Maize, Soybeans, and Canola

More than 80 percent of transgenically transformed crops planted throughout the world possess herbicide-tolerant traits (James 2009). These traits are widely deployed in soybeans as well as canola, cotton, and maize, which have been widely adopted by large- and small-scale growers in many regions of the world. Their success can be attributed to the ease of weed management options, increased yield and economic return, and soil conservation due to reduced cultivation permitted by deployment of these crops. Many specific traits, their development and nature, economic impacts, and integration with pest management practices have been reviewed in Duke and Ragsdale (2005).

Bt Protein (Insecticide) Expressing Maize, Cotton, and Potatoes

GM maize accounts for nearly one-quarter of the world's production of this commodity, or more than 0.35 million km^2 (James 2009). Management of two important pest groups (stalk and stem borers and rootworms) has been accomplished by the insertion of genetic coding that directs plant cells to produce toxic *Bt* Cry proteins that affect specific gut receptors of lepidopteran or coleopteran insects and reduce their feeding damage on stalks and ear tissue or on roots, respectively (Hellmich et al. 2008). A wide range of *Bt* Cry protein-expressing events have been incorporated in commercial maize hybrids since the introduction of this technology in 1996. In some cases, lepidopteran-targeted events are stacked into single hybrids, or both lepidopteran- and coleopteran-targeted events are combined in new maize hybrids, thus providing some degree of protection against the development of insect tolerance. Most maize growers in the world continue to rely on conventional pest management methods of insecticide application, crop rotation, and suppression by natural enemies. Their use of biotechnology is facilitated primarily by a lack of government regulation but also to some extent on returns on their investment, degree of yield protection obtained, and market concerns.

Transgenic or GE cotton varieties have been grown commercially in the United States since 1996 and now represent 49 percent of the world's production, or about 0.15 million km^2 (James 2009). About 20 percent of the area grown contains transgenic herbicide-tolerant traits. Most of this acreage includes traits that express *Bt* Cry toxins for control of bollworms, the major global pest complex, but these traits are also very active against other lepidopterous groups such as leafworms and semi-loopers, while being only somewhat active against armyworms and cutworms. Effective control of caterpillars has drastically reduced insecticide applications in *Bt* cotton fields by as much as 50 percent, allowing for more options for pest management and the possibilities of area-wide suppression of major cotton pests (Naranjo et al. 2008; Storer, Dively, and Herman 2008). However, many secondary or nontarget pests that were suppressed by collateral insecticide action have become problematic because of the wide range of cotton-feeding insect species. Sucking and rasping insects such as stink bugs, plant bugs, and thrips have posed a problem to cotton growers in Australia (Wilson, Hickman, and Deutscher 2006). Although tolerance by pests to *Bt* Cry toxins has been documented, no indication of resistance in the field has been detected except for pink bollworm in 1997. Likely, strict conformance to untreated refuge areas of cotton fields and the development of stacked and dual-action or pyramided gene constructs have reduced the rate of tolerance development.

GE potatoes that express *Bt* Cry toxins specific for control of the Colorado potato beetle and genes for resistance to *PVY* and potato leaf-roll virus were deregulated by the USDA and the Canadian Food Inspection Agency (CFIA) in 1995 for commercial use in the United States and Canada, but were removed from the market because of negative market perceptions about GE crops (Grafius and Douches 2008). The potato is the world's fourth largest crop in production behind maize, rice, and wheat and is rising in popularity because it can be grown in adverse conditions and has high nutritional value. Stacked and pyramided genes for insect and virus suppression, herbicide tolerance, nutritional supplements, and adverse soil and moisture conditions are available for potatoes, but must await consumer acceptance in world markets before commercialization.

Virus Resistance in Papaya and Squash

Both GE virus-resistant squash and papaya have been commercialized in various regions of the world. Virus-resistant GE papaya, which was commercialized in 1998, has proven an especially important innovation that eliminates infection by the ringspot virus organism and has proven critical to maintaining papaya production in many regions (see also the discussion in Chapter 5). This hybrid has essentially replaced the conventional virus-susceptible hybrids previously grown in areas such as Hawaii and other world production areas in Florida, Thailand, Taiwan, and Venezuela (Gonsalves 1998).

Future Crop Enhancements from GE Technology: Herbicide Tolerance and Bt *Protein in Rice*

Rice is the world's most widely consumed food crop and is grown on more than 1.52 million km^2. Nearly half of the world's population depends on rice as a staple source of carbohydrate (FAOSTAT 2007). An herbicide-tolerant rice hybrid was deregulated in the United States in 2006 (APHIS 2006), and evaluation of this technology is occurring in many countries. Many *Bt* Cry toxins have been incorporated into rice hybrids through GE technology and have proven effective in reducing damage from the major lepidopterous pests of rice: stem borers, and leaf rollers (Cohen et al. 2008). Although stem borers and leaf rollers usually cause only about 2–3 percent yield loss in rice, their presence causes growers to use insecticide treatments that result in the promotion of secondary pests such as plant hoppers and weevils. Use of *Bt* rice cultivars reduces the need for insecticides to control stem borers, allowing natural control and conventionally produced resistant cultivars to reduce damage by secondary pests. The greatest constraints to acceptance of commercial lines of *Bt* rice are resistance mitigation and international market recognition. In a related vein, the development of a GE rice cultivar compatible with climate conditions in developing countries called "golden rice" has great promise for vitamin A supplementation of diets for millions of people in Asia and other locations. Golden rice, which may pave the way for acceptance of *Bt* rice cultivars in the future, is explored in a case study in Chapter 10.

Fruit and Vegetable Pest Resistance

Fruits and vegetables are essential ingredients of a well-balanced diet because they supply the nutrients not found in abundance in staple crops. They provide many opportunities for genetic manipulation through GE technology because of their variety and specific requirements for growth. Because many human communities value locally produced varieties of fruits and vegetables and these varieties command a price premium, pesticides are usually used to maintain a constant supply and high quality of these commodities. A myriad of insect pests and plant pathogens could be addressed by GE technology, and solutions to these problems could best be directed at developing economies (Shelton, Fuchs, and Shotkoski 2008). China and India alone are responsible for nearly 50 percent of the world's production of fruits and vegetables (FAOSTAT 2007). *Bt*-expressing eggplant (brinjal) has been deregulated for seed production in India and will likely be the first GE food crop released for commercial use in Asia (James 2009). Only a few fruits and vegetables with GE technology traits are commercialized in the United States, including papaya, potatoes, sweet maize, and squash.

Future Plant Traits Derived from GE Technology

Alternative pesticidal traits are being developed that are not derived from *Bt* genes. Novel insecticidal compounds derived from plants such as protease inhibitors, lectins, and alpha-amylase inhibitors are currently being investigated. Future technology will incorporate compounds active against pests such as chitinases, proteases, and toxins derived from sources such as bacteria, viruses, plants, and insects (Malone, Gatehouse, and Barratt 2008).

Nutrient enhancement and pharmaceutical expression are being developed using gene transfer from plants sources that enhance nutraceuticals such as beta carotenoids (precursor to Vitamin A) in golden rice, and lycopenes in tomatoes, glucosinolates, phytoestrogens, and phenolics (Neeser and German 2004). The pharmaceutical industry has investigated GE in searching for new processes to develop and produce drug and therapeutic compounds from plant resources (Crommelin, Sindelar, and Meibohm 2007; Keyser and Müller 2004).

Drought and salt tolerance have been thoroughly investigated and incorporated into experimental and crop plants for future commercialization (Denby and Gehring 2005). Biofuel production and industrial chemical production from plant waste and animal residues, as well as industrial components such as oils, have captured the attention of chemists searching for alternative energy sources. The challenge for biotechnology will be developing new crop plants with high biomass production per hectare and that can be converted to biofuels via digestion or produce substantial amounts of biofuel precursors. The development of these novel applications of GE expands the range of ecological considerations that must be addressed, but fortunately this analysis can be accommodated within existing frameworks for environmental risk assessment (Wolt 2009).

Animal Traits Derived from GE Technology

Traits currently being investigated include more rapid growth for meat, dairy, and fiber production; enhanced nutrient content; disease resistance; biopharmaceutical production; biomedical products; and environmental control of waste production and utilization (NRC 2002a). Although none of these traits are currently commercially available, significant progress in identifying and developing risk assessment and management guidelines has been made in Canada and the United States (CVM of FDA 2008; Moreau and Jordan 2005). In addition, a broader array of crops, animals and other organisms will emerge with a mixture of traits listed earlier.

Ecological Concerns Related to Transgenic Organisms

Unintended Impacts

Unintended impacts from the deployment of crops derived from GE technology can be classified as either nontarget impacts on other organisms or the development of resistance in the target populations. Nontarget impacts can occur to beneficial insects or natural enemies, such as parasitoids and predators of important pest species (Romeis, Bartsch, et al. 2008). Impact on beneficial insects such as honey bees is evaluated in risk assessment procedures well before commercialization. To date, direct impacts on natural enemies have been insignificant because of the very specific nature of *Bt* Cry proteins. Indirect effects on natural enemies with a narrow host range have been demonstrated when the pest host insect is eliminated by the *Bt* Cry toxin, but overall increases in natural enemy populations have occurred because of reduced impacts from pesticide use.

Impacts may occur to insects related to the target pests that are of special significance or importance, such as butterflies and endangered species (Dively et al. 2004; Sears et al. 2001; Wolt, Conlan, and Majima 2005). In field situations, it is the likelihood of exposure to the toxin expressed by the GE plant that most influences the risk to that population, rather than the toxic effect when exposure takes place. In fact, toxicity to a nontarget organism may be high because of its relationship to the target organism, but exposure of more than a small fraction of the population of the nontarget organism to the toxin may be extremely low. In current practice, the low frequency of exposure of such related species to *Bt* Cry proteins has meant that risks have been considered negligible. In the case of monarch butterflies and *Bt* maize pollen, larvae are found on their host plant, common milkweed, in and around fields of maize. However, the remaining extensive areas where monarch larvae are also found reduce exposure of populations to maize pollen to extremely low levels (Dively et al. 2004; Sears et al. 2001).

Resistance of Target Populations

Cultivation of crops expressing *Bt* Cry toxins has proved extremely effective in controlling some major pest species of maize and cotton. It has dramatically reduced insecticide use in these crops and has shifted emphasis to resistance management procedures to counteract natural selection for pest resistance to *Bt* Cry proteins (Kennedy 2008). Management practices emphasize a high-dose expression of the Cry protein in the crop and the planting of a refuge – a portion of the crop without the *Bt* Cry protein trait. These practices have substantially reduced the rate of resistance development in target pest populations. In fact, there have been no reports of resistance attributed to exposure to *Bt* crops over the past 12–13 years. Recently, the trend toward stacked multiple genes inserted in the same crop varieties for control of multiple pest groups has also reduced the selection pressure on the target pest species to develop tolerance to the *Bt* Cry toxins (Ferré, van Rie, and MacIntosh 2008).

Persistence and Escape (Weediness) of Transformed Plant

Volunteer GE plants (plants of the same crop) appearing in fields after crop rotation have been a concern, especially for herbicide-tolerant traits. To prevent this situation, growers should avoid crop rotation involving herbicide-tolerant crops with tolerance for the same herbicide and use alternate herbicides for control of volunteers from the previous herbicide-tolerant crop. Where gene escape is likely to occur, two solutions have been suggested: mitigation by transgene technology or avoidance (Lu 2008). Mitigation involves constructing the transgene components such that, in the event of gene flow to a close relative in a neighboring field or to a volunteer, fitness is reduced in the offspring. Avoidance can be achieved by regulating the planting of the transgenic crop in certain geographic areas where there are close relatives of the crop or creating buffer areas where the crop cannot be sown. Although gene escape is inevitable, it is the impact of the presence of a related plant species or crop in the field that requires evaluation, not the likelihood of gene transfer.

Persistence in the Environment of Products from Transgenes

Because crop residues contain active proteins up to and after the crops are harvested, concerns about the fate and persistence of those residues (especially the *Bt* Cry proteins) have been expressed. However, no effects of *Bt* Cry proteins on soil organisms including earthworms, nematodes, collembola, mites, protozoa, or activity of various enzymes in soil have been reported (Icoz and Stotzky 2008). Some effects of *Bt* plants on microbial communities in soil have been reported, but were transient and not related to the presence of the *Bt* Cry proteins. Protein residues were not found in water samples, but were associated with suspended colloidal material or in bottom sediments (Icoz and Stotzky 2008). Active proteins from GE maize have been hypothesized to occur in streams where plant residues from neighboring fields have accumulated; detritus feeders, such as caddisfly larvae, may feed on this debris and be affected by *Bt* protein toxins (Rosi-Marshall et al. 2007). On further evaluation, however, this appears to be a negligible concern with little impact on entire populations because of the entire caddisfly population's low overall exposure in the stream ecosystem (Wolt and Peterson 2010). Future innovations, such as expression of bioprocessing enzymes in crops, may require a broader understanding of the implications of persistence in relation to residue impacts on soil ecological services (Wolt 2009).

Transfer of Genetic Material to Related Species in Habitat

Out-crossing of transgenes to closely related noncrop species is likely to occur, but the frequency of transfer is highly dependent on the plant species involved. Measuring impacts of such out-crossings is a more important concern than the gene transfer itself. Few impacts of gene transfer have been documented for current crops produced by transgene technology, likely because of the lack of close relatives of the crop species in areas where they are primarily grown or the absence of experimental opportunity (Cohen et al. 2008; Hellmich et al. 2008). Where cross-contamination may be considered an economic or choice issue, as with organic production adjacent to transgenic crops, setbacks or buffer zones are usually prescribed to limit the frequency to lower than an acceptable limit (EU 2003).

Environmental Risk Assessment (ERA) Principles

The 1992/1998 ERA framework and guidelines from EPA provide the overarching structure and approach for the assessment of ecological risks in a way that usefully informs regulatory decision making. As applied to GE crops, the ERA process must align with guiding principles for risk assessment of GE crops, such as phytosanitary standards of the International Plant Protection Convention (IPPC) and the CPB. Particularly important aspects include the following:

1. Use of a case-by-case paradigm where each event must be considered on its own with respect to the nature of the change [phenotype], intended use, and region of deployment.
2. The principle of comparability. Comparative risk assessment is used to seek a "safe as" determination for the specific GE event as compared to the unmodified comparator crop.
3. The iterative nature of the assessment (and use of tiers) assures that the data generated and the assessments made are appropriately aligned with the nature and scale of release and degree of uncertainty regarding effects on the ecosystem.

The Risk Assessment Process

A standard risk assessment procedure describes the consequence of exposure to an adverse effect through four steps: hazard identification, dose sensitivity, nature of exposure, and risk characterization (EPA 1998). For regulatory oversight of crop plants, these stages have often been designated as problem formulation, effects characterization, exposure characterization, and risk characterization (EFSA 2006).

Problem Formulation

During the problem formulation stage, the meaningful differences between the GE plant and its non-GE counterparts are identified in order to focus the ERA on the areas of greatest concern or uncertainty. Problem formulation defines the scope of the risk assessment through the generation of testable scientific hypotheses that are subsequently addressed in the analytical phase of the risk assessment (Raybould 2006; Wolt et al. 2010).

Effects Characterization

Indication of direct hazard is usually not evidenced for GE crop plants, except for those expressing toxins conferring resistance to crop pests that subsequently may have an effect on nontarget species (Hellmich et al. 2001; Romeis, Meissle, and Bigler 2006). Laboratory and field studies are typically used to establish the degree to which organisms of concern or their surrogates may be directly affected by these toxins. In addition, more involved laboratory and field studies and monitoring may be used to establish if there are indirect or secondary effects from plant-expressed toxins occurring in the environment.

Exposure Characterization

Exposure characterization describes the nature, duration, intensity, and route of exposure so that the results of toxicity testing and the overall effects characterization can be placed in context. The nature and extent of exposure to transgenic crop plants or the products of these plants are dependent on the population behavior of the organisms exposed or the nature of the ecosystem that is exposed. Often, specific and detailed information on the potential exposure of nontarget organisms or of intentional or nonintentional consumers of GE products is not well known or difficult and time consuming to obtain. In such cases, probable scenarios may be substituted for measured or calculated routes of intensiveness of exposure.

Risk Characterization

The process of evaluating risk associated with the commercial release of a transgenic crop species is the integration or product of establishing or calculating a potential effect with the likelihood of exposure. That is, the information coming from the effects and exposure characterization is integrated to provide an understanding of the degree of risk that may be present in the environment. For plants expressing a specific toxin for suppression of a pest species, this process follows a conventional risk assessment process (Rose 2006). For situations where hazard is less well defined, such as weediness, invasiveness, or gene flow, there are effective analysis tools for new plant introductions, invasive or persistent plant contaminants, or expression of plant products from GE crops that can be used (Daehler et al. 2004; Pheloung, Williams, and Halloy 1999; Stewart, Halfhill, and Warwick 2003). In such cases, emphasis on ecological considerations specific to a geographic area is important when considering gene flow to near relatives.

An Iterative, Tiered Approach

In risk assessment, a standardized approach is desirable for comparability between evaluations from various locations or under variable conditions. Although it is desirable that the risk assessment process adhere to a standardized framework, the specific techniques used for the characterization of risk must retain flexibility so they can address the wide variety of GE crop innovations that have been and are being developed. Problem formulation is key for defining the approach to be taken (Wolt et al. 2010), and in particular, the use of an iterative or tiered approach to analysis is critical (Rose 2006; Romeis, Bartsch, et al. 2008). The concept of the tiered assessment strategy described here can be extended beyond current technology to address future innovations such as GE biofeedstock crops (Wolt 2009).

A tiered approach suggests a stepwise approach to evaluation of a source of hazard. This process has been extensively used for evaluation of nontarget or environmental effects of pesticides and can be readily standardized and used for GE crops expressing a toxin or deleterious substance. Tiers move from relatively simple dose-response evaluations to more complex testing and measurement involving behaviors of organisms and effects of the ecosystem on the process. The essence of tiered testing is that if no effect can be determined in the simple and rigorous Tier I level, there would be no need to continue testing at more complex yet less controllable conditions, thus saving time and money. Tiers of assessment can also address increasingly sophisticated determination of exposure concentrations, extending from rule-of-thumb model estimates to elaborate site-specific monitoring.

Tiers of effects testing are generally described as follows:

Tier I – Laboratory oriented with evaluation of effect at 10x of the highest expected concentration in the environment of the purified substance in question.

Tier II – Laboratory tests using plant materials that express the substance at about 1x the concentration found in the field but incorporate feeding behavior.

Tier III – Long-term laboratory or semi-field (e.g., field cages) are used to provide simulated or restricted field conditions.

Tier IV – Field tests that can incorporate differences in climate or ecosystem structure across regions or large areas under field conditions.

Elements of the tier testing process for ecological effects are summarized in the following four sections.

Species Selection

For practical and scientific reasons, only a small fraction of potentially exposed terrestrial arthropods can be considered for regulatory testing. Selection of appropriate species to serve as surrogates that can be tested under laboratory conditions is therefore important (Barrett et al. 1994; Grady and Schmuck 2001; Nickson and McKee 2002). The concept of using surrogates is widely applied in regulatory toxicity testing (Candolfi, Bigler, et al. 2000; Garcia-Alonso et al. 2006; Grady and Schmuck 2001) to monitor the effects of environmental pollutants and in conservation biology to indicate the extent of anthropogenic influences, monitor population changes of other species, or locate areas of high biodiversity (Caro and O'Doherty 1999; Levin et al. 1988). Most risk assessors consider application of the surrogate concept in regulatory testing to be the best method of mitigating the practical problems associated with testing the hundreds of nontarget species that occur in cropping systems. Key species or guilds that are representative of different functional groups are known in most systems; therefore, appropriate surrogates can be selected that are relevant to the agro-ecosystem of concern.

Study Design

Surrogate species are evaluated in properly designed experiments developed to assess the impact of pesticides on nontargets (Bates et al. 2005; Candolfi, Blumel, et al. 2000). Protocols are often modified to account for the oral exposure pathway of plant-expressed insecticidal proteins. New protocols are developed to address specific research or regulatory needs associated with risk assessment (Duan et al. 2002, 2006; Hellmich et al. 2001; Lundgren and Wiedenmann 2002; Morandin and Winston 2003; Romeis, Dutton, and Bigler 2004; Stacey et al. 2006). Toxicity exposure tests (hazard assessments in Tier I tests) are usually conducted using elevated protein doses in the laboratory and follow standardized testing protocols. This procedure assures study repeatability, interpretability, and quality, and thus a high level of confidence in the reported data.

Overall Risk Assessment

Exposure characterization, which occurs parallel to toxicity testing, allows for determination of the extent to which observed effects may apply at expected levels of exposure. Evaluation of crop characteristics (growth rate and form, expression levels, and seasonal timing), crop management, and the environment where it is grown provide estimates of potential exposure of nontargets to the protein. After this information has been summarized, the predicted *hazard* is multiplied by the predicted *exposure* to calculate a risk estimate. Simple and powerful, risk estimates are based on the product of hazard and exposure values. Risk characterization should identify and explain critical uncertainties that result from temporal and spatial variability and other areas where lack of knowledge may influence the precise characterization of risk. Ideally the risk estimate can be developed in a fully quantitative fashion, but when comprehensive quantitative analysis is not feasible, the risk estimate may be qualitatively determined on the weight of scientific evidence that is available.

A nontarget risk assessment approach addresses concerns about environmental impact and the need to integrate sustainable agricultural technologies. Tiered evaluation of potential hazards with representative indicator/surrogate species provides a rigorous and effective basis for estimating risk and minimizes the likelihood of false negatives, which could result in the release of hazardous GE commodities. In addition, the tiered approach requires testing of clearly stated relevant hypotheses and thereby minimizes collection of data that are irrelevant to risk assessment.

A Case Study: **Bt** *Maize Pollen and Monarch Butterfly Risk Assessment*

Overview

The case of *Bt* maize pollen impacts on monarch butterfly was the first fully developed example of a tiered approach to risk assessment of a nontarget impact resulting from wide deployment of a GE crop; it demonstrated the combination of toxicological, ecological, and geographical data required for risk evaluation (Hellmich et al. 2001; Oberhauser et al. 2001; Pleasants et al. 2001; Stanley-Horn et al. 2001). It was largely initiated by a media frenzy that resulted from the publication of a note in the journal *Nature* (Losey et al. 1999). Its publication resulted in U.S. and Canadian regulatory agencies expressing uncertainty as to the risk posed by pollen from several *Bt* maize events to monarch butterfly larvae in and around *Bt* maize fields. The resulting research effort in the United States and Canada over the next four to five years resulted in an assessment that concluded that the risk to the North American eastern population of monarch butterflies from exposure to *Bt* maize pollen was negligible (Dively et al. 2004; Sears et al. 2001). Other examples of a tiered approach using the same principles of risk assessment followed in subsequent years for both *Bt* maize and *Bt* cotton (Carrière et al. 2004; Wolt et al. 2003).

Universality of Approach

The risk assessment approach that has been successfully applied to the case of monarch butterflies and to other concerns of GE crop impacts on nontarget organisms in recent years demonstrates and incorporates principles of risk assessment that were previously only used in toxicology and industrial accidents or spills. This procedure can be universally applied to evaluate risk of exposure to products of GE crop plants and animals stemming from the impact both on their own populations and on other populations of organisms in their broader ecohabitat (O'Callaghan et al. 2005; Romeis, Meissle, and Bigler 2006).

Paradigm Shift for Regulatory Agencies

The experience of risk assessment for the monarch butterfly established a benchmark for a regulatory approach to risk assessment of transgenic crops that in the United States has resulted in the EPA and APHIS mandating a change in reporting risk assessments for GE crops. This approach is consistent with the overall process by which GE products are regulated in the United States (McHughen and Smyth 2008). Maintaining a case-by-case approach but designing the approach to risk assessment in consultation with industry, the EPA has developed hazard testing with regard to the specificity of *Bt* or other insecticidal constituents in crop plants. In future regulatory submissions for GE plants or animals, the groundwork laid at this time should lead to a further refinement of the tiered approach for risk assessments. A continuing challenge for regulators is the harmonization of these risk assessment principles for GE crops among national and regional authorities throughout the world.

Discussion Points and Further Needs

Evaluation of Strong/Weak Points of the Bt *Maize: Monarch Case*

The monarch butterfly case study demonstrated very solid science that provided a cohesive and well-reasoned development in a risk assessment, but it was *reactive* rather than *proactive* and is not feasible as an approach to every issue that may arise for a GE crop. A new paradigm of risk assessment specifically designed to evaluate the myriad of new approaches to crop and animal biotechnology must be developed and, preferably, accepted widely in the world. An initial approach, as outlined earlier for environmental impact evaluation (Romeis, Bartsch, et al. 2008), addresses one area of concern in evaluation of biotechnology: How can regulatory agencies establish and communicate to registrants a consistent and standardized approach to risk assessment that can be used in a world economy?

Outcomes of Risk Assessment Process Relative to Regulation and Introduction of Genetically Engineered Crop Plants or Livestock

Many in the ecological community argue that there is a need for broad-scale monitoring (such as in Tier IV tests) and more of a population perspective in ERAs (NRC 2002b). There are several points to consider in evaluating this approach to risk assessment:

1. The process for development and regulatory assessment of GE crops requires that decisions regarding ecological risk be made before widespread commercial release. Only monitoring

at the commercial scale is useful. Therefore, postcommercial monitoring can only confirm the adequacy of ERA conclusions made on a smaller scale (lab/greenhouse/semi-field).
2. Risk assessment should be hypothesis-driven; monitoring is generally not.
3. Careful ERA allows for understanding of population effects (the monarch butterfly ERA is a good example), but opportunities exist for improvement (this is an ERA issue that goes beyond GE crops).

ERA can be improved by careful problem formulation and follow-up of species selection and study design as indicated previously (Wolt et al. 2010). The more difficult question is whether this approach could be universally applied under all the spatially and temporally diverse habitats represented by agricultural ecosystems throughout the world. Regulatory agencies should consider adopting a consistent and well-recognized tiered process for ERA of GE crops that can be part of the regulatory process in all world jurisdictions.

Scale of Economy for Biotechnology Research and Development

Introduction of a new transformed trait into a crop variety may cost from $30–$50 million presently. Meeting biosafety requirements for new traits transformed into crops can cost $10–$15 million alone. The cost greatly restricts small seed companies or research institutions from participating, thereby limiting the diversity of approach and innovation needed for crop and food production for the future. A concern in all situations is the need for a case-by-case ERA and the capacity of regulatory agencies in all world jurisdictions to sustainably address this approach. The driving force behind this degree of regulatory oversight is directly related to actions by detractors of GE technology and international businesses. It ultimately falls to regulatory bodies to continue to make decisions based on factual, science-based approaches to evaluation of this technology and to balance science with nongovernmental organization's (NGO's) and media-driven concerns.

References

APHIS (Animal and Plant Health Inspection Service). 2006. "USDA Deregulates Line of Genetically Engineered Rice." News release, November 24. www.aphis.usda.gov/newsroom/content/2006/11/rice_deregulate.shtml.

Barrett, K. L., N. Grandy, E. G. Harrison, S. Hassan, and P. Oomen, eds. 1994. *Guidance Document on Regulatory Testing Procedures for Pesticides with Non-Target Arthropods. From the Workshop European Standard Characteristics of Beneficials Regulatory Testing (ESCORT).* Brussels: Society of Environmental Toxicology and Chemistry Europe.

Bates, S. L., J.-Z. Zhoa, R. T. Roush, and A. M. Shelton. 2005. "Insect Resistance Management in GM Crops: Past, Present and Future." *Nature Biotechnology* 23 (1): 57–62. doi:10.1038/nbt1056.

Brookes, G., and P. Barfoot. 2006. "Global Impact of Biotech Crops: Socioeconomic and Environmental Effects in the First Ten Years of Commercial Use." *AgBioForum* 9 (3): 139–51.

Candolfi, M., F. Bigler, P. Campbell, U. Heimbach, R. Schmuck, G. Angeli, F. Bakker, et al. 2000. "Principles for Regulatory Testing and Interpretation of Semi-Field and Field Studies with Non-Target Arthropods." *Journal of Pest Science* 73 (6): 141–7.

———, S. Blumel, R. Forster, F. M. Bakker, C. Grimm, S. A. Hassan, U. Heimbach, et al., eds. 2000. *Guidelines to Evaluate Side-Effects of Plant Protection Products to Non-Target Arthropods*. Gent: IOBC/WPRS (International Organisation for Biological and Integrated Control of Noxious Animals and Plants/ West Palaearctic Regional Section).

Caro, T. M., and G. O'Doherty. 1999. "On the Use of Surrogate Species in Conservation Biology." *Conservative Biology* 13 (4): 805–14.

Carrière, Y., P. Dutilleul, C. Ellers-Kirk, B. Pedersen, S. Haller, L. Antilla, T. J. Dennehy, and B. E. Tabashnik. 2004. "Sources, Sinks, and Zone of Influence of Refuges for Managing Insect Resistance to *Bt* Crops." *Ecological Applications* 14 (6): 1615–23.

Cohen, M. B., M. Chen, J. S. Bentur, K. L. Heong, and G. Ye. 2008. "*Bt* Rice in Asia: Potential Benefits, Impact and Sustainability." In *Integration of Insect-Resistant Genetically Modified Crops within IPM Programs*, edited by J. Romeis, A. M. Shelton, and G. Kennedy, 223–48. Vol. 5 of *Progress in Biological Control*. New York: Springer.

Crommelin, D. J. A., R. D. Sindelar, and B. Meibohm, eds. 2007. *Pharmaceutical Biotechnology: Fundamentals and Applications*, 3rd ed. New York: Informa Healthcare.

CVM of FDA (Center for Veterinary Medicine of the U.S. Food and Drug Administration). 2008. *Animal Cloning: A Risk Assessment*. Rockville, MD: U.S. Department of Health and Human Services. http://www.fda.gov/downloads/AnimalVeterinary/SafetyHealth/AnimalCloning/UCM124756.pdf.

Daehler, C. C., J. S. Denslow, S. Ansari, and H.-C. Kuo. 2004. "A Risk-Assessment System for Screening out Invasive Pest Plants from Hawaii and Other Pacific Islands." *Conservation Biology* 18 (2): 360–8.

Denby, K., and C. Gehring. 2005. "Engineering Drought and Salinity Tolerance in Plants: Lessons from Genome-Wide Expression Profiling in *Arabidopsis*." *Trends in Biotechnology* 23 (11): 547–52. doi:10.1016/j.tibtech.2005.09.001.

Dively, G. P., R. Rose, M. K. Sears, R. L. Hellmich, D. E. Stanley-Horn, J. M. Russo, D. D. Calvin, and P. L. Anderson. 2004. "Effects on Monarch Butterfly Larvae (Lepidoptera: Danaidae) after Continuous Exposure to Cry1Ab-Expressing Corn During Anthesis." *Environmental Entomology* 33 (4): 1116–25.

Duan, J. J., G. Head, M. J. McKee, T. E. Nickson, J. W. Martin, and F. S. Sayegh. 2002. "Evaluation of Dietary Effects of Transgenic Corn Pollen Expressing Cry3Bb1 Protein on a Non-Target Ladybird Beetle, *Coleomegilla maculata*." *Entomologia Experimentalis et Applicata* 104 (2–3): 271–80. doi: 10.1023/A:1021258803866

———, M. S. Paradise, J. G. Lundgren, J. T. Bookout, C. Jiang, and R. N. Wiedenmann. 2006. "Assessing Nontarget Impacts of *Bt* Corn Resistant to Corn Rootworms: Tier-1 Testing with Larvae of *Poecilus chalcites* (Coleoptera: Carabidae)." *Environmental Entomology* 35 (1): 135–42. doi: 10.1603/0046-225X-35.1.135

Duke, S. O., and N. N. Ragsdale, eds. 2005. "Herbicide-Resistant Crops from Biotechnology." Special issue, *Pest Management Science* 61 (1).

EFSA (European Food Safety Authority). 2006. *Guidance Document of the Scientific Panel on Genetically Modified Organisms for the Risk Assessment of Genetically Modified Plants and Derived Food and Feed*. Parma, Italy: EFSA. http://www.efsa.europa.eu/en/scdocs/doc/99.pdf.

EPA (Environmental Protection Agency). 1998. *Guidelines for Ecological Risk Assessment. Federal Register* 63 (93): 26846–924. May 14. http://www.epa.gov/raf/publications/pdfs/ECOTXTBX.PDF.

EU (European Union). 2003. "Regulation (EC) No. 1830/2003 of the European Parliament and of the Council of 22 September 2003 Concerning the Traceability and Labelling of Genetically Modified Organisms and the Traceability of Food and Feed Products Produced from Genetically Modified Organisms and Amending Directive 2001/18/EC." *Official Journal of the European Union* 268: 24–8. http://ec.europa.eu/food/food/animalnutrition/labelling/reg_1830-2003.pdf.

FAOSTAT. 2007. "Crops production." (accessed December 16, 2010). http://faostat.fao.org/site/567/default.aspx.

Ferré, J., J. van Rie, and S. C. MacIntosh. 2008. "Insecticidal Genetically Modified Crops and Insect Resistance Management (IRM)." In *Integration of Insect-Resistant Genetically Modified Crops within IPM Programs*, edited by J. Romeis, A. M. Shelton, and G. Kennedy, 41–86. Vol. 5 of *Progress in Biological Control*. New York: Springer.

Garcia-Alonso, M., E. Jacobs, A. Raybould, T. E. Nickson, P. Sowig, H. Willekens, P. van der Kouwe, et al. 2006. "A Tiered System for Assessing the Risk of Genetically Modified Plants to Non-Target Organisms." *Environmental Biosafety Research* 5 (2): 57–65.

Gonsalves, D. 1998. "Control of Papaya Ring Spot Virus in Papaya: A Case Study." *Annual Review of Phytopathology* 36: 415–37.

Grady, N., and R. Schmuck. 2001. *Guidance Document on Regulatory Testing and Risk Assessment Procedures for Plant Protection Products with Non-Target Arthropods*. Pensacola, FL: Society of Environmental Toxicology and Chemistry.

Grafius, E. J., and D. S. Douches. 2008. "The Present and Future Role of Insect-Resistant Genetically Modified Potato Cultivars in IPM." In *Integration of Insect-Resistant Genetically Modified Crops within IPM Programs*, edited by J. Romeis, A. M. Shelton, and G. Kennedy, 195–222. Vol. 5 of *Progress in Biological Control*. New York: Springer.

Hellmich, R. L., R. Albajes, D. Bergvinson, J. R. Prasifka, Z.-Y. Wang, and M. J. Weiss. 2008. "The Present and Future Role of Insect-Resistant Genetically Modified Maize in IPM." In *Integration of Insect-Resistant Genetically Modified Crops within IPM Programs*, edited by J. Romeis, A. M. Shelton, and G. Kennedy, 119–58. Vol. 5 of *Progress in Biological Control*. New York: Springer.

______, B. D. Siegfried, M. K. Sears, D. E. Stanley-Horn, M. J. Daniels, H. R. Mattila, T. Spencer, K. G. Bidne, and L. C. Lewis. 2001. "Monarch Larvae Sensitivity to Bacillus Thuringiensis-Purified Proteins and Pollen." *Proceedings of the National Academy of Sciences of the United States of America* 98 (21): 11925–30. doi: 10.1073/pnas.211297698.

Icoz, I., and G. Stotzky. 2008. "Fate and Effects of Insect Resistant *Bt* Crops in Soil Ecosystems." *Soil Biology and Biochemistry* 40 (3): 559–86. doi:10.1016/j.soilbio.2007.11.002.

James, C. 2009. *Global Status of Commercialized Biotech/GM Crops: 2009*. ISAAA (International Service for the Acquisition of Agri-Biotech Applications) Brief 37. Ithaca, NY: ISAAA.

Kennedy, G. G. 2008. "Integration of Insect-Resistant Genetically Modified Crops within IPM Programs." In *Integration of Insect-Resistant Genetically Modified Crops within IPM Programs*, edited by J. Romeis, A. M. Shelton, and G. Kennedy, 1–26. Vol. 5 of *Progress in Biological Control*. New York: Springer.

Keyser, O., and R. H. Müller, eds. 2004. *Pharmaceutical Biotechnology: Drug Discovery and Clinical Applications*. Weinheim, Germany: Wiley.

Levin, S. A., M. A. Harwell, J. R. Kelly, and K. D. Kimball. 1988. "Ecotoxicology Problems and Approaches." In *Ecotoxicology: Problems and Approaches*, edited by S. A. Levin, M. A. Harwell, J. R. Kelly, and K. D. Kimball, 3–8. New York: Springer.

Losey, J. E., L. S. Rayor, and M. E. Carter. 1999. "Transgenic Pollen Harms Monarch Larvae." *Nature* (London) 399: 214.

Lu, B.-R. 2008. "Transgene Escape from GM Crops and Potential Biosafety Consequences: An Environmental Perspective." In *Collection of Biosafety Reviews*, vol. 4, 66–141. Trieste: ICGEB (International Center for Genetic Engineering and Biosafety).

Lundgren, J. G., and R. N. Wiedenmann. 2002. "Coleopteran-Specific Cry3Bb Toxin from Transgenic Corn Pollen Does Not Affect the Fitness of a Nontarget Species, *Coleomegilla maculata* DeGeer (Coleoptera: Coccinellidae)." *Environmental Entomology* 31 (6): 1213–18.

Malone, L. A., A. M. R. Gatehouse, and B. I. P. Barratt. 2008. "Beyond Bt: Alternative Strategies for Insect-Resistant Genetically Modified Crops." In *Integration of Insect-Resistant Genetically Modified Crops within IPM Programs*, edited by J. Romeis, A. M. Shelton, and G. Kennedy, 357–418. Vol. 5 of *Progress in Biological Control*. New York: Springer.

McHughen, A., and S. Smyth. 2008. "U.S. Regulatory System for Genetically Modified, rDNA Crop Cultivars." *Plant Biotechnology Journal* 6 (1): 2–12. doi: 10.1111/j.1467-7652.2007.00300.x.

Morandin, L. A., and M. L. Winston. 2003. "Effects of Novel Pesticides on Bumble Bee (Hymenoptera: Apidae) Colony Health and Foraging Ability." *Environmental Entomology* 32 (3): 555–63.

Moreau, P. I., and L. T. Jordan. 2005. "A Framework for the Animal Health Risk Analysis of Biotechnology-Derived Animals: A Canadian Perspective." *Revue Scientifique et Technique* 24 (1): 51–60.

Naranjo, S. E., J. R. Ruberson, H. C. Sharma, L. Wilson, and K. Wu. 2008. "The Present and Future Role of Insect-Resistant Genetically Modified Cotton in IPM." In *Integration of Insect-Resistant Genetically Modified Crops within IPM Programs*, edited by J. Romeis, A. M. Shelton, and G. Kennedy, 159–94. Vol. 5 of *Progress in Biological Control*. New York: Springer.

Neeser, J. R., and J. B. German, eds. 2004. *Biotechnology for Functional Foods and Nutraceuticals. Vol. 2 of Nutraceutical Science and Technology*, edited by F. Shahidi. New York: Marcel Dekker.

Nickson, T., and M. McKee. 2002. "Ecological Assessment of Crops Derived through Biotechnology." In *Biotechnology and Safety Assessment*, 3rd ed., edited by J. A. Thomas and R. L. Fuchs, 234–55. New York: Academic Press.

NRC (National Research Council). 2002a. *Animal Biotechnology: Science-Based Concerns*. Washington, DC: National Academy Press.

______. 2002b. *Environmental Effects of Transgenic Plants: The Scope of Adequacy of Regulation*. Washington, DC: National Academy Press.

Oberhauser, K., M. D. Prysby, H. R. Mattila, D. E. Stanley-Horn, M. K. Sears, G. Dively, E. Olsen, J. M. Pleasants, W. K. Lam, and R. L. Hellmich. 2001. "Temporal and Spatial Overlap between Monarch Larvae and Corn Pollen." *Proceedings of the National Academy of Sciences of the United States of America* 98 (21): 11913–18. doi: 10.1073/pnas.211234298.

O'Callaghan, M., T. R. Glare, E. P. J. Burgess, and L. A. Malone. 2005. "Effects of Plants Genetically Modified for Insect Resistance on Non-Target Organisms." *Annual Review of Entomology* 50 (1): 271–92.

Pheloung, P. C., P. A. Williams, and S. R. Halloy. 1999. "A Weed Risk Assessment Model for Use as a Biosecurity Tool Evaluating Plant Introductions." *Journal of Environmental Management* 57 (4): 239–51. doi:10.1006/jema.1999.0297.

Pleasants, J., R. L. Hellmich, G. P. Dively, M. K. Sears, D. E. Stanley-Horn, H. R. Mattile, J. E. Foster, P. Clark, and G. D. Jones. 2001. "Corn Pollen Deposition on Milkweeds in and Near Cornfields." *Proceedings of the National Academy of Sciences of the United States of America* 98 (21): 11919–24. doi: 10.1073/pnas.211287498.

Raybould, A. 2006. "Problem Formulation and Hypothesis Testing for Environmental Risk Assessments of Genetically Modified Crops." *Environmental Biosafety Research* 5 (3): 119–25. doi: 10.1051/ebr:2007004.

Romeis, J., D. Bartsch, F. Bigler, M. P. Candolfi, M. M. C. Gielkens, S. E. Hartley, R. L. Hellmich, et al. 2008. "Assessment of Risk of Insect-Resistant Transgenic Crops to Nontarget Arthropods." *Nature Biotechnology* 26 (2): 203–8. doi:10.1038/nbt1381.

______, A. Dutton, and F. Bigler. 2004. "*Bacillus thuringiensis* Toxin (Cry1Ab) Has No Direct Effect on Larvae of the Green Lacewing *Chrysoperla carnea* (Stephens) (Neuroptera: Chrysopidae)." *Journal of Insect Physiology* 50 (2–3): 175–83.

______, M. Meissle, and F. Bigler. 2006. "Transgenic Crops Expressing *Bacillus thuringiensis* Toxins and Biological Control." *Nature Biotechnology* 24 (1): 63–71. doi:10.1038/nbt1180.

______, A. M. Shelton, and G. Kennedy, eds. 2008. *Integration of Insect-Resistant Genetically Modified Crops within IPM Programs. Vol. 5 of Progress in Biological Control.* New York: Springer.

Rose, R. I. 2006. "Tier-Based Testing for the Effects of Proteinaceous Insecticidal Plant-Incorporated Protectants on Non-Target Arthropods in the Context of Regulatory Risk Assessments." Proceedings of the meeting "Ecological Impact of Genetically Modified Organisms," Llieda, Spain, June 1–3, 2005. *IOBC/WPRS Bulletin* 29 (5): 143–50. http://www.iobc-wprs.org/pub/bulletins/bulletin_2006_29_05.pdf.

Rosi-Marshall, E. J., J. L. Tank, T. V. Royer, M. R. Whiles, M. Evana-White, C. Chambers, N. A. Griffiths, J. Pokelsek, and M. L. Stephen. 2007. "Toxins in Transgenic Crop Byproducts May Affect Headwater Stream Ecosystems." *Proceedings of the National Academy of Sciences USA* 104 (41): 16204–8. doi: 10.1073/pnas.0707177104.

Sears, M. K., R. L. Hellmich, B. D. Siegfried, B. D. Pleasants, J. M. Stanley-Horn, K. S. Oberhauser, and G. P. Dively. 2001. "Impact of *Bt* Corn Pollen on Monarch Butterfly Populations: A Risk Assessment." *Proceedings of the National Academy of Sciences of the United States of America* 98 (1): 11937–42.

Shelton, A. M., M. Fuchs, and F. A. Shotkoski. 2008. "Transgenic Vegetables and Fruits for Control of Insects and Insect-Vectored Pathogens." In *Integration of Insect-Resistant Genetically Modified Crops within IPM Programs*, edited by J. Romeis, A. M. Shelton, and G. Kennedy, 249–72. Vol. 5 of *Progress in Biological Control*. New York: Springer.

Stacey, D., G. Graser, M. Mead-Briggs, and A. Raybould. 2006. "Testing the Impact on Non-Target Organisms of Insecticidal Proteins Expressed in Transgenic Crops." Proceedings of the meeting "Ecological Impact of Genetically Modified Organisms," Llieda, Spain, June 1–3, 2005. *IOBC/WPRS Bulletin* 29 (5): 165–72. http://www.iobc-wprs.org/pub/bulletins/bulletin_2006_29_05.pdf.

Stanley-Horn, D. E., G. P. Dively, R. L. Hellmich, H. R. Mattila, M. K. Sears, R. Rose, L. C. Jesse, J. E. Losey, J. J. Obrycki, and L. Lewis. 2001. "Assessing the Impact of Cry1Ab-Expressing Corn Pollen on Monarch Butterfly Larvae in Field Studies." *Proceedings of the National Academy of Sciences of the United States of America* 98 (21): 11931–6. doi: 10.1073/pnas.211277798.

Stewart, C. N., Jr., M. D. Halfhill, and S. I. Warwick. 2003. "Transgene Introgression from Genetically Modified Crops to Their Wild Relatives." *Nature Reviews Genetics* 4 (10): 806–17.

Storer, M. P., G. P. Dively, and R. A. Herman. 2008. "Landscape Effects of Insect-Resistant Genetically Modified Crops." In *Integration of Insect-Resistant Genetically Modified Crops within IPM Programs*, edited by J. Romeis, A. M. Shelton, and G. Kennedy, 272–302. Vol. 5 of *Progress in Biological Control*. New York: Springer.

Wilson, L., M. Hickman, and S. Deutscher. 2006. "Research Update on IPM and Secondary Pests." In *Proceedings of the 13th Australian Cotton Research Conference*, 249–58.

Thirteenth Australian Cotton Research Conference, Broadbeach, Queensland, August 8–10, 2006.

Wolt, J. D. 2009. "Advancing Environmental Risk Assessment for Transgenic Biofeedstock Crops." *Biotechnology for Biofuels* 2 (1): 27. doi:10.1186/1754-6834-2-27.

———, C. A. Conlan, and K. Majima. 2005. "An Ecological Risk Assessment of Cry1F Maize Pollen Impact to Pale Grass Blue Butterfly." *Environmental Biosafety Research* 4 (4): 243–51.

———, P. Keese, A. Raybould, J. W. Fitzpatrick, M. Burachik, A. Gray, S. S. Olin, J. Schiemann, M. Sears, and F. Wu. 2010. "Problem Formulation in the Environmental Risk Assessment for Genetically Modified Plants." *Transgenic Research* 19 (3): 425–36. doi:10.1007/s11248-009-9321-9.

———, and R. K. D. Peterson. 2010. "Prospective Formulation of Environmental Risk Assessments: Probabilistic Screening for Cry1A(b) Maize Risk to Aquatic Insects. *Ecotoxicology and Environmental Safety* 73:1182–8.

———, R. K. D. Peterson, P. Bystrak, and T. Meade. 2003. "A Screening Level Approach for Non-Target Insect Risk Assessment: Transgenic *Bt* Corn Pollen and the Monarch Butterfly (Lepidoptera: Danaidae)." *Environmental Entomology* 32 (2): 237–46.

9

Organic Agriculture as an Alternative to a GE-Based System

Erin Silva

> Organic agriculture is a holistic production management system which promotes and enhances agroecosystem health, including biodiversity, biological cycles, and soil biological activity. It emphasizes the use of management practices in preference to the use of off-farm inputs, taking into account that regional conditions require locally adapted systems. This is accomplished by using, where possible, cultural, biological and mechanical methods, as opposed to using synthetic.
>
> *(Codex Alimentarius 2010, 2)*

Risk assessment is often accompanied by the evaluation of alternatives. Organic agriculture is increasingly recognized worldwide as a viable, sustainable alternative to conventional farming systems. To comply with these standards, organic farmers must adopt practices that inherently lead to greater sustainability of the system, such as optimization of nutrient additions, crop rotation and diversification, enhancement of biodiversity, and the implementation of proactive disease and pest control strategies. Also key in the compliance to the organic standards is the prohibition of the use of crop varieties and other materials such as soil inoculants that have been developed using GE technology.

Organic agriculture continues to expand worldwide. In 2008, organically managed area across the globe totaled 0.35 million km^2, farmed by almost 1.4 million producers in 154 countries (Willer and Kilcher 2010). Australia holds the largest certified organic surface area with 0.12 million km^2, followed by Argentina with 0.028 million km^2 (Willer and Kilcher 2010). This area represents a significant growth in organically managed area worldwide, with a 9 percent increase experienced in all geographic regions of the world from 2007–8 (Willer and Kilcher 2010). Global organic sales reached \$50.9 billion in 2008, double the \$25 billion recorded in 2003 (Willer and Kilcher 2010). Organic agriculture in the United States has followed the same path as these global trends. As of 2008, certified organic area in the United States reached more than 0.019 million km^2 (USDA 2009). U.S. sales of organic food and beverages grew exponentially from $1 billion in 1990 to $24.8 billion in 2009, reaching 3.7 percent of overall purchases in 2009; organic fruits and vegetables represented an even greater fraction of the market, comprising 11.4 percent of all U.S. fruit and vegetable sales (OTA 2010).

In the United States, organic farming practices are dictated by the National Organic Program (NOP). As part of the USDA's Agricultural Marketing Service, this agency sets forth the regulations to be followed by organic farmers throughout the United States. In addition, all foods entering the United States and labeled as organic must adhere to these standards. Similar to the Codex Alimentarius definition at the beginning of this chapter, NOP defines organic agriculture as production "that is managed in accordance with the Organic Foods Production Act (OFPA) of 1990 and regulations in Title 7, Part 205 of the Code of Federal Regulations to respond to site-specific conditions by integrating cultural, biological, and mechanical practices that foster cycling of resources, promote ecological balance, and conserve biodiversity" (NOP 2010).

Can Organic Agriculture Feed the World?

Agriculture's ability to feed the world's growing population will be greatly challenged as we move forward into this century. As described in earlier chapters in this book, current food production levels will need to increase greatly to feed the world's growing population. Increasing evidence is accumulating that the adoption of organic agricultural practices can have beneficial impacts on global environmental sustainability and human health. However, critics often debate the utility of the adoption of organic agriculture, stating that, despite positive environmental impacts, organic agriculture will not be able to meet the needs of a growing world population. Despite these assertions from critics, data point to the high probability that organic agriculture methods can support farming systems capable of producing the yields necessary to feed the world's growing population using a comparable growing area. Badgley et al. (2006) estimated the potential of a global organic food supply using ratios calculated from 293 published comparisons of average organic/conventional yield ratios using two different models. In Model 1, organic/conventional yield ratios are derived from studies conducted in developed countries where production uses of modern technology are applied to the entire worldwide agricultural land base. Essentially, Model 1 assumes that, if converted to organic production, the low-intensity agriculture present in much of the developing world would have the same or slightly reduced potential yields that have been reported for the developed world, assuming that agriculture in the developing world will follow the same technological trajectories as those of the developed world. Under the assumptions of this model, for many crop categories (grains, sweeteners, tree nuts, oil crops, fruits, meat, animal fats, milk, and eggs), the estimated organic food supply is similar in magnitude to the current global conventional food supply, with yield ratios ranging from 0.93–1.06. Starchy roots, legumes, and vegetables exhibited the potential for a slight decline in yield potential under current organic management strategies using available varieties, with yield ratios of 0.82–0.89.

Under the assumptions of Model 2 in Badgley et al. (2006), yield ratios derived from studies in the developed world of food production are applied to the entire agricultural base of the developed world, and the yield ratios derived from studies in the developing world are applied to the entire agricultural production base of the developing world, with the sum of these separate estimates providing the global estimate. This model shows that yields produced by transitioning to an entirely global organic food supply actually exceed the current food supply in all crop categories, with higher estimates resulting primarily from the high average yield ratios (1.80) of organic versus current methods of production in the developing world. The results of Model 2 more realistically reflect the impact that organic management's agroecological practices can have on a regional basis in the developing world in which small local farms, stressful environmental conditions, and lack of mechanization predominate; these characteristics describe situations where organic agriculture can offer a significant advantage over the large-scale industrial models of the developed world.

These production trends appear in the United States as well. Several research studies performed on land managed organically for a significant amount of time illustrate that organic yields are comparable to conventional yields. A study initiated in Iowa in 1977 using both conventional and organic systems as treatments provides one example of yield comparability. Compared to conventional continuous corn and conventional corn-soybean rotations, organic corn yields were 401.7 kg/ha higher and 539.8 kg/ha lower, respectively (Duffy, Liebman, and Pecinovsky 2002). Organic soybean outperformed conventional soybean with respect to yield for both feed-grade varieties (188.3 kg/ha) and food-grade varieties (87.4 kg/ha).

Additionally, Delate and Cambardella (2004), in a study also conducted in Iowa, showed no significant differences in yield between organic and conventional corn and soybean production. In another study conducted in the Upper Midwest, Posner, Baldock, and Hedtcke (2008) concluded that organically managed and similar low-input forage crop systems can yield as much, or more, dry matter as their conventionally managed counterparts, with quality sufficient to produce as much milk as the conventional systems. They also found that organically managed corn, soybean, and winter wheat can produce about 90 percent as well as their conventionally managed counterparts. In the western United States, results from the first eight years of a study conducted at University of California-Davis show that the organic and low-input systems had yields comparable to the conventional systems in all crops that were tested (tomato, safflower, corn, and bean), with some instances yielding higher than conventional systems (Clark 1999).

Although comparisons of conventional and organic yields may not necessarily include transgenic varieties of corn and soybean in conventional systems, there is a lack of peer-reviewed, published data to suggest that the use of transgenic varieties would significantly change the results of the aforementioned studies or result in substantial yield increases, particularly yield increases that would lead to the 70 percent increase in food production required by the year 2050 as projected by FAO (2009). Crop yields can be described by two measurements: intrinsic yield and operational yield. Intrinsic yield, or potential yield, describes the highest yield that can be achieved under production in ideal conditions (Gurian-Sherman 2009). Operational yields are obtained under field conditions, when environmental factors, such as pests and stress, result in yields considerably lower than intrinsic yield. Thus far, the transgenic crops that are currently available contain genes that improve operational yield by reducing losses from pests and stress (Gurian-Sherman 2009). Protecting crops from pests and stress through other methods, including organically approved production techniques, can achieve the same results as GE. In the past 20 years during which the use of transgenic crops has become widespread, no currently available transgenic varieties have been shown to enhance the intrinsic yield of any crops; rather, the intrinsic yield increases observed in corn and soybean during the 20th century have primarily been due to successes in traditional breeding (Gurian-Sherman 2009).

Thus far, results of peer-reviewed, third-party, replicated research comparing the yields of transgenic versus conventionally produced crops have not been impressive. In a study comparing several transgenic lines of *Bt* soybean in replicated field trials in 2003–7, soybean 100-seed weights and harvested yields were similar between the *Bt* and non-*Bt* entries each year of the study (McPherson and McCrae 2009). Although not a food crop, gluphosinate-tolerant varieties of cotton have not been adopted by growers on a wide scale because of poor agronomic performance. When investigated under replicated trial conditions, results indicated that the gluphosinate-tolerant variety did not perform as well when yield potential was high, validating grower concerns with this herbicide technology in cotton production (Balkcom et al. 2010).

Experiments investigating the performance of *Bt* corn have shown similar results. In a study conducted in Canada, *Bt* hybrids of corn had no higher yield potentials and thus added no yield benefits over their near-isolines (Ma and Subedi 2005). Furthermore, in all cases, there was no indication that the *Bt* hybrids produced higher yield, took up more nitrogen (N), or partitioned more N into the grain than their non-*Bt* near-isolines, with two full-season hybrids producing lower N concentration in grains. Thus, not only do *Bt* hybrids fail to increase yield but they also fail to improve quality, as total grain N or crude protein content in corn is an important quality trait for food and feed (Ma and Subedi 2005). Further evidence that transgenic corn varieties do not significantly benefit yield is found in a study by Barry et al. (2000), which was conducted jointly by the USDA and the University of Missouri.

Barry et al. (2000) compared the performance of transgenic hybrids for insect control and yield and found significant differences among the yields of the various hybrids. As a group, the *Bt* corn hybrids produced significantly lower yields than the non-*Bt* corn hybrids (11.4 kg versus 12.4 kg per plot).

The yield potential of organic versus conventional agricultural is particularly favorable under stressful environmental conditions. A number of studies have shown that, under drought conditions, crops under organic management produce higher yields than comparable conventionally produced crops. Several studies have shown that organic crops can outyield conventional crops by 70–90 percent under severe drought conditions (Lockeretz, Shearer, and Kohl 1981; Lotter, Liebhardt, and Seidel 2003; Smolik, Rickerl, and Dobbs 1995; Stanhill 1990). Data collected by the Rodale Institute (1999) show that in 1999, during one of the worst droughts on record in Pennsylvania, yields of organic soybeans were 2.0 Mg/ha compared to only 1.1 Mg/ha from conventionally grown soybeans. This characteristic of resiliency under harsh environmental conditions makes organic agriculture an important tool in maintaining agricultural sustainability during times of climatic variability, particularly for the agriculture in developing countries that especially suffers from environmental extremes.

Beyond yield, the economic impact of expanding organic acreage and its effect on organic agriculture's ability to feed the world's growing population must be considered. As cited by Pimentel et al. (2005), on average, organic systems require approximately 15 percent more labor (Granatstein 2010; Sorby 2002), but the increase in labor input may range from 7 percent (Brumfield, Rimal, and Reiners 2000) to 75 percent (Karlen, Duffy, and Colvin 1995; Nguyen and Haynes 1995). These increased labor inputs are responsible for much of the increased production costs of organic agriculture. However, many developing countries still rely on hand labor, with on-farm labor providing a significant job base for much of the population. For example, in Tanzania, as in most countries of Sub-Saharan Africa, the hand hoe is still used for land preparation on 80 percent of the cultivated land (Mrema, Baker, and Kahan 2008). Results of a study analyzing the benefits of substituting manual labor preparation in Nigeria with modern tractor mechanization indicates that, although hand labor still predominates, agriculture was found to be more profitable, environmentally sound, and efficient under the prevailing smallholder farming system with hand labor (Eziakor 1990).

Conversely, the introduction of transgenic crops to the agricultural landscape has not necessarily resulted in increased profitability for the farmer. Jost et al. (2008) analyzed the profitability of cotton production in Georgia in the United States and found that it was associated most closely with yields and not the transgenic technologies. Similarly, Cox, Hanchar, and Shield (2009) investigated the profitability of stacked transgenic corn hybrids and showed that stacking the *Bt* corn borer trait with the glyphosate-resistant trait resulted in only three of eight site-rotation comparisons showing a profit and two site-rotation comparisons showing a profit loss when compared with the near-isolines. Data regarding the use of transgenic soybean have shown a slight yield reduction and are insufficient to validate an increase in profits (Marra, Pardey, and Alston 2002). Thus, increased adoption of transgenic crops worldwide would not definitively translate to economic benefits, especially in light of the static yields described earlier in this chapter.

Organic Agriculture as Part of a Sustainable Agriculture Model

Impact of Organic Agriculture on Sustaining Ecological Health

Organic agriculture has been shown to have multiple ecological benefits that foster sustainability in agriculture systems. Increasing numbers of research studies illustrate the roles that organic agriculture can play in mitigating climate change, increasing carbon sequestration, improving the quality of our soil and water resources, and benefiting human health globally.

One of the primary criticisms of organic agriculture centers on the supposition that organic practices consume higher amounts of energy than conventional agriculture because of the high levels of mechanization used for weed management in developed countries. However, research indicates a 15 to 45 percent reduction in energy per unit yield as compared to conventionally managed systems (Gomiero, Paoletti, and Pimentel 2008). This reduction is attributable to several factors: (1) lack of input of synthetic N fertilizers, which account for more than 50 percent of the total energy input in conventional agriculture; (2) low input of other mineral fertilizers (e.g., phosphorus and potassium) in organic systems; and (3) restrictive use of synthetic pesticides and herbicides under organic management (Cormack 2000; Gomiero et al. 2008; Haas, Wetterich, and Köpke 2001; Hoeppner et al. 2006; Lampkin 2002; Lockeretz et al. 1981; Pimentel et al. 2005; Refsgaard, Halberg, and Kristensen 1998). The energy savings provided by forgoing synthetic fertilizers and pesticides exceeds the higher energy demand for machinery (Cormack 2000).

Prohibitions on the use of synthetic fertilizers in organic agriculture can also decrease the contribution of greenhouse gas emissions created by agricultural production worldwide. The energy used for the chemical synthesis of nitrogen fertilizers used in conventional agriculture systems represents up to 0.4–0.6 Pg of CO_2 emissions and accounts for much as 10 percent of direct global agricultural emissions and approximately 1 percent of total anthropogenic greenhouse gas (GHG) emissions (Eggleston et al. 2006; Scialabba and Müller-Lindenlauf 2010). These emissions are avoided in organic systems. Prohibition of the use of synthetic fertilizers in organic agriculture also reduces the nitrous oxide emissions from fertilizer application and the energy demand for fertilizer manufacturing, thus potentially lowering the direct global agricultural GHG emissions by about 20 percent (Scialabba and Müller-Lindenlauf 2010). In addition to the prohibitions on synthetic fertilizers, which decrease GHG emissions, restrictions on burning and farming on recently cleared land also give organic agriculture an advantage with respect to its impact on climate change. CH_4 and N_2O from biomass burning account for 12 percent of the global agricultural GHG emissions (Scialabba and Müller-Lindenlauf 2010). In organic agriculture, preparation of land by burning vegetation is restricted to a minimum, and International Federation of Organic Agriculture Movements (IFOAM) organic standards ban the certification of primary ecosystems that have recently been cleared or altered (EAC 2007; IFOAM 2002; Regional Organic Task Force 2008). Thus, because of the regulations and principles supporting organic agricultural systems, organic producers have an inherent potential to reduce GHG emissions.

A frequent misconception of organic agriculture involves the use of manure as an organic fertilizer. It is important to note that animal manure is not the only source of fertility on organic farms; legume-based fertility programs are very common among organic producers. To comply with the pasturing requirement of organic agriculture, as well as the average smaller herd size of an organic livestock operation, if the manure used on the production fields is generated on-farm, it is likely to be solid (versus liquid) manure. This distinction is very important, because manure handled as a solid product contributes much less than liquid manure to methane emissions from manure application. Liquid management systems are increasing in number on larger, automated confinement operations (Edison Electric Institute 2009) that are not allowed under organic management. These lagoon-based systems lose 40 times or more methane than the systems used on most organic farms. Conversely, application of solid manure resulted in substantially lower N_2O emission (0.99 kg N_2O–N per ha per year) than the application of liquid manure (2.83 kg N_2O–N per ha per year) or mineral fertilizer (2.82 kg N_2O–N per ha per year; Gregorich et al. 2005).

Organic agriculture's reliance on leguminous cover crops and manure as fertility inputs does not prevent global crop production from increasing to feed a growing population. Badgley et al. (2008) estimated that globally, 140 Tg of N is fixed by the use of additional leguminous crops as fertilizer, which is 58 Tg greater than the 82 Tg of synthetic N in use in 2001. At the long-term Farm Systems Trial at the Rodale Institute in Pennsylvania, the organic system maintains comparable yields to the conventional averages when legume cover crops are used as the only source of nitrogen fertility (Pimentel et al. 2005). Breeding and selection for cover crop varieties and green manures could further expand the potential of N-fixing crops by increasing the rates of N fixation. The potential of N fixation in agricultural systems could also be increased by the identification of improved strains of N-fixing symbionts, the establishment of recommendations to decrease the amount of N lost from legume-based production systems through leaching, and the overall increase in the planted acreage of legume cover crops (van Groeningen et al. 2004).

The impact of organic agriculture on environmental sustainability is further increased by its impact on carbon sequestration. Recent research has begun to illuminate organic agriculture's beneficial effects on maintaining carbon pools in the soil due to its reliance on such management practices as cover cropping and additions of green manures and compost (Barthès et al. 2004; Fliessbach et al. 2007; Freibauer et al. 2004; Küstermann, Kainz, and Hülsbergen 2008; Pimentel et al. 2005). In the United States, field trials in Pennsylvania comparing three distinct maize/soybean agro-ecosystems – two organic (animal-based and legume-based) and one conventional – showed a fivefold higher carbon sequestration in the organic system in comparison with conventional management. For fertility in the animal-based organic system, aged cattle manure was applied at a rate of 5.6 Mg per ha^2 (dry), two years out of every five immediately before plowing the soil for corn; additional N was supplied by the plow-down of legume-hay crops (Pimentel et al. 2005). Crops in the rotation included corn, soybeans, corn silage, wheat, and red clover-alfalfa hay, as well as a rye cover crop before corn silage and soybeans. In the organic legume-based fertility system, a hairy vetch winter cover crop was incorporated before corn planting as a green manure, adding an average of 49 kg (or 140 kg per ha for any given year with a corn crop) of total N to this system per ha per year. The rotation in this system included hairy vetch (winter cover crop used as a green manure), corn, rye (winter cover crop), soybeans, and winter wheat. The conventional cropping system, based on synthetic fertilizer and herbicide use, represented a typical cash-grain, row-crop farming system and used a simple five-year crop rotation (corn, corn, soybeans, corn, soybeans) that reflected commercial conventional operations in the region and throughout the Midwest. Fertilizer and pesticide applications for corn and soybeans followed the Pennsylvania State University Cooperative Extension recommendations. The system did not have cover crops during the nongrowing season (Pimentel et al. 2005). Analyzing results over the 14-year period of this study, the authors estimated that the adoption of organic agriculture practices in the maize/soybean grown region in the United States would increase soil carbon sequestration by 13 to 30 Tg per year (Drinkwater, Wagoner, and Sarrantonio 1998; Hepperly, Douds, Jr., and Seidel 2006; Pimentel et al. 2005).

Other studies conducted in the United States show similar results with respect to organic agriculture's potential to sequester carbon. Results of a four-year management study of an apple orchard in Washington State showed that microbial biomass carbon was greater under organic than conventional management (Glover, Andrews, and Reganold 2000). In Reganold et al. (2010), 13 organic and conventional field pairs of strawberries (5 in 2004 and 8 in 2005) were analyzed for differences in soil properties. The organic fields had been certified organic by USDA for at least five years, providing sufficient time for the organic farming practices to influence soil properties. The selection of farms was based on grower interviews and on-farm field examinations to ensure that all soil-forming factors, except management, were the same for each field pair. Conventional and organic fields were then paired, with fields in each pair chosen based on similarities in microclimate; soil profile, type, and classification; and strawberry variety. Results showed that organically managed surface soils in strawberry production contained significantly greater total carbon (21.6 percent more) than their conventional counterparts.

Organic agriculture can also contribute to global sustainability through its positive impacts on groundwater quality. In the research of Oquist, Strock, and Mulla (2007), conducted by the University of Minnesota, the effects of alternative farming practices (including organic agriculture) on groundwater quality were compared with those of conventional farming practices. Results showed that, under the environmental conditions of Minnesota, alternative practices reduced mean daily and annual losses of nitrogen and phosphorus in subsurface drainage, especially during years when precipitation was average or above average. Oquist et al. (2007) concluded that alternative farming practices have the potential to reduce agricultural contributions to surface water pollution, including the reduction of Upper Midwest nitrate-nitrogen (nitrate N) contributions to the Mississippi River and hypoxia in the Gulf of Mexico. Additionally, Michalak (2004) conducted a Rodale Institute study in Pennsylvania comparing water quality from alternative versus conventional farming practices in the United States; results showed that nitrate N concentrations above regulatory levels were measured more frequently under conventional practices than under alternative farming practices. Kramer et al. (2006) produced similar results in Washington, indicating that the use of organic fertilizers in orchards significantly reduced harmful nitrate leaching and enhanced denitrifier efficiency; this higher efficiency allowed for lower potential relative rate of N_20 in the organically farmed soils than in the conventionally farmed soils.

Impact of Organic Agriculture on Sustaining Human Health

Organic agriculture has been shown to have the potential to create food systems that can sustain human health through several different mechanisms. Decreased pesticide residues found on organic products is one means through which organic agriculture accomplishes this result. According to EPA (2007, para. 1), "laboratory studies show that pesticides can cause health problems, such as birth defects, nerve damage, cancer, and other effects that might occur over a long period of time.... Some pesticides also pose unique health risks to children."

In its report, *Pesticides in the Diets of Infants and Children* (NRC 1993), the National Research Council asserts that "dietary intake of pesticides represents the major source of exposure for infants and children." A review of three independent datasets measuring amounts of pesticide residue on food indicated that organically grown foods have fewer and generally less pesticide residues than conventionally grown foods: organic foods typically contained pesticide residues only one-third as often as conventionally grown foods (Baker et al. 2002).

Research directly measuring the amount of pesticide metabolites in children further illustrates the reduction in pesticide load provided by organic agriculture (Lu et al. 2008). In the Seattle, Washington, area, 23 children aged 3–11 years who consumed diets of only conventionally grown products in 2003–4 were assessed for amounts of pesticide metabolites present in their urine. These same children switched to organic diets for five consecutive days in the summer and fall, and their urinary metabolites were measured for malathion, chlorpyrifos, and other organophosphate (OP) pesticides twice daily for a period of 7, 12, or 15 consecutive days. Substituting organic fresh fruits and vegetables for corresponding conventional food items reduced the median urinary metabolite concentrations to nondetected or close to nondetected levels for malathion and chlorpyrifos at the end of the five-day organic diet intervention period in both summer and fall seasons.

In addition to contributing potential health benefits through lowered pesticide residue loads on agricultural products, organically produced foods have been shown to have a potentially beneficial impact on human health through higher levels of nutritionally valuable compounds. Lairon (2009), in a review article of available scientific literature, found that organic products tended to have higher nutrient density and higher levels of health-promoting compounds: plant products contained more dry matter, some minerals (Fe, Mg), and antioxidant micronutrients (phenols, resveratrol), and animal organic products contained more polyunsaturated fatty acids. Additionally, Mitchell et al. (2007) found that tomatoes grown under organic management over a 10-year period consistently had significantly more flavonoids than their conventional counterparts. Studies conducted by the USDA have shown that catsup produced by organic food companies was higher in lycopene and total carotenoid contents, total antioxidant activity, and lipophilic antioxidant activities than comparable conventionally produced varieties (Ishida and Chapman 2004). This additional nutrient density becomes increasingly important in ensuring that the nutritional needs of people are met in areas where food may be scarce or food shortages common.

Breeding for Organic Agriculture: An Alternative to GMO-Based Breeding Technology

Organic agricultural systems have unique needs with respect to variety types and availability. Because of the restrictions and prohibitions on chemical pesticides and fertilizers, organic agricultural systems have a much greater need to adapt and meld to local environmental and ecosystem conditions. The reduction in environmental variability in conventional agriculture through the widespread use of chemical inputs allows individual cultivars to be successful over large geographic areas (Wolfe et al. 2008). Because organic agriculture has fewer tools to quickly alleviate abiotic and biotic stresses on the systems, varieties adapted to perform well under varied environmental conditions are extremely important (Wolfe et al. 2008). Furthermore, because organic farming practices are tailored to site-specific conditions, there is a wide diversity in practices among individual farms. Therefore, species and varieties need to be chosen for their adaptation to local soil and climate and their resistance to local pests and diseases.

Organic farmers, under both U.S. and EU regulations, are required to use organically produced seed and stock. Regulations also prohibit the use of crops developed using GE technology. However, organic agriculture still primarily depends on varieties originally bred for conventional agriculture using traditional breeding methods (Löschenberger et al. 2008). These varieties often do not possess the unique characteristics of organic varieties, such as nutrient-use efficiency, weed suppression, high levels of disease resistance, tolerance to mechanical weed control, tolerance to abiotic stress, insect resistance, and qualities desired by the organic consumer (Lammerts van Bueren, Jones, et al. 2010).

Centralized breeding strategies employed by conventional agriculture are not well suited to meet the challenges of variety development across these wide ranges of environmental conditions. It is widely recognized that conventional plant breeding has been more beneficial to farmers in high-potential environments or those who can modify their environment to suit new cultivars than to the poorest and low-resource farmers. Low-resource farmers lack financial capital or other resources to acquire outside inputs or newer varieties to replace their traditional, well-known, and reliable varieties (Byerlee and Husain 1993; Eyzaguirre and Iwanaga 1996; Trutmann 1996). Unfortunately, conventional breeding programs have focused little on the needs of organic agriculture (Reid et al. 2010). Crop varieties developed for conventional agriculture are primarily bred under homogeneous conditions at a few breeding sites or nurseries. Using these methods, identifying superior genotypes that will perform well under low-input conditions and in highly heterogenic environments is often more difficult (Haugerud and Collinson 1990). Although organic agriculture is not necessarily low input, it is inarguably highly heterogenic in nature.

Partly as a result of breeding under a centralized model, varieties bred for conventional agriculture differ in performance under conventional versus organic management. Trials that compare varieties under organic and conventional conditions have found differences sufficiently large enough to warrant breeding under specific conditions (Wolfe et al. 2008). Several studies have reported differences in the performance of wheat cultivars in organic and conventional management systems, with some cultivars being better suited to organic management in northern North America (Carr et al. 2006; Mason et al. 2007; Nass, Ivany, and MacLeod 2003). Przystalski et al. (2008) concluded that selection of cultivars should be conducted under conditions closely mimicking commercial organic farms and including traits important to organic farmers, because specific cultivars exhibited crossover interactions between management systems despite high genetic correlations between systems. In Washington State, the direct selection of wheat varieties for yield under organic management has resulted in genotypic ranking different from conventional management (Murphy et al. 2007). However, to date there are only few varieties that are specifically bred for organic and low-input systems in developed countries; more than 95 percent of organic agriculture is based on crop varieties that were bred for the conventional high-input sector with selection in conventional breeding programs (Lammerts van Bueren, Struik, and Jacobsen 2002).

Participatory Plant Breeding

Participatory plant breeding (PPB) is a unique approach to providing suitable varieties for diverse organic systems without using GE technology, which is prohibited in organic systems. PPB involves several partners (e.g., farmers, traders, consumers, breeders, researchers) in the crop selection process and is based on the interaction of skills and knowledge of each partner (Wolfe et al. 2008). It uses these partners' knowledge to develop crop varieties suited to particular agro-ecological zones, with the partners working together to set breeding objectives, generate genetic variability, make selections, evaluate experimental varieties, and generate and disseminate seeds (Dawson and Goldberger 2008).

In practice, three kinds of participation are usually distinguished: consultative (information sharing), collaborative (task sharing), and collegial (sharing responsibility, decision making, and accountability; Desclaux and Hédont 2006; Sperling et al. 2001). The type of participation may determine whether the breeding activity is centralized or decentralized. Although PPB is usually decentralized, it can also be carried out in centralized research stations where farmers are invited to visit, give their opinions, and practice selection among plants being grown at the station. PPB can also be divided into two categories: complete participatory breeding, in which farmers and scientists collaborate continuously throughout the breeding process, and participatory varietal selection, in which initial stages of the breeding process are performed exclusively by scientists and farmer participation is restricted to evaluating finished cultivars (Morris and Bellon 2004).

The practice of PPB is relatively new and has been developing along several different modes of operation. Sperling et al. (2001) reviewed 65 long-term examples of PPB that began during the 1990s and showed the various technical and organizational strategies. Ideally, farmers involved in PPB were researchers alongside the plant breeders: setting priorities for the breeding process, making crosses, screening germplasm, testing selections in multiple environments, and leading the seed multiplication and distribution process. However, the roles of farmers can be quite diverse, including involvement in the following activities:

1. Setting breeding targets
2. Generating (or accessing) variation through crossing (or using collections)
3. Selecting in segregating populations

4. Variety testing and characterization
5. Interacting with seed systems (release, popularization/marketing/diffusion, seed production, distribution; modified from Schnell 1982, in Sperling et al. 2001).

Farmer–researcher collaboration can vary at each of these stages and is one of the factors useful in comparing PPB models.

PPB can be further classified depending on the leadership of the projects. Formally led PPB programs are led by researchers who invite farmer participation in the formal research (Sperling et al. 2001). These programs are often conducted in an academic research setting, thus requiring appropriate experimental design and replication of results. Farmer collaborators are expected to augment the formal research program through defining varietal preferences and setting research priorities. Generally, formally led PPB programs also involve strong linkages to formal variety release and seed production systems. Finally, scientists involved in formally led programs are usually expected by the scientific community to extrapolate their methods beyond the individual community with which they work and are encouraged to show the advantages of PPB compared to formal breeding approaches (Weltzien et al. 1999).

On the other end of the spectrum are farmer-led PPB projects. In these projects, although researchers or other professionals function as facilitators in a supporting role, farmers establish breeding objectives, bear the main responsibility for both the labor and cost of conducting experiments, select materials for seed multiplication, and disseminate both seed and information (Sperling et al. 2001). The objective of farmer-led PPB is to provide varieties or populations that suit the specific local environment and local preferences, and any broader applicability beyond local circumstances is fortuitous (Sperling et al. 2001). Farmer-led PPB tends to work for a specific client group or groups that have no obligation either to disseminate information for wider geographical extrapolation or to feed products such as varieties into external formal release and seed systems (McGuire, Manicad, and Sperling 1999).

PPB appears to increase the probability and the speed of adoption of new varieties. There are now several examples indicating that PPB improves breeding efficiency, leads to more acceptable varieties, accelerates adoption, and promotes genetic diversity (Ashby and Lilja 2004; Ceccarelli and Grando 2005; Morris and Bellon 2004; Weltzien et al. 2003). These achievements are partially due to the ability of the participatory program to adapt to emerging priorities (Mangione et al. 2006). Additionally, implementing participatory programs does not appear to increase the demand for breeders' services, although the number of people required as technicians and to provide production labor increases significantly. Other benefits of participatory trials include the generation of more representative biological and agronomic data, involvement of more farmers, and greater ranges of environmental conditions. These benefits facilitate the assembly of information on farmers' preferences that is not available through centralized breeding programs.

Participatory Plant Breeding Case Study: The Organic Seed Partnership

One of the largest participatory plant breeding efforts in the United States is the Organic Seed Partnership. Initially begun in 2001 as the Public Seed Initiative at Cornell University with funding from the USDA Initiative for Future Agricultural Food Systems, the project, centered in the northeastern United States, began with a series of regional organic breeding roundtable sessions that included organic farmers, breeders, and seed companies. This group identified critical organic breeding needs that then guided six on-farm organic breeding projects in which breeders collaborated with six growers. The initial three-year phase accomplished three primary goals: providing workshops for growers, facilitating on-farm trialing of germplasm available from the originating university, and providing seed-processing machinery.

After experiencing success on a regional level, the project was expanded to a national scope. To remain true to the principles of PPB, regional hubs were selected to maintain local relevance: Cornell University represented the Northeast; New Mexico State University represented the Southwest; Oregon State University represented the Pacific Northwest; Alcorn State University represented the Southeast; and West Virginia State University represented Appalachia. Goals of this second phase of the project included providing a wide array of improved crop varieties through the formal seed distribution mechanisms; enhancing organic farmers' selection and breeding capacity; expanding trialing, variety evaluation, and selection activities; and creating curricula in support of organic research objectives. Farmers were given the roles of objective setters, germplasm providers/breeders, germplasm evaluators, seed growers, and variety distributors.

The ultimate goal of the Organic Seed Partnership was to create a strong national network aimed at developing and delivering improved vegetable varieties selected for superior performance in organic systems. To achieve this goal, new varieties and improved capacity to produce large quantities of commercial-grade seed needed to be developed. This necessitated the involvement of several key partners – small businesses (e.g., farmer/breeders and regionally focused, smaller seed companies), nonprofit organizations, public sector research institutions including universities, and the USDA – to develop a large germplasm base. The list of seed company supporters included some of the world's largest producers as well as small seed companies that served the northeastern organic market. Appropriate procedures were developed to manage the transfer of genetic materials between breeders to the trialing networks to preserve the originators' rights.

The Organic Seed Partnership further developed into what is currently the Northern Organic Vegetable Improvement Cooperative (NOVIC). Drawing on some of the same collaborators as the Organic Seed Partnership (Cornell University, Oregon State University) and introducing some new relationships (University of Wisconsin, Organic Seed Alliance), this project focuses more on the needs of the broad geographical area of the northern United States. The project creates a collaborative network across the northern United States to guide organic variety evaluation and improvement. To best draw on both the expertise of the hub institutions and the needs of the organic farmers, efforts focus on the participatory breeding and trialing of specific target crops (sweet corn, carrots, broccoli, snap peas, and winter squash). Long-term goals are to enhance compliance with the NOP requirement for use of certified organic seed and to create a robust multistate network among researchers and farmers working on organic vegetable breeding and trialing.

The NOVIC project aims to enhance communication among members of the organic farming community to maximize the participatory aspects of the project. One of the ways the project seeks to accomplish this goal is through the creation of a relational database of all published organic variety trialing results to advise organic farmers and researchers; this database will facilitate the identification of needs and resources in plant improvement for organic systems, and it will provide insight into the factors common to organic systems that influence varietal adaptation. In addition, field days and participatory plant breeding workshops engage and empower organic producers with skills and information. Thus, this project has the potential to create strong grassroots, regionally specific, and relevant alternatives to the industry-based variety development and seed production model that currently predominates in the United States.

Conclusion

Organic agriculture offers a viable, sustainable alternative to conventional and GE-based agricultural systems. Many peer-reviewed research studies demonstrate that the development and availability of transgenic crops thus far have not resulted in increases in crop yields over traditionally produced conventional varieties. Instead, about half of the yield advances witnessed in the 20th century can be attributed to traditional plant breeding (Gurian-Sherman 2009), with the other half attributed to improvements in production, such as irrigation, mechanization, and fertilizer use (Duvick 2005).

The body of research illustrating organic agriculture's beneficial impacts in improving ecological and environmental sustainability beyond the current conventional production practices continues to grow. In addition, the importance of the role of organic agriculture in sustaining and improving human health worldwide continues to be clarified.

Estimates of the yields of a completely organic world food system show adequate levels of production to feed the world's population as of 2010 (Badgley et al. 2006). These yields are accomplished by an agricultural production system that does not rely on the biotechnology and GMOs present in many conventional food systems. By avoiding these technologies, organic agriculture avoids the concerns of safety and social equity that surround the use of biotechnology and GMO-based systems.

Although yield estimates using current organic production figures show respectable levels of global production, the productivity of organic farming systems can further be enhanced using variety development models that avoid biotechnology. Research both in the United States and overseas continues to refine organic production methods, increasing efficiency and yield through better nutrient and weed management, cover cropping, and variety selection. PPB, an alternative to transgenic breeding strategies, allows for variety development to be better tailored to organic production systems and the needs of local farmers. Not only does this method of variety development produce enhanced varieties but it also returns power and control to the local farmers and local communities, further enhancing the long-term sustainability of local agriculture worldwide.

References

Ashby, J. A., and N. Lilja. 2004. "Participatory Research: Does it Work? Evidence from Participatory Plant Breeding." *Proceedings of the 4th International Crop Science Congress, New Directions for a Diverse Planet*, Brisbane, Queensland, Australia, September 26–October 1.

Badgley, C., J. Moghtader, E. Quintero, E. Zakem, M. Chappell, K. Avilés-Vázquez, A. Samulon, and I. Perfecto. 2006. "Organic Agriculture and the Global Food Supply." *Renewable Agriculture and Food Systems* 22 (2): 86–108. doi:10.1017/S1742170507001640.

Baker, B. P., C. M. Benbrook, E. Groth, III, and K. L. Benbrook. 2002. "Pesticide Residues in Conventional, Integrated Pest Management (IPM)-Grown and Organic Foods: Insights from Three U.S. Data Sets." *Food Additives and Contaminants* 19 (5): 427–66. doi:10.1080/02652030110113799.

Balkcom, K., A. Price, E. Santen, D. Delaney, D. Boykin, F. Arriaga, J. Bergtold, T. Kornecki, and R. Raper. 2010. "Row Spacing, Tillage System, and Herbicide Technology Affects Cotton Plant Growth and Yield." *Field Crops Research* 117 (2/3): 219–25. doi:10.1016/j.fcr.2010.03.003.

Barry, B. D., L. L. Darrah, D. L. Huckla, A. Q. Antonio, G. S. Smith, and M. H. O'Day. 2000. "Performance of Transgenic Corn Hybrids in Missouri for Insect Control and Yield." *Journal of Economic Entomology* 93 (3): 993–9. doi:10.1603/0022-0493-93.3.993.

Barthès, B., A. Azontonde, E. Blanchart, C. Girardin, C. Villenave, S. Lesaint, R. Oliver, and C. Feller. 2004. "Effect of a Legume Cover Crop (*Mucuna pruriens* var. *utilis*) on Soil Carbon in an Ultisol under Maize Cultivation in Southern Benin." *Soil Use and Management* 20 (2): 231–9. doi:10.1111/j.1475-2743.2004.tb00363.x.

Brumfield, R., A. Rimal, and S. Reiners. 2000. "Comparative Cost Analyses of Conventional, Integrated Crop Management, and Organic Methods." *HortTechnology* 10 (4): 785–93.

Byerlee, D., and T. Husain. 1993. "Agricultural Research Strategies for Favoured and Marginal Areas: The Experience of Farming System Research in Pakistan." *Experimental Agriculture* 29 (2): 155–71. doi:10.1017/S0014479700020603.

Carr, P., H. Kandel, P. Kandel, R. Porter, R. Horsley, and S. Zwinger. 2006. "Wheat Cultivar Performance on Certified Organic Fields in Minnesota and North Dakota." *Crop Science* 46 (5): 1963–71. doi:10.2135/cropsci2006.01-0046.

Ceccarelli, S., and S. Grando. 2005. "Decentralized-Participatory Plant Breeding." In *In the Wake of the Double Helix: From the Green Revolution to the Gene Revolution*, edited by R. Tuberosa, R. L. Phillips, and M. Gale, 145–56. Proceedings of an International Congress, University of Bologna, Italy, May 27–31, 2003. Bologna: Avenue Media.

Clark, S. 1999. "Crop-Yield and Economic Comparisons of Organic, Low-Input, and Conventional Farming Systems in California's Sacramento Valley." *American Journal of Alternative Agriculture* 14 (3): 109–21. doi:10.1017/S0889189300008225.

Codex Alimentarius. 2010. *Guidelines for the Production, Processing, Labelling and Marketing of Organically Produced Foods*. CAC/GL 32 – 1999. Adopted 1999; revised

2001, 2003, 2004, 2007; amended 2008, 2009, 2010. Rome: Codex Alimentarius. http://www.codexalimentarius.net/download/standards/360/cxg_032e.pdf.
Cormack, W. 2000. *Energy Use in Organic Farming Systems*. MAFF (Ministry of Agriculture, Fisheries, and Food) project code OF0182. London: MAFF. http://orgprints.org/8169/1/OF0182_181_FRP.pdf.
Cox, W., J. Hanchar, and E. Shield. 2009. "Stacked Corn Hybrids Show Inconsistent Yield and Economic Responses in New York." *Agronomy Journal* 101 (6): 1530–7. doi:10.2134/agronj2009.0297.
Dawson, J. C., and J. R. Goldberger. 2008. "Assessing Farmer Interest in Participatory Plant Breeding: Who Wants to Work with Scientists?" *Renewable Agriculture and Food Systems* 23 (3): 177–87. doi:10.1017/S1742170507002141.
Delate, K., and C. Cambardella. 2004. "Agroecosystem Performance during Transition to Certified Organic Grain Production." *Agronomy Journal* 96 (5): 1288–98. doi:10.2134/agronj2004.1288.
Desclaux, D., and M. Hédont, eds. 2006. *Proceedings of ECO-PB Workshop: "Participatory Plant Breeding: Relevance for Organic Agriculture?"* Workshop held in Domaine de la Besse (Camon, Ariège), France, June 11–13. Paris: ITAB (Institut Technique en Agriculture Biologique). http://www.eco-pb.org/09/proceedings_060613.pdf.
Drinkwater, L. E., P. Wagoner, and M. Sarrantonio. 1998. "Legume-Based Cropping Systems Have Reduced Carbon and Nitrogen Losses." *Nature* 396 (6708): 262–5. doi:10.1038/24376.
Duffy, M., M. Liebman, and K. Pecinovsky. 2002. *Organic vs. Conventional Farming Systems*. Report ISRF02-13. Ames: Iowa State University, Northeast Research and Demonstration Farm. http://www.ag.iastate.edu/farms/02reports/ne/OrganicConvSystems.pdf.
Duvick, D. N. 2005. "The Contribution of Breeding to Yield Advances in Maize (*Zea mays* L.)." *Advances in Agronomy* 86:83–145.
EAC (East African Community). 2007. *East African Organic Products Standard*. Report EAS 456:2007. Arusha: EAC. http://www.unep-unctad.org/CBTF/events/dsalaam2/EAS%20456-2007_Organic%20products%20standard_PRINT.pdf.
Edison Electric Institute. 2009. "Animal Waste Methane Energy Recovery." Last revised December 11. http://www.uspowerpartners.org/Topics/SECTION6Topic-AnimalWasteMethane.htm.
Eggleston, S., L. Buendia, K. Miwa, T. Ngara, and K. Tanabe, eds. 2006. *Agriculture, Forestry and Other Land Use*. Vol. 4 of *IPCC Guidelines for National Greenhouse Gas Inventories*. Hayama, Japan: IGES (Institute for Global Environmental Strategies). http://www.ipcc-nggip.iges.or.jp/public/2006gl/vol4.html.
EPA (Environmental Protection Agency). 2007. "Pesticides and Food: Health Problems Pesticides May Pose." http://www.epa.gov/pesticides/food/risks.htm.
Eyzaguirre, P., and M. Iwanaga. 1996. "Farmers' Contribution to Maintaining Genetic Diversity in Crops and its Role within the Total Genetic Resources System." In *Participatory Plant Breeding*, edited by P. Eyzaguirre and M. Iwanaga, 9–18. Proceedings of a Workshop on Participatory Plant Breeding, July 26–9, 1995, Wageningen, The Netherlands. Rome: IPGRI.
Eziakor, I. G. 1990. "Comparative Analysis of the Effectiveness of Manual versus Mechanized Tillage among Third World Smallholders: A Case Study in Bauchi State of Nigeria." *Agriculture, Ecosystems & Environment* 31 (4): 301–12. doi:10.1016/0167-8809(90)90229-7.
FAO (Food and Agriculture Organization). 2009. "2050: A Third More Mouths to Feed." News release, September 23. http://www.fao.org/news/story/0/item/35571/icode/en.
Fliessbach, A., H. Oberholzer, L. Gunst, and P. Mader. 2007. "Soil Organic Matter and Biological Soil Quality Indicators after 21 Years of Organic and Conventional

Farming." *Agriculture, Ecosystems and Environment* 118 (1–4): 273–84. doi:10.1016/j.agee.2006.05.022.

Freibauer, A., M. Rounsevell, P. Smith, and J, Verhagen. 2004. "Carbon Sequestration in the Agricultural Soils of Europe." *Geoderma* 122 (1): 1–23.

Glover, J. D., P. K. Andrews, and J. P. Reganold. 2000. "Systematic Method for Rating Soil Quality of Conventional, Organic, and Integrated Apple Orchards in Washington State." *Agriculture, Ecosystems & Environment* 80 (1/2): 29–45. doi:10.1016/S0167-8809(00)00131-6.

Gomiero, T., M. Paoletti, and D. Pimentel. 2008. "Energy and Environmental Issues in Organic and Conventional Agriculture." *Critical Reviews in Plant Sciences* 27 (4): 239–54. doi:10.1080/07352680802225456.

Granatstein, D. 2010. *Tree Fruit Production with Organic Farming Methods.* Wenatchee, WA: Center for Sustaining Agriculture and Natural Resources. Accessed December 13. http://organic.tfrec.wsu.edu/OrganicIFP/OrganicFruitProduction/OrganicMgt.PDF.

Gregorich, E. G., P. Rochette, A. J. VandenBygaart, and D. A. Angers. 2005. "Greenhouse Gas Contributions of Agricultural Soils and Potential Mitigation Practices in Eastern Canada." *Soil and Tillage Research* 83 (1): 53–72.

Gurian-Sherman, D. 2009. *Failure to Yield: Evaluating the Performance of Genetically Engineered Crops.* Cambridge: UCS (Union of Concerned Scientists). http://www.ucsusa.org/assets/documents/food_and_agriculture/failure-to-yield.pdf.

Haas, G., F. Wetterich, and U. Köpke. 2001. "Comparing Intensive, Extensified and Organic Grassland Farming in Southern Germany by Process Life Cycle Assessment." *Agriculture, Ecosystems and Environment* 83 (1–2): 43–53. doi:10.1016/S0167-8809(00)00160-2.

Haugerud, A., and M. Collinson. 1990. "Plants, Genes and People: Improving the Relevance of Plant Breeding in Africa." *Experimental Agriculture* 26 (3): 341–62. doi:10.1017/S0014479700018500.

Hepperly, P., D. Douds, Jr., and R. Seidel. 2006. "The Rodale Farming System Trial 1981 to 2005: Long Term Analysis of Organic and Conventional Maize and Soybean Cropping Systems." In *Long-Term Field Experiments in Organic Farming*, edited by J. Raupp, C. Pekrun, M. Oltmanns, and U. Köpke, 15–32. Bonn: ISOFAR (International Society of Organic Agricultural Research).

Hoeppner, J. W., M. H. Hentz, B. G. McConkey, R. P. Zentner, and C. N. Nagy. 2006. "Energy Use and Efficiency in Two Canadian Organic and Conventional Crop Production Systems." *Renewable Agriculture and Food Systems* 21 (1): 60–7. doi:10.1079/RAF2005118.

IFOAM (International Federation of Organic Agricultural Movements). 2002. *Basic Standards for Organic Production and Processing.* Approved by the IFOAM General Assembly, Victoria, Canada, August.

Ishida, B. K., and M. H. Chapman. 2004. "A Comparison of Carotenoid Content and Total Antioxidant Activity in Catsup from Several Commercial Sources in the United States." *Journal of Agricultural and Food Chemistry* 52 (26): 8017–20. doi:10.1021/jf040154o.

Jost, P., D. Shurley, S. Culpepper, P. Roberts, R. Nichols, J. Reeves, and S. Anthony. 2008. "Economic Comparison of Transgenic and Nontransgenic Cotton Production Systems in Georgia." *Agronomy Journal* 100 (1): 42–51. doi:10.2134/agrojnl2006.0259.

Karlen, D. L., M. D. Duffy, and T. S. Colvin. 1995. "Nutrient, Labor, Energy, and Economic Evaluations of Two Farming Systems in Iowa." *Journal of Production Agriculture* 8 (4): 540–6.

Kramer, S. B., J. P. Reganold, J. D. Glover, B. J. M. Bohannan, and H. A. Mooney. 2006. "Reduced Nitrate Leaching and Enhanced Denitrifier Activity and Efficiency in Organically Fertilized Soils." *Proceedings of the National Academy of Sciences of the United States of America* 103 (12): 4522–7. doi:10.1073/pnas.0600359103.

Küstermann, B., M. Kainz, and K.-J. Hülsbergen. 2008. "Modeling Carbon Cycles and Estimation of Greenhouse Gas Emissions from Organic and Conventional Farming Systems." *Renewable Agriculture and Food Systems* 23 (1): 38–52. doi:10.1017/S1742170507002062.

Lairon, D. 2009. "Nutritional Quality and Safety of Organic Food: A Review." *Agronomy for Sustainable Development* 30 (1): 33–41. doi:10.1051/agro/2009019.

Lammerts van Bueren, E. T., S. S. Jones, L. Tamm, K. M. Murphy, J. R. Myers, C. Leifert, and M. M. Messmer. 2010. "The Need to Breed Crop Varieties Suitable for Organic Farming, Using Wheat, Tomato and Broccoli as Examples: A Review." *NJAS: Wageningen Journal of Life Sciences*. doi:10.1016/j.njas.2010.04.001.

Lammerts van Bueren, E. T., P. C. Struik, E. Jacobsen. 2002. "Ecological Concepts in Organic Farming and their Consequences for an Organic Crop Ideotype." *NJAS: Wageningen Journal of Life Sciences* 50 (1): 1–26.

Lampkin, N. 2002. *Organic Farming*, 2nd revised ed. Suffolk, UK: Old Pond.

Lockeretz, W., G. Shearer, and D. H. Kohl. 1981. "Organic Farming in the Corn Belt." *Science* 211 (4482): 540–7. doi:10.1126/science.211.4482.540.

Löschenberger, F., A. Fleck, H. Grausgruber, H. Hetzendorfer, G. Hof, J. Lafferty, M. Marn, A. Neumayer, G. Pfaffinger, and J. Birschitzky. 2008. "Breeding for Organic Agriculture: The Example of Winter Wheat in Austria." *Euphytica* 163 (3): 469–80. doi:10.1007/s10681-008-9709-2.

Lotter, D.W., W. Liebhardt, and R. Seidel. 2003. "The Performance of Organic and Conventional Cropping Systems in an Extreme Climate Year." *American Journal of Alternative Agriculture* 18 (3): 146–54.

Lu, C., D. B. Barr, M. A. Pearson, and L. A. Waller. 2008. "Dietary Intake and its Contribution to Longitudinal Organophosphorus Pesticide Exposure in Urban/Suburban Children." *Environmental Health Perspectives* 116 (4): 537–42. doi:10.1289/ehp.114-a572b.

Ma, B. L., and K. D. Subedi. 2005. "Development, Yield, Grain Moisture and Nitrogen Uptake of *Bt* Corn Hybrids and Their Conventional Near-Isolines." *Field Crops Research*. 93 (2–3): 199–211.

Mangione, D., S. Senni, M. Puccioni, S. Grando, and S. Ceccarelli. 2006. "The Cost of Participatory Barley Breeding." *Euphytica* 150 (3): 289–306. doi:10.1007/s10681-006-0226-x.

Marra, M. C., P. G. Pardey, and J. M. Alston. 2002. "The Payoffs to Transgenic Field Crops: An Assessment of the Evidence." *AgBioForum* 5 (2): 43–50.

Mason, H. E., A. Navabi, B. Frick, J. O'Donovan, and D. Spaner. 2007. "The Weed Competitive Ability of Canada Western Red Spring Wheat Cultivars Grown under Organic Management." *Crop Science* 47 (3): 1167–76. doi:10.2135/cropsci2006.09.0566.

McGuire, S., G. Manicad, and L. Sperling. 1999. *Technical and Institutional Issues in Participatory Plant Breeding: From a Perspective of Farmer Plant Breeding*. Working Document 3. Cali, Columbia: CGIAR (Consultative Group for International Agricultural Research) PRGA (Participatory Research and Gender Analysis for Technology Development and Institutional Innovation) Program.

McPherson, R. M., and T. C. McCrae. 2009. "Evaluation of Transgenic Soybean Exhibiting High Expression of a Synthetic *Bacillus thuringiensis cry1A* Transgene for Suppressing Lepidopteran Population Densities and Crop Injury." *Journal of Economic Entomology* 102 (4): 1640–8. doi:10.1603/029.102.0431.

Michalak, P. 2004. *Water, Agriculture, and You*. Kutztown, PA: Rodale Institute. http://www.rodaleinstitute.org/files/water_booklet.pdf.

Mitchell, A. E., Y.-J. Hong, E. Koh, D. M. Barrett, D. E. Bryant, R. F. Denison, and S. Kaffka. 2007. "Ten-Year Comparison of the Influence of Organic and Conventional

Crop Management Practices on the Content of Flavonoids in Tomatoes." *Journal of Agricultural and Food Chemistry* 55 (15): 6154–9. doi:10.1021/jf070344+.
Morris, M., and M. Bellon. 2004. "Participatory Plant Breeding Research: Opportunities and Challenges for the International Crop Improvement System." *Euphytica* 136 (1): 21–35. doi:10.1023/B:EUPH.0000019509.37769.b1.
Mrema, G. C., D. Baker, and D. Kahan. 2008. *Agricultural Mechanization in Sub-Saharan Africa: Time for a New Look.* Rome: FAO (Food and Agriculture Organization). http://www.fao.org/Ag/ags/publications/docs/AGSF_OccassionalPapers/OP22-web.pdf.
Murphy, K. M., K. G. Campbell, S. R. Lyon, and S. S. Jones. 2007. "Evidence of Varietal Adaptation to Organic Farming Systems." *Field Crops Research* 102 (3): 172–7. doi:10.1016/j.fcr.2007.03.011.
Nass, H., J. Ivany, and J. MacLeod. 2003. "Agronomic Performance and Quality of Spring Wheat and Soybean Cultivars under Organic Culture." *American Journal of Alternative Agriculture* 18 (3): 164–70. doi:10.1079/AJAA200348.
Nguyen, M., and R. Haynes. 1995. "Energy and Labour Efficiency for Three Pairs of Conventional and Alternative Mixed Cropping (Pasture-Arable) Farms in Canterbury (New Zealand)." *Agriculture, Ecosystems and Environment* 52 (2–3): 163–72. doi:10.1016/0167-8809(94)00538-P.
NOP (National Organic Program). 2010. http://www.ams.usda.gov/AMSv1.0/nop.
NRC (National Research Council). 1993. *Pesticides in the Diets of Infants and Children.* Washington, DC: National Academy Press.
Oquist, K. A., J. S. Strock, and D. J. Mulla. 2007. "Influence of Alternative and Conventional Farming Practices on Subsurface Drainage and Water Quality." *Journal of Environmental Quality.* 36 (4): 1194–204. doi:10.2134/jeq2006.0274.
OTA (Organic Trade Association). 2010. *Industry Statistics and Projected Growth* (last updated June 16). http://www.ota.com/organic/mt/business.html.
Pimentel, D., P. Hepperly, J. Hanson, D. Douds, and R. Seidel. 2005. "Environmental, Energetic and Economic Comparison of Organic and Conventional Farming Systems." *Bioscience* 55 (7): 573–82.
Posner, J. L., J. O. Baldock, and J. L. Hedtcke. 2008. "Organic and Conventional Production Systems in the Wisconsin Integrated Cropping Systems Trials: I. Productivity 1990–2002." *Agronomy Journal* 100 (2): 253–60. doi:10.2134/agrojnl2007.0058.
Przystalski, M., A. Osman, E. Thiemt, B. Rolland, L. Ericson, H. Østergård, L. Levy, et al. 2008. "Comparing the Performance of Cereal Varieties in Organic and Nonorganic Cropping Systems in Different European Countries." *Euphytica* 163 (3): 417–33. doi:10.1007/s10681-008-9715-4.
Refsgaard, K., N. Halberg, and E. S. Kristensen. 1998. "Energy Utilization in Crop and Dairy Production in Organic and Conventional Livestock Production Systems." *Agricultural Systems* 57 (4): 599–630. doi:10.1016/S0308-521X(98)00004-3.
Reganold, J. P., P. K. Andrews, J. R. Reeve, L. Carpenter-Boggs, C. W. Schadt, J. R. Alldredge, C. F. Ross, N. M. Davies, and J. Zhou. 2010. "Fruit and Soil Quality of Organic and Conventional Strawberry Agroecosystems." *PLoS ONE* 5 (9): e12346. doi:10.1371/journal.pone.0012346.
Regional Organic Task Force. 2008. *Pacific Organic Standard.* Nouméa, New Caldonia: Secretariat of the Pacific Community.
Reid, T. A., R.-C. Yang, D. F. Salmon, A. Navabi, and D. Spaner. 2010. "Realized Gains from Selection for Spring Wheat Grain Yield Are Different in Conventional and Organically Managed Systems." *Euphytica.* doi:10.1007/s10681-010-0257-1.
Rodale Institute. 1999. "100-Year Drought Is No Match for Organic Soybeans: Scientific Trials at the Rodale Institute Give Hope to Farmers Everywhere." November 8. http://www.rodaleinstitute.org/19991109/fst.

Schnell, F. W. 1982. "A Synoptic Study of the Methods and Categories of Plant Breeding." *Zeitschrift für Pflanzenzüchtung* 89 (1): 1–18.

Scialabba, N. E. -H., and M. Müller-Lindenlauf. 2010. "Organic Agriculture and Climate Change." *Renewable Agriculture and Food Systems* 25 (2): 158–69. doi:10.1017/S1742170510000116.

Smolik, J. D, D. H. Rickerl, and T. L. Dobbs. 1995. "The Relative Sustainability of Alternative, Conventional, and Reduced-Till Farming Systems." *American Journal of Alternative Agriculture* 10 (1): 25–35.

Sorby, K. 2002. *Toward More Sustainable Coffee?* Background paper to the World Bank Agricultural Technology Note 30. Washington, DC: World Bank.

Sperling, L., J. A. Ashby, M. E. Smith, E. Weltzien, and S. McGuire. 2001. "A Framework for Analyzing Participatory Plant Breeding Approaches and Results." *Euphytica* 122 (3): 439–50. doi:10.1023/A:1017505323730.

Stanhill, G. 1990. "The Comparative Productivity of Organic Agriculture." *Agriculture, Ecosystems and Environment* 30 (1–2): 1–26. doi:10.1016/0167-8809(90)90179-H.

Trutmann, P. 1996. "Participatory Diagnosis as an Essential Part of Participatory Breeding: A Plant Protection Perspective." In *Participatory Plant Breeding*, edited by P. Eyzaguirre and M. Iwanaga, 37–43. Proceedings of a Workshop on Participatory Plant Breeding, July 26–9, 1995, Wageningen, The Netherlands. Rome: IPGRI.

USDA (United States Department of Agriculture). 2009. "Organic Production." http://www.ers.usda.gov/Data/Organic/.

van Groeningen, J. W., G. J. Kasper, G. L. Velthof, A. van den Pol–van Dasselaar, and P. J. Kuikman. 2004. "Nitrous Oxide Emissions from Silage Maize Fields under Different Mineral Fertilizer and Slurry Application." *Plant and Soil* 263 (1): 101–11. doi:10.1023/B:PLSO.0000047729.43185.46.

Weltzien, E., M. E. Smith, L. S. Meitzner, and L. Sperling. 1999. *Technical and Institutional Issues in Participatory Plant Breeding: From the Perspective of Formal Plant Breeding: A Global Analysis of Issues, Results and Current Experience*. Working Document 3. Cali, Columbia: CGIAR (Consultative Group for International Agricultural Research) PRGA (Participatory Research and Gender Analysis for Technology Development and Institutional Innovation) Program.

______. 2003. *Technical and Institutional Issues in Participatory Plant Breeding: From the Perspective of Formal Plant Breeding*. PPB Monograph No. 1. Cali, Columbia: CGIAR (Consultative Group for International Agricultural Research) PRGA (Participatory Research and Gender Analysis for Technology Development and Institutional Innovation) Program. http://www.prgaprogram.org/descargas/plant_breeding/monographs/Technical%20and%20Institutional%20Issues%20in%20Participatory%20Plant%20Breeding%20from%20the%20Perspective%20of%20Formal.pdf.

Willer, H., and L. Kilcher, eds. 2010. *The World of Organic Agriculture – Statistics and Emerging Trends 2010*. Bonn: IFOAM (International Foundation of Organic Agriculture Movements).

Wolfe, M. S., J. P. Baresel, D. Desclaux, I. Goldringer, S. Hoad, G. Kovacs, F. Löschenberger, T. Miedaner, H. Østergård, and E. T. Lammerts van Bueren. 2008. "Developments in Breeding Cereals for Organic Agriculture." *Euphytica* 163 (3): 323–46. doi:10.1007/s10681-008-9690-9.

10

A Case Study of Rice from Traditional Breeding to Genomics

Rice – Food for the Gods

Pamela Ronald

With a special prayer, Iban farmers of Borneo beseeched the rice (*padi*) spirits to stand watch over their community's crop so it would thrive and resist ever-present diseases (Hamilton 2003):

> Oh sacred padi,
> You the opulent, you the distinguished,
> Our padi of highest rank;
> Oh sacred padi,
> Here I am planting you;
> Keep watch o'er your children,
> Keep watch o'er your people,
> Over the little ones, over the young ones,
> Oh do not be laggard, do not be lazy,
> Lest there be sickness, lest there be ailing;
> You must visit your people, visit your children.
> You who have been treated by Pulang Gana;
> Oh do not neglect to give succour,
> Oh do not tire, do not fail in your duty.
> *Iban prayer, quoted in Freeman (1970, 154–5)*

The farmers had good reason to pray. The survival of the rice ensured the survival of the people. If the rice crop failed, so would the people. Because of its importance to the survival of so many Asian cultures, rice maintains a near-divine status in Asia. The grain is viewed as a sacred food given to humans, which continues to sustain the Earth's people in a profound manner (Hamilton 2003).

Today, rice provides more of the calories consumed by humans than any other food. It is the staple food for more than half of the global population. In Asia alone, more than 2 billion people obtain 60 to 70 percent of their calories from rice and its products. When all developing countries are considered together, rice provides 27 percent of daily energy intake and 20 percent of their dietary protein.

Rice is the most rapidly growing food source in Africa and is of significant importance to food security in an increasing number of low-income food-deficit countries. The production of rice and its associated postharvest operations employ nearly 1 billion people in rural areas of developing countries. About four-fifths of the world's rice is grown by small-scale farmers in low-income countries. China and India together are responsible for more than half of global output. Developed countries account for about 5 percent of global rice production. Thailand, Vietnam, United States, India, and Pakistan are the top exporters of rice.

Given the cultural and nutritional importance of rice, its history is a worthwhile case study of the evolution of crop genetic research. This chapter traces the origin of rice and describes the rapid expansion in the number of new rice varieties as well as the knowledge gained from modern genetic research.

When and Where Did Farmers First Start Growing Rice?

There are two widely cultivated subspecies of rice (*Oryza sativa*): ssp. *indica* (long-grain rice) and ssp. *japonica* (short-grain rice). *Japonica* rice varieties were introduced into the Yangtze River valley in China about 8,000 years ago, suggesting that rice may be the oldest crop in continuous use (Hamilton 2003; Huke and Huke 1990; Khush et al. 2003). *Japonica* varieties were introduced to Korea and then to Japan in the third century BCE (Huke and Huke 1990). California farmers started growing *japonica* varieties as a result of the Gold Rush and the need to supply food to the thousands of Asian immigrants arriving in the state. In South Carolina and Georgia, farmers began growing rice as a major crop for the colonists in the 1700s. Today, 80 percent of the U.S. rice acreage is located in the southeastern United States. *Indica* rice originated in the eastern Himalayan foothills, from where it spread to the rest of the Indian subcontinent, Sri Lanka, and the lowlands of southeast Asia and then to China (Khush et al. 2003). In some areas of the world, such as the river valleys and low-lying areas in China, rice was likely cultivated widely as a wetland crop. In Southeast Asia, by contrast, rice was originally produced under dry land conditions in the uplands, and only recently did it come to occupy the vast river deltas (Huke and Huke 1990).

Since the time that rice was first cultivated along the shores of the Yangtze River, more than 120,000 *O. sativa* varieties are estimated to have been developed through conventional breeding; the gene bank at the International Rice Research Institute (IRRI) alone contains 100,000 varieties. Many of these traditional varieties are called "landraces." Landraces are selected by farmers for their adaptations to specific locations in areas where local subsistence agriculture has long prevailed. Often, they have been handed down from one generation to the other, following family or community custom. These seeds carry prized genes for disease resistance and other agronomic characteristics that were used in the development of modern varieties.

During the Song dynasty in early 11th-century China, new varieties of faster ripening rice plants imported from the kingdom of Champa – an area now part of Vietnam – allowed double cropping in the fertile Yangtze delta. The double cropping vastly improved the yields of the rice and shifted the economic center of China from the north to the south, where it has remained ever since (Hamilton 2003). The successes gained through these traditional genetic modifications continue to motivate breeders today.

The Green Revolution

Outputs were dramatically increased again during the Green Revolution of the 1960s and 1970s. One of the difficulties with rice production is that the top of a mature rice plant becomes quite heavy with grain. Wind or rain can then cause the plant to fall over into the mud, thereby ruining the grain. One of the breakthroughs that led to the Green Revolution was the development of semi-dwarf varieties that did not fall over as easily. These varieties carry a mutation in a gene that makes the rice plant shorter. For example, in Japan a semi-dwarf variety was developed that increased rice production threefold between 1900 and 1965. However the *japonica* subspecies of rice produced high yields only in temperate climates such as that of Japan or California. Scientists at IRRI in the Philippines wanted to achieve similar results with the *indica* subspecies of rice that grew well in the warmer and more humid conditions of South and Southeast Asia and makes up more than three-fourths of the total rice traded in the world. Rice breeders at IRRI crossed the Indonesian *indica* variety Peta, which was higher yielding and relatively drought resistant, to Dee-geo-woo-gen, a semi-dwarf landrace developed in late-19th-century Taiwan, to create IR8, released in 1966.

IR8 had superior characteristics to both Peta and Dee-geo-woo-gen. It was semi-dwarf because of the presence of the sd-1 gene, high yielding (i.e., producing twice the amount of grain as other varieties in response to nitrogen fertilizers and appropriate irrigation), and ripened quickly. Its introduction dramatically increased per-acre output and, in some places, allowed three crops per year because it matured in 125 days rather than 210 days. It also flowered independently of the length of the day and was resistant to a devastating fungal disease called rice blast. IR8 was called "miracle rice."

Breeders usually build on a variety that they have developed, constantly improving it to develop superior new varieties or raise the standards. For example, breeders at IRRI, as well as at rice centers in other countries, have continued to modify IR8 and other varieties by using various techniques to incorporate genes from diverse varieties, landraces, and wild rice species from around the world. California breeders used IR8 to introduce the semi-dwarf gene sd-1 into their locally adapted *japonica* varieties. In India alone, there are 800 new varieties all carrying sd-1 (Khush et al. 2003). The most widely planted rice variety today is IR64, which was derived from 22 landraces from seven countries.

The breeders were tremendously successful in achieving their goals, and their work highlighted the power of genetic modification. Currently, 90 percent of the Asian harvest comes from modern varieties, but yield increases on farms were never as great as those observed at research stations such as IRRI. Production at IRRI remained twice as high as that in most farmers' fields. Poor water control, inadequate nitrogen fertilizers, insects, and disease all contributed to lower yields. Because the modern varieties were developed specifically to maximize the plants' responsiveness to water and nitrogen fertilizers, controlling the crops' water and having adequate fertilizer were even more important for the new varieties. However, controlling water and having sufficient fertilizer require money and community organization – things many farmers and communities lacked. Therefore many farmers who lacked capital were not able to realize higher yields from the miracle rice. In addition, the improved rice varieties displaced some of the local landraces, which led to reduced genetic diversity.

The loss of diversity was, and is, problematic for several reasons. First, the landraces were adapted to the local environments and often were better able than the modern varieties to respond to unique local stresses such as flooding. For example, around 0.114 km^2 of rice lands in South and Southeast Asia are subject to uncontrolled flooding. These floods can occur after local and remote heavy rain and may completely submerge the crop for several days. Varieties well adapted to such areas are tolerant of both deepwater and submergence. Seedlings of submergence-tolerant varieties can withstand complete submergence for 1–10 days. Deepwater varieties can elongate 10–12 cm in one day to keep their leaves above water and can grow in 1–3 m of stagnant water for many months. Such varieties can even thrive in daily tidal fluctuations that sometimes may also cause complete submergence. Although the average yields of traditional landraces are relatively small – only about 1.4 Mg/ha – the rice crops planted in these areas support more than 100 million people. Some farmers in these areas replaced such traditional varieties with modern varieties, hoping to achieve the high yields gained at research stations, which were sometimes as high as 10 Mg/ha. However, because the plants were less well adapted to this environment and flood control was not available, the yields were often lower than the yields of the local varieties.

Wide planting of a single variety also reduces the rice's adaptability to diseases. In the mid-1970s, there was a large increase in the incidence of bacterial leaf blight, a serious disease of rice. Whereas the displaced landraces had some resistance to this disease, the new varieties lacked resistance. In addition, dense plantings and the addition of nitrogen fertilizers seemed to make the plants more susceptible.

Finally, lack of genetic diversity can also reduce the diversity in taste and other attributes of rice. Because different cultures have different preferences regarding rice's taste, texture, color, and stickiness, one rice variety cannot please everyone. Dry, flaky rice is eaten in South Asia and the Middle East. Moist, sticky varieties of rice are preferred in Japan, Taiwan, the Republic of Korea, Egypt, and northern China. Furthermore, red rice is the grain of choice in parts of southern India. In its natural state (before milling and polishing remove the outer layer), rice comes in many different colors, including brown, red, purple, and even black. These colorful rice varieties are often prized for their health benefits.

Although the Green Revolution led to stronger and higher yielding rice varieties for some farmers in some environments, it also had some unintended and undesirable consequences. The reduction in genetic diversity was often costly because the new varieties were not well adapted to local conditions and had increased susceptibility to disease. In some cases the farmers incurred additional costs in buying fertilizer and strengthening their irrigation systems to sustain the rice. In addition, runoff into nearby streams caused by the overuse of fertilizer degraded the environment in some places. Thus, many of the advances made during the Green Revolution were not considered sustainable, and new approaches became necessary.

Efforts to Develop Rice Varieties That Are Genetically Diverse, Are Locally Adapted, and Require Fewer Inputs

Many genetic techniques are now being used to reintroduce genetic diversity into cultivated rice varieties and to develop varieties that do not require as much nitrogen fertilizer, pesticides, or other inputs. One of the time-tested techniques is called *embryo rescue*. It is used by breeders who want to cross two different species of rice. For example, a wild species from Mali called *O. longistaminata* was found to contain a valuable gene for resistance against a bacterial blight indigenous to Africa and Asia. Because two distinct species cannot be crossed by pollen-transfer techniques, embryo rescue was performed.

The process of embryo rescue began with placing the pollen from *O. longistaminata* on the female part of the *O. sativa* flower (called a genetic cross). Then the embryo was dissected from the kernel, placed in laboratory dishes, and coaxed to germinate by adding hormones, nutrients, and other chemicals to the dish. The resulting hybrid carried half the genes from *O. sativa* and half the genes from *O. longistaminata*, including the gene for resistance against bacterial blight, Xa21. This hybrid was then used as a parent for additional breeding work. Breeders at IRRI and other research stations have now introduced Xa21 into *O. sativa* varieties using these advanced breeding approaches. Song et al. (1995) isolated the gene in 1995 and genetically engineered rice for disease resistance.

Another widely used technique to generate genetic diversity – and therefore agronomically useful traits – is mutagenesis. During this process, seeds are put in a mutagenic and highly carcinogenic solution that induces random changes in the DNA. The seeds are then planted, and usually about half die. The seeds from the surviving plants are collected and then planted. This mutant population can then be surveyed for new traits by breeders. Induced mutation techniques are commonly practiced in traditional breeding and can result in the same mutations that result from spontaneous mutations occurring in nature.

Induced and spontaneous mutations have created many different traits within the crop, including both stress tolerance and grain characteristics. For example, one of these spontaneous mutants gave rise to the precious sticky rice *mochi* (also called glutinous or sweet rice) of Thailand. Sticky rice is an important culinary and cultural component throughout East Asia and is used in festival foods and desserts. In upland regions of Southeast Asia, it is a staple food in many homes. Ten percent of the rice traded each year is sticky rice. The development of sticky rice is one example of plant breeders' modification of plants in response to local cultural preferences.

The precise origin of sticky rice remains obscure because it is not found in the archeological record. Laotian Buddhist legend places the origin of sticky rice at around 1,100 years ago, although Chinese folklore indicates that it was in existence around the time of the death of the poet Qu Yuan more than 2,000 years ago. A recent study by two North Carolina State University geneticists using modern genetic techniques suggested that sticky rice most likely originated at a single time in Southeast Asia (Olsen and Purugganan 2002). Here is how they traced the origin of sticky rice: It is widely known that sticky rice lacks the starch amylose, which constitutes up to 30 percent of the total starch in nonsticky rice endosperm. The lack of amlyose is due to a mutation in a gene called waxy, which encodes an enzyme required for amylose synthesis.

The researchers knew that different rice varieties all carry very similar genes, but in a significant proportion of cases (this varies from one species to the next), a given gene in one variety will be slightly different from its counterpart in a very closely related variety. That is the nature of genetic diversity – the genes are different in diverse rice varieties. They also knew that all breeders used the exact same variety as a breeding parent and then picked those seedlings that carried a single trait of interest encoded by a single gene. Thus the new varieties would all contain the same gene with the same sequence, and all would be derived from the same parent. Gene differences can be observed by looking carefully at the sequence of a gene from many varieties. The researchers hypothesized that if a single breeder 2,000 years ago developed a really good sticky rice variety, and then generously shared it with other breeders, the new varieties would all contain exactly the same waxy gene. That gene would have a single origin: breeder #1's sticky rice. To test this idea they looked at the sequence of the waxy gene of 105 landraces and found that all those that were sticky carried nearly the same sequence in the waxy gene, including a mutation that knocked out production of amylose. This result suggests that the early breeders of sticky rice liked the adhesive quality conferred by that single mutation and preserved that particular trait through breeding.

What We Are Learning Today: Discovering the Hidden Gems

The most precious things are not jade and pearls, but the five grains.

The five grains in this Chinese proverb are most likely rice, wheat, millet, sorghum, and maize. The grains are considered precious because so many people depend heavily on them for their daily caloric intake. Together, these cereal grains account for up to 60 percent of the calories consumed by people in the developing world. Hidden deep within these cereal grains, scientists found more precious gems still – valuable genetic information. In some ways, the genetic information in rice is the most precious of all. Rice is in the same plant family as other cereals and is therefore closely related to them. Rice also has much less genetic information to sort through than other grains. For example, wheat contains 40 times the genetic information as rice. Therefore finding the gems in rice is easier, just as finding a needle in a small haystack is easier than in a large haystack. Collaborations between public and private research laboratories have resulted in a publicly available sequence of the rice genome. This was a large international project and demonstrated the importance of rice to different cultures and the communal nature of scientific work. From the sequence, biologists can predict the location and sequence of many of the estimated 40,000–60,000 rice genes. The genomic sequence provides a rich resource for understanding biological processes of plants.

From the sequence, scientists have already learned some astounding facts. Before rice genome sequencing, many researchers assumed that most plants would have very similar genes, which would only differ slightly from animal genes. We now know that rice not only contains a greater number of genes than do humans but that also many of the genes are unique and have not previously been found. Many of these genes likely control functions in the rice plant that do not occur in the other plants, such as leaf structure, submergence tolerance, grain size, and flavor. Many of the genes that occur in rice and other plants do not occur in humans at all. Perhaps this is because plants cannot run away and need very specific recognition and signaling systems to respond to environmental stresses.

The functional roles for most of the rice genes have not been determined, but once a gene has been identified and sequenced, uncovering the function becomes somewhat easier. First, biologists can develop hypotheses about what role the gene plays in a plant's production. Often, this is done by studying what the gene does when its expression is altered in rice or another species. For instance, if a yeast gene has a clear function and then a gene that has a very similar sequence is discovered in rice, it is likely that the rice gene has the same function as the yeast gene. For example, the rice genome sequence reveals several genes that encode phosphate transporters similarly to yeast genes and are important for the uptake of phosphate (Goff et al.

2002). Therefore, the rice genes possibly are involved in the uptake of phosphate from soils. Genes controlling disease resistance, tolerance to environmental stresses such as drought or cold, or synthesis of essential vitamins have also been identified by such comparative analysis (DellaPenna 2001). This information allows biologists to form hypotheses regarding which genes govern specific biochemical and metabolic pathways, such as the development of grain size, number of leaves, and synthesis of nutrients.

Experiments can also be designed to determine whether a gene of interest has the function that is predicted from the analysis of its sequence. For example, rice plants carry genes that seem similar to genes in yeast involved in moving phosphate in and out of the cell. Breeders can now work with molecular geneticists to identify a set of rice varieties that carry different sequences for phosphate transporters. These varieties can then be tested for their ability to uptake phosphate from the soil. Without the sequence information, breeders would not have been able to identify or test this set of rice varieties. Rice varieties exhibiting useful traits such as high uptake of phosphate from the soil can be used as a breeding parent in the development of a new variety.

By overexpressing a gene using GE, scientists can gain powerful insights into its role. For example, if a gene is suspected to govern disease resistance, biologists can load up a plant with that gene – overexpression – and monitor whether the plant's resistance to disease is altered. Altered resistance suggests that the suspected gene does indeed have a role in disease resistance. Because overexpression of a gene can cause diverse effects, additional experimentation is needed to determine the gene's precise function. Conversely, the gene can be "knocked out" (taken out of a plant) to learn if the disease tolerance decreases. Such sequence-guided testing holds great promise in biology, allowing researchers to "turn up" desirable traits in plants and "turn off" the unhelpful qualities either by traditional breeding or GE.

Genes are expressed ("turned on") when an organism is carrying out one of the gene's important functions. For example, when rice plants are exposed to low temperatures, entire sets of genes are turned on and turned off. In the past, scientists were only able to study the expression of one gene at a time. Now, using the new method of microarray technology, the expression of all of the predicted genes in the rice genome (ca. 50,000) can be assessed in a single afternoon. In this method, high-density arrays of small stretches of DNA, each uniquely corresponding to a single gene, are placed on a single glass slide. Such arrays, also called "DNA chips," can be used to identify genes expressed at a particular point in time (Jung, An, and Ronald 2008). For example, a rice variety that is exposed to a disease against which it has no resistance will express a different set of genes than a variety that carries resistance to this disease. Comparison of the expression of each of the 50,000 genes in these two varieties discloses which genes work together to cause the rice plant's resistance. Once identified, these genes can be introduced into normally susceptible varieties through traditional breeding or GE.

Knowing the sequence of specific genes also allows scientists to take advantage of the natural genetic variation of existing crop species. In rice, the germplasm comprises more than 90,000 traditional rice varieties and wild species collected from a broad range of geo-climates and held in trust in the IRRI gene bank in the Philippines. Each variety holds unique attributes that have allowed it to survive in different habitats: Some are tall varieties that can float through floods, some can stay submerged and go into "hibernation," some are resistant to fungal diseases, some are drought tolerant, etc. These attributes may help other varieties survive in other environments. To date, this wealth of information has remained largely untapped because of the difficulty identifying the genes that are most critical to crop production.

Now, breeders are able to sort through the collection of 90,000 rice varieties to identify those that carry genes of interest. Those that do can be tested to see if they do indeed carry a trait that is helpful to crop production. As described previously, different varieties carry similar genes with slightly different variations, and each variant likely has different levels of relative usefulness. For example, a farmer in Africa may like the taste of her grain, and she may be reasonably satisfied with the yield in a good year. Unfortunately, in years of severe droughts, she may lose her crop. To be able to introduce genes from a variety in the gene bank that confers drought resistance would be of great value to this farmer. Similarly, the introduction of genes that confer tolerance to flooding would also benefit poor farmers in Asia where flooding from monsoon rains inundates approximately 20 percent of lowland rice paddies each year.

After sorting through 20,000 rice varieties using a combination of traditional breeding and modern genetic techniques, scientists were able to identify a gene called Sub1 from an Indian variety, which confers tolerance to 14 days of submergence. Recently we carried out fine mapping of this locus and isolated the gene using a positional cloning strategy (Xu et al. 2000, 2006).

Xu et al. (2006) demonstrated that the Sub1a1-1 allele conferred complete submergence tolerance in transgenic rice. The breeding team at IRRI, led by David Mackill, has introduced the gene using precision breeding into cultivars favored by growers in South and Southeast Asia (Xu et al. 2006). Sub1 varieties have already outperformed conventional varieties in farmers' fields in Bangladesh and India, yielding two- to fivefold greater amounts after flooding, and have been adopted at unprecedented rates.

For farmers in California, weed management is the most difficult part of production. One effective weed-control strategy centers on the use of field flooding to suppress weeds directly and to give the crop a competitive advantage. Pregerminated rice seeds are broadcast from airplanes into shallow paddies. This pregermination method requires that farmers use high seeding rates to ensure good stand establishment, but the shallowness of paddies does not adequately discourage weed development.

Although organic growers, who do not use herbicides, control weeds by gradually raising the water level, this approach is not always effective because the plant will die if completely submerged for more than a few days. Development of California rice varieties that can be submerged for long periods of time would benefit both conventional and organic production. Because the Sub1 gene has now been introduced into the background of a widely grown California rice *japonica* cultivar, this variety can now be tested for its effectiveness in weed suppression schemes.

Identification and characterization of Sub1 provide an excellent example of how collaborations between breeders and geneticists can lead to the development of agronomically useful varieties. With the availability of the rice genome sequence, the pace of identifying and characterizing such useful genes will continue to accelerate. The goal of the International Rice Functional Genomics Consortium is to identify the function of every rice gene within 15 years. Once such a catalog is complete, a greater challenge will be to determine the functionality of these genes in the whole plant and in a particular environment. This can be done most efficiently through a combination of modern genetics and traditional breeding. A sustainable rice farming system must incorporate new genes from diverse sources to ensure resistance to disease as well as enabling reduced inputs of fertilizers and pesticides while maintaining productive and healthy rice farms. Thus the challenge ahead for the plant research community is to design efficient ways to tap into the wealth of rice genome sequence information to address production constraints in an environmentally sustainable manner. We now have more information about the rice plant than the breeders of the Green Revolution could have ever dreamed. We have the opportunity to use this information to benefit the maximum number of rice consumers.

Bio-Fortified

Many crops are now vitamin fortified. It is well established that vitamin A deficiency (VAD) is a public health problem in more than 100 countries, especially in Africa and Southeast Asia, hitting hardest young children and pregnant women. Worldwide, more than 124 million children are estimated to be vitamin A deficient. Many of these children go blind or become ill from diarrhea, and nearly eight million preschool-aged children die each year as the result of this deficiency. The World Health Organization (WHO) estimates that improved vitamin A nutritional status could prevent the deaths of 1.3–2.5 million late-infancy and preschool-aged children each year. The heartache of losing a child to a preventable disease is not one commonly encountered in the developed world.

To combat VAD, WHO has proposed an arsenal of nutritional "well-being weapons," which include a combination of breastfeeding and vitamin A supplementation coupled with long-term solutions such as the promotion of vitamin A–rich diets and fortification. In response to this challenge, the Rockefeller Foundation supported a group of scientists who attempted to fortify rice plants with higher levels of carotenoids, the precursors to vitamin A. They introduced a gene from daffodils (which produce many carotenoids to create the pigment that gives the flower its yellow color) and two genes from a bacterium into rice using GE (Ye et al. 2000). The resulting GE rice grains were golden and carotenoid-rich. In a sense, the resulting nutritionally enhanced rice is similar to vitamin D–enriched milk, except the fortification is put there by a different process – GE – which some people view with suspicion. It is also similar to adding iodine to salt, a process credited with drastically reducing iodine-deficiency disorders in infants. Today 94 percent of households in Kazakhstan use iodized salt, and the UN is expected to certify the country officially free of iodine-deficiency disorders.

Eating carotenoid-fortified rice is expected to directly benefit children once it is released in 2012. Preliminary results from human feeding studies suggest that the carotenoids in golden rice can be properly metabolized into the vitamin A that is needed by children (Tang et al. 2009). Other studies also support the idea that widespread consumption of golden rice would reduce VAD and save thousands of lives (Stein, Sachdev, and Qaim 2006). The positive effects of golden rice are predicted to be most pronounced in the lowest income groups at a fraction of the cost of the current supplementation programs (Stein et al. 2006). If confirmed, this relatively low-tech, sustainable, publicly funded, consumer-oriented effort can complement other approaches, such as the development of home gardens with vitamin A–rich crops such as beans and pumpkins.

Conclusion

Many consider rice to be the ideal crop plant for scientific research, and several genome-sequencing projects are now rapidly advancing our understanding of rice genetics. We are gaining tools to develop improved varieties to enhance sustainable farming practices around the world, and we are gaining new respect for this sacred plant.

References

DellaPenna, D. 2001. "Plant Metabolic Engineering." *Plant Physiology* 125 (1): 160–3.

Freeman, D. 1970. *Report on the Iban*. Athlone: Berg.

Goff, S. A., D. Ricke, T. Lan, G. Presting, R. Wang, M. Dunn, et al. 2002. "A Draft Sequence of the Rice Genome (*Oryza sativa* L. ssp. *japonica*)." *Science* 296 (5565): 92–100.

Hamilton, R. 2003. *The Art of Rice: Spirit and Sustenance in Asia*. Los Angeles: University of California.

Huke, R. E., and E. H. Huke. 1990. *Rice: Then and Now*. Manila: IRRI (International Rice Research Institute).

Jung, K., G. An, and P. Ronald. 2008. "Towards a Better Bowl of Rice: Assigning Biological Function to 60,000 Rice Genes." *Nature Reviews Genetics* 9: 91–101.

Khush, G. S., D. S. Brar, P. S. Virk, S. X. Tang, S. S. Malik, G. A. Busto, Y. T. Lee, et al. 2003. *Classifying Rice Germplasm by Isozyme Polymorphism and Origin of Cultivated Rice*. Discussion Paper 46. Los Baños, Philippines: IRRI (International Rice Research Institute).

Olsen, K. M., and M. D. Purugganan. 2002. "Molecular Evidence on the Origin and Evolution of Glutinous Rice." *Genetics* 162 (2): 941–50.

Song, W.-Y., G.-L. Wang, L.-H. Zhu, C. Fauquet, and P. Ronald. 1995. "A Receptor Kinase-Like Protein Encoded by the Rice Disease Resistance Gene Xa21." *Science* 270: 1804–6.

Stein, A. J., H. P. Sachdev, and M. Qaim. 2006. "Potential Impact and Cost-Effectiveness of Golden Rice." *Nature Biotechnology* 24: 1200–1. doi:10.1038/nbt1006-1200b.

Tang, G., J. Qin, G. G. Dolnikowski, R. M. Russell, and M. A. Grusak. 2009. "Golden Rice Is an Effective Source for Vitamin A." *American Journal of Clinical Nutrition* 89 (6): 1776–83. doi:10.3945/ajcn.2008.27119.

Xu, K., X. Xu, P. Ronald, and D. J. Mackill. 2000. "A High-Resolution Linkage Map in the Vicinity of the Rice Submergence Tolerance Locus Sub1." *Molecular and General Genetics* 263 (4): 681–9.

———., X. Xu, T. Fukao, P. Canlas, S. Heuer, J. Bailey-Serres, A. Ismail, P. Ronald, and D. Mackill. 2006. "Sub1A Is an Ethylene-Responsive-Factor Gene That Confers Submergence Tolerance to Rice." *Nature* 442: 705–8. doi:10.1038/nature04920.

Ye, X., S. Al-Babili, A. Klöti, J. Zhang, P. Lucca, P. Beyer, and I. Potrykus. 2000. "Engineering the Provitamin A (Beta-Carotene) Biosynthetic Pathway into (Carotenoid-Free) Rice Endosperm." *Science* 287 (5451): 303–5. doi: 10.1126/science.287.5451.303.

11

Case Study

Healthy Grown Potatoes and Sustainability of Wisconsin Potato Production

Alvin J. Bussan, Deana Knuteson, Jed Colquhoun, Larry Binning, Shelley Jansky, Jiming Jiang, Paul D. Mitchell, Walter R. Stevenson, Russell Groves, Jeff Wyman, Matt Ruark, and Keith Kelling

Earlier chapters have described some of the challenges that genetically engineered crops have faced – many of them rooted in public opinion. However, as this case study demonstrates, even foods produced specifically in a manner to minimize negative impacts on the environment can face resistance from consumers and others along the marketing chain.

The Wisconsin Eco-Potato collaboration developed after several meetings between representatives of the Wisconsin Potato and Vegetable Growers Association (WPVGA) and the World Wildlife Fund (WWF). The representatives had first met at the National Potato Council meeting and identified a common interest in the development of environmentally friendly farming systems. Collaboration activities evolved as an experiment in large-scale, reduced-pesticide agriculture that adopted production practices identified as desirable by consumers (Lynch et al. 2000). The Memorandum of Understanding (MOU) signed between WWF and WPVGA in 1997 included the following goals (Sexson 2006):

1. Increase IPM adoption.
2. Decrease use of high-risk pesticides (both human and environmental risk).
3. Improve biodiversity.
4. Inform public policy.
5. Create marketplace incentives – resulting in the Healthy Grown Potato Program.

The University of Wisconsin–Madison (UW) joined the collaboration in 1999. The International Crane Foundation, Michael Fields Institute, Defenders of Wildlife, and other NGOs joined the collaboration after the initial MOU was signed between the WPVGA and WWF.

The development of the Healthy Grown Potato standard emerged from the fifth goal of the Wisconsin Eco-Potato Collaboration. It determined the need for unbiased evaluation of individual growers as a key component of the Healthy Grown standard. The integrity of the Healthy Grown brand relies on certification by Protected Harvest

(Soquel, CA), an independent entity with no grower or retailer alignment. The third-party certifier monitors current management standards and scrutinizes established standards to ensure that goals of decreased environmental impacts are met. Third-party certification by Protected Harvest involves field inspection of individual farms and evaluation of production and pest management records. Potato growers keep detailed management records that are facilitated by database software developed for tracking field-specific practices for reduced use of and reliance on high-risk pesticides in potato.

Healthy Grown Potato is a voluntary program. Growers choose to adopt the production system in the hope that consumers would voluntarily pay higher prices. In turn higher prices would encourage growers to adopt management schemes with higher potential costs and risks. Conceptually, the enhanced price would be shared across multiple components of the value chain and would reward growers, packing sheds, distributors, and retailers certified by Protected Harvest as meeting the standard criteria. Conversely, costs of implementing Healthy Grown could be recovered by increased market share for those growers certifying potatoes under the standard. Furthermore, marketplace incentives might encourage other growers to adopt integrated pest management (IPM) systems, advancing agriculture from the early adaptors to the broader community. Expanding adoption voluntarily was highly desirable in contrast to using regulation as a means to lead to shifts in production systems. For example, EPA is currently discussing restrictions on the use of metam sodium (EPA 2008). The use of fertilizers now requires nutrient management plans as mandated by the Wisconsin State Legislature and the USDA Natural Resource Conservation Service (WI DATCP 2009).

Standards of Practice for Healthy Grown Potatoes

Initial Healthy Grown standards were drafted to promote adoption of IPM and minimize use of high-risk pesticides, especially those posing greatest environmental and applicator risk. Healthy Grown standards defined by Protected Harvest prohibit a list of pesticides from being used within certified fields. Inclusion on the do-not-use list was based on high toxicity scores, which were calculated based on human health risk and potential environmental effects of a pesticide (Benbrook et al. 2002). On the list were pesticides from the organophosphate, carbamate, and organochlorine chemical families (Table 11.1). The WWF urged that pesticides suspected of causing endocrine disruption in wildlife and humans be restricted as well (Lyons 1999). In addition to the non-use list, the total pesticide use on a Healthy Grown potato field is limited by accumulated toxicity of all pesticides applied. The maximum accumulated toxicity scores for Healthy Grown potato fields are 900 and 1,200 toxicity units for short (<90 day) and long (>90 day) season crops, respectively (Sexson et al. 2004).

Table 11.1. *List of do-not-use pesticides within the Healthy Grown potato production standard*

Common name
Aldicarb
Azinphos-methyl
Carbofuran
Carbaryl
Diazinon
Disulfoton
Endosulfan
Methamidophos
Oxamyl
Paraquat
Permethrin
Phorate

Source: University of Wisconsin (2008).

Finally, Protected Harvest/Healthy Grown standards include resistance management strategies because having a limited list of allowable pesticides increases the likelihood of resistance development (Stevenson 2004). Healthy Grown Potato relies on neonicotinoid insecticides and strobilurin fungicides. Insensitivity of Colorado potato beetle to neonicotinoid insecticides has emerged in potato production areas across the United States (Alyokhin et al. 2006; Mota-Sanchez et al. 2006). Early blight resistance to strobilurin fungicides has been documented in many areas as well (Ma, Felts, and Micailides 2003; Pasche, Wharam, and Gudmestad 2004; Rosenzweig et al. 2008). The IPM and pesticide protocols were developed through research and based on efficacy, economic viability, and environmental risk reductions (Stevenson et al. 1994). Reduced-risk practices were as reliable, efficacious, and economical as conventional practices (Sexson 2002; Stevenson et al. 1994).

In addition to monitoring pesticide use, Healthy Grown potato growers must implement and adopt advanced IPM practices. UW personnel drafted an IPM workbook with management recommendations for every aspect of the production system (Table 11.2; Sexson et al. 2004), based on increasingly advanced IPM practices. Failure to implement IPM practices such as crop rotation of three years or longer leads to fields automatically failing certification. The lowest score is assigned to growers using current best management strategies. Growers adopting more advanced IPM strategies receive higher scores. More advanced strategies remain under development. New research-based information contributes toward the development of standards, and growers must adopt novel management systems. Growers complete yearly self-assessments of IPM adoption, with their score representing the summation of points assigned to each practice used in their season-long production system. Growers must reach minimum IPM scores or the field fails certification.

Table 11.2. *IPM survey categories across different phases of the growing season*

Growing season phase	Management season component
Preplant	Resistance management in rotational years Pest management in rotational crops Soil sampling Potato early dying management Field selection Cultivar selection
Planting	Seed preparation Field preparation Cultural pest management options Resistance management Planting process Fertility
In-season	General IPM Scouting Disease management Insect management Weed management Plant nutrition Irrigation Resistance management Biological control
Harvest	Pest management decisions Environmental conditions
Postharvest	Storage conditions
Farm operations and sustainability	Pest management decisions General operations Energy Workers

Source: Sexson et al. (2004).

Healthy Grown has evolved to include other management system components. Habitat improvement standards were incorporated into the Protected Harvest/Healthy Grown standards in 2007 (Anchor 2007; Vander Zanden, Olden, and Gratton 2006; Zedler et al. 2007). Healthy Grown is one of the first sustainable agriculture programs to require habitat and plant community management standards in noncropland (Knuteson 2007a). Minimal inputs enhance native species within plant communities that maintain the greatest biodiversity. Habitat improvement standards address the management of noncropland to enhance plant community composition and improve ecosystem function. Habitat improvement standards address invasive plant management, restoration of native plant communities, and habitat improvement for wildlife. They target the restoration of natural communities by assessing habitat types and quality from presettlement archives of the Wisconsin landscapes (Knuteson 2007b). These restoration practices improve habitat for threatened species and desirable wildlife. Crop- and noncroplands within farms are evaluated as an ecosystem within the larger landscape to capitalize on the multiple resources available.

Future standards are under development for reduced fumigation, improved water quality and maintenance of water quantity, and enhanced socioeconomic metrics. On-farm research is essential for ensuring that standard practices meet desired outcomes and to assess the costs and risks of implementing the system for the grower.

Data Collection and Management

Farmers enrolled fields into the Healthy Grown program from 2001 through 2008. Each field within a farm was identified and evaluated for certification separately and served as an experimental unit. Fields were only evaluated during the potato production year of the rotation, except for IPM practices related to crop rotation and frequency. Potato management practices certified by Protected Harvest had to comply with state and federal regulations with regard to pesticide and fertilizer use. Noncropland within the farm was enrolled for certification under the natural community standard. Fields were not evaluated for yield or quality or social, labor, or economic factors.

Individual farmers were responsible for data collection and entry into database management software, and Protected Harvest was responsible for verification. Software was developed by SureHarvest, distributed by WPVGA, and Healthy Grown growers were trained by UW. Growers completed surveys at the end of each growing season to quantify IPM practices across their farms (Sexson et al. 2004). Growers were provided a list of IPM practices with scores from 1 to 4 across multiple categories (Table 11.2; Sexson 2006). IPM scores were summed across all categories. Pesticide use was entered for each field, including date of application, product, rate, and target pest. Application of products from the do-not-use list automatically resulted in failed certification. All other products were assigned a toxicity value based on the application rate (Benbrook et al. 2002). Protected Harvest quantified the accumulated toxicity score for all applied products in a given field. If total accumulated toxicity scores for the field were less than the maximum for short or long season crops, then the field was certified as Healthy Grown.

Healthy Grown Farms identified noncropland for certification by Protected Harvest under the natural community standard from 2006 to 2008. Ecologists from UW and the International Crane Foundation evaluated habitat type with site inspection and historical vegetation records of Wisconsin (Knuteson 2007a; Zedler et al. 2007). Management plans were developed to enhance plant communities and return vegetation to the state prior to disturbance by cropping or the introduction of invasive species. Farmers implemented practices targeted by management plans with training

Table 11.3. *Total farms and acres enrolled in Healthy Grown potato program, certification success, and market success of certified potatoes in Wisconsin*

			Certified Healthy Grown[1]			Sold as Healthy Grown	
Year	Growers	Total Hectares	Hectares	Fields	% Passed	Hectares	% of Certified Potatoes
2001	12	3,774	2,028	53	54	17	1
2002	11	4,717	2,157	61	46	75	5
2003	11	3,174	1,886	48	59	127	6.7
2004	11	2,160	1,999	59	93	27	1.4
2005	11	2,747	2,229	58	81	23	1.2
2006	11	2,421	1,882	58	35		
2007	11	2,627	1,991	60	76		
2008	7	2,308	1,223	53	77		
2008	7	3,275	0				

[1] Certified by Protected Harvest.

and assistance by UW and International Crane Foundation personnel. Protected Harvest certified and quantified acres managed in accordance with management plans, and farms recorded the hours required to complete prescribed practices.

The database is retained by the WPVGA and SureHarvest. Data for this evaluation were made available to researchers from UW in anonymous form with no designation of farm or field. Total acres enrolled and certified as Healthy Grown were summed for each year from 2001 to 2008. Toxicity and IPM scores for certified fields were evaluated for means and standard error by field and year. Entire potato industry data were generated by assigning toxicity scores to the National Agricultural Statistics Service (NASS) data on pesticide use in Wisconsin in 2003 and 2005. Total acres enrolled in natural community standard were summarized by habitat type and practice implemented for 2006 to 2008.

Healthy Grown Adoption

Certification by Protected Harvest began in 2001 after all participants agreed on the standards. In the first year of the program 12 farms participated, 11 farms participated from 2002 until 2007, and 7 farms participated in 2008 (Table 11.3). Growers entered 3,774 ha into the program in 2001, with 4,717 ha entered in 2002. After the first two years of the program, enrollment decreased to between 2,100 and 3,500 ha. The high dose of certification and lack of market acceptance of the product led to this reduction. Not all potato fields entered into the Healthy Grown potato program passed certification. The proportion of the fields that met minimum standards ranged from 35 to 93 percent. Rates of certification of less than 60 percent in three of seven years are indicative of the rigorous nature of the certification standards. The variability in

Table 11.4. *Environmental impact of Healthy Grown potato program as quantified by BioIPM score and toxicity unit accumulation of certified acres*

Year	BioIPM score[1]	Toxicity units[2]
2001	208	1,111
2002	237	1,052
2003	237	872
2004	241	925
2005	270	924
2006	282	753
2007	263	796
2008	317	787

[1] IPM scores were calculated from grower surveys on adopted practices across a range of categories.
[2] Toxicity scores were calculated by summing toxicity values for all pesticides applied within each potato field.

grower success in meeting certification criteria was partially due to the variability in the incidence of pests (i.e., late blight and early blight). Years with conditions conducive to the development of disease or with episodic insect pests required more intensive management and pesticide use (Stevenson et al. 1995, 1996). Despite certification rates as low as 35 percent, Healthy Grown growers successfully produced almost 2,000 ha every year with annual production of at least 800,000 T.

Healthy Grown Performance Metrics

Average toxicity scores of potato fields was 1,100 per field in 2001, about 100 units less than the maximum allowed (Table 11.4). Since the program was initiated, the average accumulated toxicity scores of all pesticides decreased every year in certified fields. However, toxicity scores of non-Healthy Grown potato fields also decreased. For example, toxicity scores for pesticide use in Wisconsin decreased from 2,156 in 2003 to 1,560 in 2008 (NASS 2004, 2009). Toxicity scores of fields certified Healthy Grown were nearly 50 percent of the industry average over this time. Failed certification was mostly due to application of pesticides on the do-not-use list, although a few fields were rejected because of toxicity scores that exceeded the maximum limit.

The Healthy Grown potato program benefits from the use of labeled rescue treatments if pest populations or management issues develop. These potatoes can then still be sold in commercial markets. Other management systems disallow or restrict products for use as rescue treatments, which is of particular concern in organic potatoes because limited pest management products are available (OMRI 2010). Most of the toxicity units in potatoes are accumulated by fungicide applications (Benbrook et al. 2002). Potatoes in Wisconsin require 10 to 12 fungicide applications to manage early and late blight (Stevenson et al. 1995, 1996). Insecticides contribute fewer toxicity units than fungicides, and herbicides contribute the least (Benbrook et al. 2002). A major threat to Healthy Grown potatoes is pesticide resistance by key pests (Georghiou 1986; Stevenson 2004). Healthy Grown pest management programs rely on newer pesticides with lower potential environmental and human risks, but higher risk of resistance development. Resistance management is critical for maintaining efficacy of low-risk pesticides (Sexson et al. 2004).

Another important metric relates to the adoption of IPM. Healthy Grown producers achieved IPM scores of just over 200 in 2001 (Table 11.4). IPM scores increased 30 to 40 points in Healthy Grown fields from 2002 to 2004. Every grower increased IPM scores over the six to seven years of the program, with variability in scores from farm to farm. Healthy Grown farmers have achieved their goal of increasing IPM adoption during the first seven years of the program: By 2005, IPM adoption had increased 30 to 40 percent from the beginning of the program in 2001. Preliminary survey data indicated that corresponding increases in IPM adoption were seen in non-Healthy Grown potatoes but not to the extent witnessed in certified potatoes (data not shown). Evaluation of the yearly IPM scores provided sufficient data for targeted educational efforts that contributed to increase adoption of IPM practices over the life of the program.

Natural community standards were implemented in 2006 to quantify management and improvements to noncropland. Noncropland managed under the natural community standard was 69, 130, and 62 ha by six, ten, and seven farms in 2006, 2007, and 2008, respectively. In 2007, noncrop ecosystems managed by Healthy Grown potato farmers included savannah (30 ha), prairie (4.7 ha), oak/pine barrens (75 ha), and sedge meadow (20 ha; Table 11.5). Ecosystems were managed by burning, removal of undesirable brush species, and establishment of native species or improved biodiversity. All ecosystems were also mowed, with the exception of the prairie. Invasive plants, especially garlic mustard and spotted knapweed, were managed in the prairie and sedge meadow. Native vegetation was established on oak/pine barrens. Desirable native plant species appeared to have increased in all restored areas based on preliminary observational data. Healthy Grown farmers spent 564 hours implementing plans for meeting natural community standards. Time reported does not include time required in the generation of the plan.

Costs of the Healthy Grown Program

The Healthy Grown program was developed to meet perceived consumer preference for foods produced with environmental considerations, such as organic vegetables, which receive price premiums. A Healthy Grown program goal was to produce a potato management system with minimal impact on the environment. In exchange, growers

Table 11.5. *Ecosystem types and treatments applied on participating farms for the 2006–2007 season*

Ecosystem type	Area treated (ha)	Prescribed burn		Mechanical cutting		Invasive species control		Establish native vegetation	
		Hours[1]	Hectares	Hours	Hectares	Hours	Hectares	Hours	Hectares
Savanna	30.4	52	14.2	70	20.2	0	0	0	0
Prairie	4.7	16	4.7	0	0	20	4.7	0	0
Oak /Pine Barrens	75.1	74	20.8	80	31.6	0	0	200	27.5
Sedge Meadow	20.2	32	6.1	10	4	10	10.1	0	0
Total	130.4	174	45.8	160	55.8	30	14.8	200	27.5

[1] Hours and hectares affected are listed.

would receive price premiums that would offset adoption and certification costs and create production incentives. The Healthy Grown program costs $60 to $90 per ha for implementation. The program costs were higher partially because several key plant protection products were under patent protection and were more expensive than conventional management programs with no restriction on product use (Sexson 2002). In addition, Healthy Grown has increased management requirements for meeting standards and implementing IPM practices (Sexson et al. 2004). Initial research suggested that Healthy Grown potato yields were more variable than commercially managed potatoes, but risks declined with grower experience. The cost per potato of implementing Healthy Grown is nearly $0.003/kg.

In 2007, as mentioned earlier Healthy Grown potato growers spent 564 hours conducting land management practices to meet natural community standards at a minimal cost of more than $5,000 per farm. The true cost was likely five to ten times that level because growers volunteered time and ecologists were hired on grant funds. Certification of Healthy Grown is also a substantial cost. Estimated costs for 2009 include a $5,000 program technology fee charged to the Wisconsin potato industry that is covered by grants. In addition, each farmer pays a $1,600 participation fee and $0.50/ha charge. Total certification cost is predicted to be $19,700 or $2.80/ha, in addition to the management costs. Other costs include development of packaging, segregation of certified potatoes in storage and packaging, and tracking the finished product.

The cost of developing Healthy Grown potato program also includes the research into management systems, development of the program, and education of farmers on program implementation. Research addressing the Wisconsin Eco-Potato collaboration goals and that related to Healthy Grown program development was initiated in the early 1980s. A succession of potato management software was released including Potato Disease Management, the Wisconsin Irrigation Scheduling Program, Potato Crop Management, and WISDOM (Stevenson et al. 1995, 1996). Since the initiation of the Wisconsin Eco-Potato Collaboration in 1997, it has received more than 20 federal, regional, state, and UW research and extension grants worth more than $3 million (Table 11.6). The Wisconsin Potato Improvement Board has also invested $150,000 to $200,000 in sustainable potato production research grants on an annual basis. The UW College of Agriculture and Life Sciences and Cooperative Extension Service have invested salary and fringe benefits of 15 to 20 researchers involved in the research that generated the Healthy Grown standards. Other costs included establishment of pesticide toxicity scores.

These investments have affected all potato production in Wisconsin – not just certified fields. Survey data have demonstrated shifts in management practices by non-Healthy Grown growers, suggesting benefits to the entire industry. Expansion of Healthy Grown and other sustainable agriculture programs will require continued and

Table 11.6. *Grant funding supporting research and extension programs needed for development and implementation of Healthy Grown potato program in Wisconsin*

Source	Title	Time span	Amount	Work plan – grant purpose
USDA/CSREES/Fund for Rural America	Leveraging community and industry-wide resources to foster biointensive IPM	1997–2001	$100,000	IPM measurement (Nowak, Benbrook), UW communications/coordination area-wide research (Wyman), cover crop research (MacGuidwin) – *research, industry survey, survey development*
USDA/CSREES/IPM	Biointensive IPM establishment and measurement in irrigated vegetables	1998–9	$100,000	IPM measurement (Benbrook), area-wide research (Wyman), cover crop research (MacGuidwin) – *research*
USDA/CSREES Pest Management Alternatives Program	FQPA mitigation strategies for potatoes	2000–2	$149,476	Coloma Farms reduced risk insect & disease research (Wyman, Stevenson) – *research*
USDA/CSREES/ IPM/RAMP	Novel tactics for risk reduction, resistance management, and profit through pest suppression	2001–3	$372,245 for WI	Wisconsin potato research/Florida Glades Crop Care/Benbrook advancing IPM and resistance management; $822,065 total – *research*
USDA/CSREES/ IPM/Methyl Bromide Transition	Using soil solarization to reduce inoculum potential of *Verticillium dahliae* and nematodes in vegetable production systems	2003–6	$100,000	Soil solarization as an alternative to chemical fumigation (MacGuidwin, Sexson) – *research*
EPA/Pesticide Stewardship Program	WPVGA Stewardship Grant	1998–9	$75,000	Survey of growers, Collaboration coordination) – *documentation, outreach*
EPA/American Farmland Trust	Program Plan for Implementing IPM through public–private partnerships; Year 1	2000–1	$150,000	Targeted growers maintain or improve their biointensive IPM scores, reduce or maintain pesticide toxicity reductions. Collaboration coordination – *outreach, management, standards development*
EPA/American Farmland Trust	Program Plan for Implementing IPM through public–private partnerships; Year 2	2001–2	$150,000	Targeted growers maintain or improve their biointensive IPM scores, reduce or maintain pesticide toxicity reductions – *outreach, management, market development*

EPA/American Farmland Trust	Program Plan for Implementing IPM through public–private partnerships; Year 3	2002–3	$150,000	25 growers maintain or improve their biointensive IPM scores, reduce or maintain pesticide toxicity reductions. Economic analysis. BioIPM handbook. Collaboration, coordinating – *Outreach, management, market development*
WDATCP Agricultural Development & Diversification (ADD)	Adding Value through Environmental Marketing: opportunities for WI potato growers	2000–2	$49,700	WPVGA contracted consulting for branding and eco-label (Dlott, Barstow) – *market development*
USDA Congressional	Potato Pest Management: value-added marketing opportunities	2001	$182,400	Establish standard and structure for eco-labeling – *standard and market development*
USDA Congressional	Potato Pest Management	2002	$192,160	Eco-label, ecosystem. and soil quality work – *research, outreach, and management*
USDA Congressional	Potato Pest Management	2003	$185,160	Grower education, ecosystem, coordination – *outreach and management*
USDA Congressional	Potato Pest Management	2004	$164,413	Grower education, ecosystem, coordination – *outreach and management*
UW Consortium Grant	Ecosystem Restoration on WI Potato Farms, with UWSP	2003–5	$46,000	Ecosystem work, part of Ted and student help – *research*
USDA NRI	Managed Ecosystem Program	2005–8	$450,000	Promoting biodiversity in agricultural landscapes: integrating natural habitats with ecologically sound agricultural management – *research, standard development*
USDA North Central Region IPM Center	Potato IPM Working Group for North Central U.S.	2005	$25,000	Extend potato IPM evaluation and materials; monitor resistance in key pests – *research and outreach*
USDA Congressional	Potato Pest Management	2006	$169,517	Grower education, ecosystem, coordination – *outreach and management*
HATCH multistate research formula fund	Pesticide use & risk reduction in vegetable & fruit crops in WI	2007	$146,186	Grower education, ecosystem, coordination – *outreach and management*
USDA Congressional	Potato Pest Management	2008	$131,833	Grower education, coordination, reduced ecosystem – *outreach and management*

increasing investment in agricultural research and extension to ensure success and accomplishment of identified goals.

Market Challenges for Healthy Grown Potatoes

All potatoes grown and certified as Healthy Grown have been intended for the fresh market. Processing (chips or fries) companies have yet to adopt a value-added product based on the Healthy Grown concept, although increasing market pressures by retailers may create demand for chip or processed potato products that are sustainably produced. Potato varieties grown for Healthy Grown include russets such as Russet Norkotah, Silverton, Russet Burbank, and Gold Rush; red-skinned varieties such as Red Norland and Dark Red Norland; and specialty potatoes, such as Yukon Gold and Sierra Gold. Certification does not require information on yield or quality, so the productivity of Healthy Grown potatoes has not been documented. However, little to no difference in yield was observed between Healthy Grown and potatoes managed under conventional production systems in field-scale research (Sexson 2002).

Healthy Grown potato sales have not met expectations. The challenges marketing Healthy Grown potatoes have been related to a lack of consumer education about the program, little produce buyer interest, and store shelf allocation. Healthy Grown potatoes were initially sold at prices similar to standard potato products, but price incentives of $0.01 to $0.02 per kg have been received more recently. The value of Healthy Grown potatoes appears to have covered production costs, but the total cost of packaging must also be accounted for in the price incentive. Sale of Healthy Grown potatoes reached 5 to 6.7 percent of the certified potato production in 2002 and 2003, with little to no price incentives (Table 11.3). Sales declined to just over 1 percent of certified Healthy Grown potatoes in 2004 and 2005 sold under the Healthy Grown label. Increased demand for sustainably produced or locally grown fruits and vegetables may offer opportunities for expanded sales of Healthy Grown potatoes. The collaboration may need to partner with retailers and distributors to generate marketing to consumers about the merits of Healthy Grown. Packing sheds continue to work with produce buyers to expand market share. Healthy Grown was introduced as a new branded product in 2000 and remained that way until 2008. Healthy Grown is now focusing on co-branding strategies with existing, privately owned labels to increase sales and marketability.

Healthy Grown growers continue to develop new standards toward more sustainable production systems. The primary emphasis to date has been on environmental protection and economic sustainability through a shift in pesticide use, implementation of IPM, and natural community standards. Healthy Grown growers are collaborating with researchers in the development of achievable standards for fumigation and nutrient management. Recent discussions focused on water conservation, energy use, and carbon crediting. In addition to environmental standards, social sustainability goals have become a top priority for Healthy Grown potatoes. Economic and social standards relate to an improved quality of life for potato growers by increasing, or at least maintaining, per acre and per unit profitability. In 2008, a farm operations and sustainability module was added to the standard. This section encompasses general operations, use of energy, and workers. Additional future social standards will document labor management efforts related to fair salary or wages, health and other fringe benefits, and opportunities for labor investment into farm enterprises for full-time employees. Temporary or seasonal employee standards include assuring comfortable housing, state-defined worker protection, fair wages, and fringe benefits.

Value of Biotechnology to the Healthy Grown Potato Program

GE potatoes were developed in the late 1980s, with commercial field-scale evaluation conducted during the mid-1990s (Magretta 1997). Initial traits conferred pest resistance as a means to reduce pesticide use in potato production. Traditionally fungicides are applied ten or more times and insecticides three or more times per year to manage early blight and late blight, Colorado potato beetle, potato leafhopper, and other pests. Furthermore, potato quality is defined by suitability for fresh market or processing, and adoption of new varieties has been slow. The primary fresh market and processing potato varieties in Wisconsin are Russet Norkotah and Russet Burbank, which were released 20 and 100 years ago, respectively. Slow development and transition to improved varieties due to specific end use make potato an ideal crop in which GE can incorporate enhanced traits. However, GE potatoes with pest resistance traits were never adopted even though multiple traits were approved by the USDA (Athayde 2001). Acceptance of GE potatoes by food service industries, consumers, and foreign markets will be required before GE varieties return to the production system. GM potatoes are not allowed in the current Healthy Grown standard, and no GM potatoes are grown in the United States.

Pest and herbicide resistance traits could have substantial value to the potato industry. Insect resistance in NewLeaf potatoes was genetically transformed with the *Bt* gene conferring resistance to Colorado potato beetle and potato tuber moth (Coombs et al. 2003; Douches et al. 2001). Colorado potato beetle resistance would reduce in-season insecticide use 75 to 80 percent, with further reductions in pesticide use possible if potato leafhopper and aphid resistance traits were also integrated into the potato. Insect-resistant potatoes would decrease toxicity scores of Healthy Grown potato attributed to insecticides by 65 to 80 percent. However, the *Bt* potato would not reduce toxicity scores accumulated due to insecticides by 100 percent because toxin produced by the *Bt* and other genes would also accumulate toxicity units, but with lower scores. Herbicide resistance would have less impact on toxicity scores because the potato crop would still need to be treated with an herbicide. Glyphosate-resistant potatoes would lower toxicity scores by 50 percent for weed management compared to non-herbicide-resistant potatoes, but this would only reduce total scores for the crop by 5 percent.

Fungicides contribute more to toxicity scores than all other pesticides combined, making up 80 percent of the total. *RB* is a broad-spectrum resistance gene to late blight that was isolated from *Solanum bulbocastanum* and inserted into Katahdin and other potato varieties (Bradeen et al. 2009; Song et al. 2003). Katahdin transformed with *RB* has demonstrated robust resistance to multiple strains of late blight in field and greenhouse trials (Bradeen et al. 2009; Halterman et al. 2008; Kramer et al. 2009; Kuhl et al. 2007). *RB*-transformed potato would still require fungicide application to avoid yield losses due to early blight (Stevenson et al. 2007). *RB*-transformed potato varieties with early blight resistance would reduce the need for fungicides further, but genes conferring early blight resistance have yet to be discovered. Defender potato variety has moderate to high level of resistance to late blight and early blight through conventional breeding (Novy et al. 2006). Defender treated with fungicides every third week of the growing season had a similar yield to potatoes treated every seven days. Planting Defender could allow for 67 percent reduction in pesticide use while maintaining production (Stevenson et al. 2007). Planting early and late blight-resistant potato varieties would reduce toxicity scores by up to 90 percent (Kirk et al. 2005). Unfortunately, Defender has many agronomic, postharvest qualities, and other production issues that limit adoption by growers. Transformation of potatoes with the *RB* gene, early blight resistance, and *Bt* could reduce toxicity scores by more than 95 percent. Other disease-resistance genes such as Potato virus Y, *Verticillium*, common scab, and nematode resistance would further reduce the need for pesticides or improve productivity or crop recovery (Brown et al. 2006; Celebi-Toprak, Slack, and Jahn 2002; Flis et al. 2005; Novy et al. 2004; Sato et al. 2006; Szajko et al. 2008).

The Healthy Grown program has not established standards for nutrient management or fumigation. No suitable alternative IPM has been found. Questions are incorporated in the current potato production portion of the standard that push growers toward the most applicable Best Management Practices (BMPs) for fumigation alternatives and nutrient management. Fumigation is applied to 85 percent of the Wisconsin potato crop. Potato varieties have been identified with tolerance to early dying that would eliminate the need for fumigation (Jansky 2009). Field-scale studies have confirmed that yields of tolerant varieties were similar under fumigated and nonfumigated conditions under low to moderate early dying pressure (LeMere 2007). Identification of early dying resistance genes and transformation of currently suitable varieties could eliminate the need for fumigation and allow development of appropriate standards. Recently, a major gene conferring *Verticillium* wilt resistance has been reported in potato (Bae, Halterman, and Jansky 2008). Commonly grown potato varieties have a high nutrient demand. Many varieties expressing tolerance to early dying have been shown to have lower nitrogen fertilizer requirements. Bannock Russet produced optimum yield at 150 kg N/ha, whereas Russet Burbank required over 200 kg N/ha to optimize yield (LeMere 2007). Early dying resistance and improved nitrogen use efficiency appear to be linked, but whether the same gene is related to each characteristic is unknown.

The Healthy Grown model demonstrates how pest resistance traits incorporated through GE could lower pesticide needs and toxicity scores. The economic advantage to the grower, packer, or other parts of the potato industry is uncertain. Transformed varieties will certainly incur license fees for seed and other potential costs. In addition, pest-resistant potatoes will likely result in increased production through improved efficiency, which could influence the supply–demand relationships. Potato productivity has increased over the last 10–15 years while consumption has declined. Potato growers often sold crops at a loss during the early 2000s, resulting in the loss of a number of farms and a reduction in potato acres. Analysis of the economic impact of GE potatoes must consider not only savings in pesticide costs but also impacts on per unit price and profitability. The potential reduction in labor and reduced exposure of farm laborers to pesticides must also be considered when determining the sustainability of pest-resistant potato varieties developed through GE.

Consumer attitudes toward GE crops with agronomic or pest management traits have ranged from nonchalant to opposition. Potato offers an example of a crop that can be transformed for improved human nutrition. The average U.S. citizen consumes close to 50 kg of potato per year, and potato is still the most widely consumed vegetable worldwide. Potatoes provide vitamin C, potassium, and carbohydrates with minimal fats until fried or covered with condiments. Transgenic potatoes with enhanced levels of vitamin E have been developed (Crowell, McGrath, and Douches 2007). More than 60 percent of potatoes in the United States are consumed as chips or French fries, and almost all potatoes are eaten with salt, butter (or substitute), or sour cream. GE potato with altered flavor, starch composition, sugar composition, or other nutraceutical aspects could have a positive impact on human nutrition. Potato clones have been identified with "resistant" starch that acts as a fiber on consumption and decreases the glycemic index (Jansen et al. 2001). Recently transformed potatoes at UW have lower reducing sugars and acrylamide levels in fried potato. Genes identified in potato with nutraceutical aspects could change the functional role of potato in human nutrition (Crowell et al. 2007; Ducreux et al. 2008; Hovenkamp-Hermelink et al. 1987; Jobling 2004; Visser et al. 1991).

Summary

The Healthy Grown program promoted adoption of sustainable production practices based on consumer preference. The initial plan included a retail price incentive that could be shared throughout the supply chain. Today's increased focus on sustainably produced foods provides an opportunity for Healthy Grown, with economic advantages derived through increased market share rather than value-added price.

GE is a potential means for establishing new standards such as reduced reliance on fumigation or alternative nutrient management practices. GE provides critical tools for developing new potato end products with improved nutritional value. Systematic evaluation of potato and other agricultural production systems will be critical for identifying new goals for improving agricultural sustainability. GE, technological development, grower commitment, collaborations, and targeted research will be critical for the continued evolution of sustainable agricultural systems.

Healthy Grown is an example of a management system derived to reach sustainability goals. It was only achieved through cooperation and collaboration between growers, environmental groups, and researchers to develop standards based on science. For this example, sustainability has been broken down into three components: economic, environmental, and social concerns. The success of implementing Healthy Grown derives from its identification of goals for addressing these three components. Even though successes have been achieved in meeting production standards, challenges remain in the marketing of Healthy Grown. The primary lesson from the Healthy Grown program is the importance of close working relationships with retailers and end users. Partnerships with retailers, processors, and others in the value chain are critical to gaining access in the marketplace. Consumer preference is only part of the equation regarding purchase decisions. Retailers, distributors, and processors must manage risk, turnover, and competition with other stores in produce purchasing decisions. Produce buyers and other retail managers regulate availability to consumers. The future of Healthy Grown's success will require partnerships with produce buyers and raw product managers in the retail and processing sectors. Similarly, GE crops must be acceptable to consumers, retailers, food service, processors, and other key partners. Focusing on traits with consumer benefits rather than production benefits could improve market acceptance.

References

Alyokhin, A., G. Dively, M. Patterson, M. Mahoney, D. Rogers, and J. Wollam. 2006. "Susceptibility of Imidacloprid-Resistant Colorado Potato Beetles to Non-Neonicotinoid Insecticides in the Laboratory and Field Trials." *American Journal of Potato Research* 83 (6): 485–94.

Anchor, T. 2007. "Standards for the Management of Natural Areas on Wisconsin Vegetable Farms." Invited talk at the School of Natural Resources, University of Missouri–Columbia, May 3.

Athayde, M. 2001. "Monsanto Drops Its Biotech Potato." *World Watch* 14: 5–7.

Bae, J., D. Halterman, and S. H. Jansky. 2008. "Development of a Molecular Marker Associated with Verticillium Wilt Resistance in Diploid Interspecific Potato Hybrids." *Molecular Breeding* 22 (1): 61–9.

Benbrook, C. M., D. L. Sexson, J. A. Wyman, W. R. Stevenson, S. Lynch, J. Wallendal, S. Diercks, et al. 2002. "Developing a Pesticide Risk Assessment Tool to Monitor Progress in Reducing Reliance on High-Risk Pesticides." *American Journal of Potato Research* 79: 183–99.

Bradeen, J. M., M. Iorizzo, D. S. Mollov, J. Raasch, L. C. Kramer, B. P. Millett, S. Austin-Phillips, et al. 2009. "Higher Copy Numbers of the Potato RB Transgene Correspond to Enhanced Transcript and Late Blight Resistance Levels." *Molecular Plant-Microbe Interactions* 22 (4): 437–46.

Brown, C. R., H. Mojtahedi, S. James, R. G. Novy, and S. Love. 2006. "Development and Evaluation of Potato Germplasm Resistant to Columbia Root-Knot Nematode (*Meloidogyne chitwoodi*)." *American Journal of Potato Research* 83: 1–8.

Celebi-Toprak, F., S. A. Slack, and M. M. Jahn. 2002. "A New Gene, Nytbr, for Hypersensitivity to Potato Virus Y from *Solanum tuberosum* Maps to Chromosome IV." *Theoretical and Applied Genetics* 104: 669–74.

Coombs, J. J., D. S. Douches, W. Li, E. J. Grafius, and W. L. Pett. 2003. "Field Evaluation of Natural, Engineered, and Combined Resistance Mechanisms in Potato for Control of Colorado Potato Beetle." *Journal of the American Society for Horticultural Science* 128: 219–24.

Crowell, E. F., M. J. McGrath, and D. S. Douches. 2007. "Accumulation of Vitamin E in Potato (*Solanum tuberosum*) Tubers." *Transgenic Research* 17: 205–17.

Douches, D. S., T. J. Kisha, W. Li, W. L. Pett, and E. J. Grafius. 2001. "Effectiveness of Natural and Engineered Host Plant Resistance in Potato to the Colorado Potato Beetle (*Leptinotarsa decemlineata* (Say))." *HortScience* 36: 967–70.

Ducreux, L. J. M., W. L. Morris, I. M. Prosser, J. A. Morris, M. H. Beale, F. Wright, T. Shepherd, et al. 2008. "Expression Profiling of Potato Germplasm Differentiated in Quality Traits Leads to the Identification of Candidate Flavour and Texture Genes." *Journal of Experimental Botany* 59: 4219–31.

EPA (Environmental Protection Agency). 2008. "Chloropicrin, Dazomet, Metam Sodium/Potassium, and Methyl Bromide; Amendments to Reregistration Eligibility Decisions." EPA-HQ-OPP-2008-0518, July 16. *Federal Register* 73 (137): 40871–3.

Flis, B., J. Hennig, D. Strzelczyk-Żyta, C. Gebhardt, and W. Marczewski. 2005. "The Ry-fsto Gene from *Solanum stoloniferum* for Extreme Resistance to Potato Virus Y Maps to Potato Chromosome XII and Is Diagnosed by PCR Marker GP122718 in PVY Resistant Potato Cultivars." *Molecular Breeding* 15 (1): 95–101. doi: 10.1007/s11032-004-2736-3.

Georghiou, P. G. 1986. "The Magnitude of the Resistance Problem." In *Pesticide Resistance Strategies and Tactics for Management*. Washington, DC: National Academy Press.

Halterman, D., L. C. Kramer, S. M. Wielgus, and J. Jiang. 2008. "Performance of Transgenic Potato Containing the Late Blight Resistance Gene RB." *Plant Disease* 92: 339–43.

Hovenkamp-Hermelink J. H. M., E. Jacobsen, A. S. Ponstein, R. G. F. Visser, G. H. Vos-Scheperkeuter, E. W. Bijmolt, J. N. de Vries, B. Witholt, and W. J. Feenstra. 1987. "Isolation of an Amylose-Free Mutant of the Potato (*Solanum tuberosum* L.)." *Theoretical and Applied Genetics* 75 (1): 217–21. doi: 10.1007/BF00249167.

Jansen, G., W. Flamme, K. Schuler, and M. Vandrey. 2001. "Tuber and Starch Quality of Wild and Cultivated Potato Species and Cultivars." *Potato Research* 44 (2): 137–46.

Jansky, S. 2009. "Verticillium Wilt Resistance in U.S. Potato Breeding Programs." *American Journal for Potato Research* 86 (6): 504–12.

Jobling, S. 2004. "Improving Starch for Food and Industrial Applications." *Current Opinion in Plant Biology* 7 (2): 210–18.

Kirk, W. W., F. M. Abu-El Samen, J. B. Muhinyuza, R. Hammerschmidt, D. S. Douches, C. A. Thill, H. Groza, and L. Thompson. 2005. "Evaluation of Potato Late Blight Management Utilizing Host Plant Resistance and Reduced Rates and Frequencies of Fungicide Applications." *Crop Protection* 24 (11): 961–70.

Knuteson, D. 2007a. "Ecologists Take to Potato Fields and Edges." *Badger Common'Tater* 59: 17–19.

———. 2007b. “Potato Growers Restore Native Plant Communities.” *Badger Common’Tater* 59: 22–3.

Kramer, L. C., M. J. Choudoir, S. M. Wielgus, P. B. Bhaskar, and J. Jiang. 2009. “Correlation between Transcript Abundance of the RB Gene and the Level of the RB-Mediated Late Blight Resistance in Potato.” *Molecular Plant-Microbe Interactions* 22 (4): 447–55.

Kuhl, J. C., K. Zarka, J. Coombs, W. W. Kirk, and D. S. Douches. 2007. “Late Blight Resistance of RB Transgenic Potato Lines.” *Journal of the American Society for Horticultural Science* 132 (6): 783–9.

LeMere, M. 2007. “Effect of Fumigation, Supplemental Nitrogen, and Cultivar Resistance on Severity of Early Blight and Early Dying and Crop Quality and Yield in Four Potato Cultivars.” MS Thesis, University of Wisconsin–Madison.

Lynch, S., D. L. Sexson, C. Benbrook, M. Carter, J. Wyman, P. Nowak, J. Barzen, S. Diercks, and J. Wallendal. 2000. “Working the Bugs Out: A Continuing Effort in Wisconsin Models a Promising Pathway toward Addressing Both Public and Producer Concerns over Pesticide Risk and Pest Control.” *Choices* 15 (3): 29–32.

Lyons, G. 1999. “Endocrine Disrupting Pesticides.” *Pesticide News* 46: 16–19.

Ma, Z., D. Felts, and T. J. Micailides. 2003. “Resistance to Azoxystrobin in Alternaria Isolates from Pistachio in California.” *Pesticide Biochemistry and Physiology* 77 (2): 66–74.

Magretta, J. 1997. “Growth through Global Sustainability.” *Harvard Business Review* 75: 79–88.

Mota-Sanchez, D., R. M. Hollingworth, E. J. Grafius, and D. D. Moyer. 2006. “Resistance and Cross-Resistance to Neonicotinoid Insecticides and Spinosad in the Colorado Potato Beetle, *Leptinotarsa decemlineata* (Say) (Coleoptera: Chrysomelidae).” *Pest Management Science* 62 (1): 30–7.

NASS (National Agricultural Statistics Service). 2004. *Agricultural Chemical Usage: 2003 Field Crops Summary*. Washington, DC: USDA. http://usda.mannlib.cornell.edu/usda/nass/AgriChemUsFC//2000s/2004/AgriChemUsFC-05-20-2004.pdf.

———. 2009. *Agricultural Chemical Usage 2008 Field Crops Summary*. Washington, DC: USDA. www.usda.gov/nass/.

Novy, R. G., J. M. Alvarez, D. L. Corsini, A. Nasruddin, E. B. Radcliffe, and D. W. Ragsdale. 2004. “Resistance to PVY, PLRV, PVX, Green Peach Aphid, Colorado Potato Beetle, and Wireworm in the Progeny of a Tri-Species Somatic Hybrid.” *American Journal of Potato Research* 81: 77–8.

———, S. L. Love, D. L. Corsini, J. J. Pavek, J. L. Whitworth, A. R. Mosley, S. R. James, et al. 2006. “Defender: A High-Yielding, Processing Potato Cultivar with Foliar and Tuber Resistance to Late Blight.” *American Journal of Potato Research* 83: 9–19.

OMRI (Organic Materials Review Institute). 2010. “OMRI Products List, Web Edition.” Last modified November 8. http://www.omri.org/sites/default/files/opl_pdf/complete_company.pdf.

Pasche, J. S., C. M. Wharam, and N. C. Gudmestad. 2004. “Shift in Sensitivity of *Alternaria solani* in Response to QoI Fungicides.” *Plant Disease* 88: 181–7.

Rosenzweig, N., G. Olaya, Z. K. Atallah, S. Cleere, C. Stanger, and W. R. Stevenson. 2008. “Monitoring and Tracking Changes in Sensitivity to Azoxystrobin Fungicide in *Alternaria solani* in Wisconsin.” *Plant Disease* 92: 555–60.

Sato, M., K. Nishikawa, K. Komura, and K. Hosaka. 2006. “Potato Virus Y Resistance Gene, Rychc, Mapped to the Distal End of Potato Chromosome 9.” *Euphytica* 149: 367–72.

Sexson, D. 2002. “Economic Analysis of BioIPM Programs in Wisconsin Potatoes.” *Proceedings of the Wisconsin Crop Management Conference*, Madison, WI, January 17. http://www.soils.wisc.edu/extension/wcmc/2002proceedings/Sexson-Conf-2002.pdf.

———. 2006. *Companion Documentation for the Eco-Potato Standards.* WWF/WPVGA/UW (World Wildlife Fund/Wisconsin Potato and Vegetable Growers Association/University of Wisconsin) Collaboration. http://ipcm.wisc.edu/LinkClick.aspx?fileticket=%2fpiedC5K%2fp4%3d&tabid=87&mid=643.

———, T. Connell, A. J. Bussan, K. Kelling, W. R. Stevenson, and J. Wyman. 2004. *Potato IPM Workbook.* Madison: University of Wisconsin.

Song, J., J. M. Bradeen, S. K. Naess, J. A. Raasch, S. M. Wielgus, G. T. Haberlach, J. Liu, et al. 2003. "Gene RB Cloned from *Solanum bulbocastanum* Confers Broad Spectrum Resistance to Potato Late Blight." *Proceedings of the National Academy of Sciences of the United States of America* 100: 9128–33.

Stevenson, W. R. 2004. "Role of Extension in Management of Pest Resistance." In *Management of Pest Resistance: Strategies Using Crop Management, Biotechnology and Pesticides,* edited by B. J. Jacobsen and S. R. Matten, 116–17. CAST (Council of Agricultural Science and Technology) Special Publication No. 24. Ames, IA: CAST.

———, L. K. Binning, T. R. Connell, J. A. Wyman, and D. Curwen. 1996. *Integrated Pest Management WISDOM – Professional Software for Agricultural Systems, Version 1.31.06.* Madison: University of Wisconsin Press.

———, D. Curwen, K. A. Kelling, J. A. Wyman, L. K. Binning, and T. R. Connell. 1994. "Wisconsin's IPM Program for Potato: The Developmental Process." *HortTechnology* 4: 90–5.

———, R. V. James, D. A. Inglis, D. A. Johnson, R. T. Schotzko, and R. E. Thornton. 2007. "Fungicide Spray Programs for Defender – A New Potato Cultivar with Resistance to Late Blight and Early Blight." *Plant Disease* 91: 1327–36.

———, J. A. Wyman, K. A. Kelling, and L. K. Binning. 1995. "Prescriptive Crop and Pest Management Software for Farming Systems Involving Potatoes." In *Ecology and Modeling of Potato Crops under Conditions Limiting Growth,* edited by A. J. Haverkort and D. K. L. MacKerron, 291–304. Norwell, MA: Kluwer Academic Publishers.

Szajko, K., M. Chrzanowska, K. Witek, D. Strzelczyk-Żyta, H. Zagorska, C. Gebhardt, J. Hennig, and W. Marczewski. 2008. "The Novel Gene Ny-1 on Potato Chromosome IX Confers Hypersensitive Resistance to Potato Virus Y and Is an Alternative to Ry Genes in Potato Breeding for PVY Resistance." *Theoretical and Applied Genetics* 116 (2): 297–303.

University of Wisconsin. 2008. *Wisconsin Eco-Potato Standard.* Madison: University of Wisconsin. http://ipcm.wisc.edu/LinkClick.aspx?fileticket=%2bbe09M%2fHXHg%3d&tabid=87&mid=643.

Vander Zanden, J., J. Olden, and C. Gratton. 2006. "Food Web Approaches in Restoration Ecology." In *Foundations of Restoration Ecology,* edited by D. A. Falk, M. A. Palmer, and J. B. Zedler, 165–89. Washington, DC: Island Press.

Visser, R. G. F., I. Somhorst, G. J. Kuipers, W. J. Feenstra, and E. Jacobsen. 1991. "Inhibition of the Expression of the Gene for Granule-Bound Starch Synthase in Potato by Antisense Constructs." *Molecular and General Genetics* 225 (2): 289–96.

WI DATCP (Wisconsin Department of Agriculture Trade and Consumer Protection). 2009. *Soil Water and Resource Management Program.* Chapter ATCP 50, 329–62. Madison: Wisconsin Department of Agriculture.

Zedler, P. H, C. Gratton, D. Knuteson, N. Mathews, J. Barzen, T. Anchor, H. Gaines, M. V. Knight, and L. Nye. 2007. "Conservation and Restoration in an Ecological Landscape: An Eco-Label Focused Approach." Paper presented at the Ecological Society of America/Society for Ecological Restoration Meeting, San Jose, CA, August 5–10.

12

Precautionary Practice of Risk Assessment

Caroline (Cal) Baier-Anderson and Michelle Mauthe Harvey

> The outstanding scientific discovery of the twentieth century is not television, or radio, but rather the complexity of the land organism. The last word in ignorance is the man who says of an animal or plant: "What good is it?" If the land mechanism as a whole is good, then every part is good, whether we understand it or not. If the biota, in the course of aeons, has built something we like but do not understand, then who but a fool would discard seemingly useless parts? To keep every cog and wheel is the first precaution of intelligent tinkering.
>
> *Leopold (1966, 190)*

Every day we are asked to make decisions regarding safety and risk based on limited information. Over the years we have developed risk assessment methods to help us assemble, organize, and formally evaluate the information that is available, but ultimately our decisions will require that we also balance what we know with what we don't know. Where the bar is set is ultimately a philosophical question, because we will never have all the science that we want or need.

The Environmental Defense Fund (EDF) is a science-based organization founded by scientists more than 40 years ago. Ever since, our biologists, chemists, engineers, and physicists have used science to help cut through political logjams. Whether we provide original research or enlist the expertise of top specialists in the field, we use cutting-edge science and creative thinking to solve tough environmental problems.

We are also about markets. Market competition has always been one of the most powerful engines of American innovation; the EDF long ago found ways to harness that engine for environmental progress. However, when science and markets appear to conflict, threatening harm to people and planet, we pause. We suggest that the strategic use of the precautionary principle can represent the appropriate risk management response, particularly when the scientific data indicate concern but lack sufficient information to permit high-confidence quantitative assessment.

The EDF believes that the degree of precaution applied in any given circumstance should be inversely proportional to the maturity of the science but commensurate with the magnitude and reversibility of the potential adverse impacts. Ultimately, we recognize that the context in which decisions are made include the social, economic, and community implications as well as the scientific, leaving open opportunities for more nuanced responses to risk. Keeping in mind that there are questions we may never be able to answer to our complete satisfaction, the nanotechnology case study presented later in this chapter provides an example of the EDF's approach to balancing potential risk with precaution.

Understanding the Precautionary Principle

Some still try to frame the false dichotomy: "science or precaution." Contrary to misconceptions, the intent of the precautionary principle is *not* to "prevent or restrict actions that raise even conjectural threats of harm to human health or the environment" (Miller and Conko 2001, 302). Rather, the precautionary principle starts with scientific review and comes into play when that review results in an unquantifiable possibility of unacceptable levels of harm. If the probability of harm is quantifiable, the prevention principle applies and risk management follows, with acceptable levels of risk determined and actions taken to maintain risk at a lower level (UNESCO 2005).

The precautionary principle finds its roots in European legislation of the early 1970s (UNESCO 2005), but it emerged prominently during the 1992 Rio Conference, also known as the "Earth Summit." Principle #15 of the Rio Declaration states,

> In order to protect the environment, the precautionary approach shall be widely applied by States according to their capabilities. Where there are threats of serious or irreversible damage, lack of full scientific certainty shall not be used as a reason for postponing cost-effective measures to prevent environmental degradation.
>
> *United Nations (1992)*

Six years later, in January 1998, the Science and Environmental Health Network (SEHN) hosted a conference at Wingspread in Racine, Wisconsin, and issued this statement on the precautionary principle:

> When an activity raises threats of harm to human health or the environment, precautionary measures should be taken even if some cause and effect relationships are not fully established scientifically.
>
> In this context the proponent of an activity, rather than the public, should bear the burden of proof.
>
> The process of applying the precautionary principle must be open, informed and democratic and must include potentially affected parties. It must also involve an examination of the full range of alternatives, including no action.
>
> *SEHN (1998, para. 5–7)*

In 2000, the European Union (EU) communication on the precautionary principle stated,

> The precautionary principle applies where scientific evidence is insufficient, inconclusive or uncertain and preliminary scientific evaluation indicates that there are reasonable grounds for concern that the potentially dangerous effects on the environment, human, animal or plant health may be inconsistent with the high level of protection chosen by the EU (Commission of the European Communities 2000, 8).

In 2005, the World Commission on the Ethics of Scientific Knowledge and Technology (COMEST) developed this working definition of the precautionary principle (COMEST 2005, 14):

> When human activities may lead to morally unacceptable harm that is scientifically plausible but uncertain, actions shall be taken to avoid or diminish that harm. Morally unacceptable harm refers to harm to humans or the environment that is
>
> 1. threatening to human life or health, or
> 2. serious and effectively irreversible, or
> 3. inequitable to present or future generations, or
> 4. imposed without adequate consideration of the human rights of those affected.

The judgment of plausibility should be grounded in scientific analysis. Analysis should be ongoing so that chosen actions are subject to review. Uncertainty may apply to, but need not be limited to, causality or the bounds of the possible harm. Actions are interventions that are undertaken before harm occurs that seek to avoid or diminish the harm. Actions should be proportional to the seriousness of the potential harm, with consideration of their positive and negative consequences and an assessment of the moral implications of both action and inaction. The choice of action should be the result of a participatory process.

"Morally unacceptable harm." "Scientifically plausible but uncertain." "Grounded in scientific analysis." "Subject to review." "Proportional to the seriousness of the potential harm." "Participatory." These phrases are the hallmarks of precaution.

Risk Assessment Contexts

The process of assessing risk as the probability of an adverse outcome is an everyday activity, although generally the risk calculations – whether to drive to the supermarket in a snowstorm or allow your child to walk to school alone for the first time – are conducted in an informal manner. More formal risk assessments are used as tools to inform regulatory decisions, and the process of collecting and analyzing data for the purposes of quantitative risk assessment has evolved over time as science and society have changed. Many environmental health and safety laws require risk assessments to be conducted or imply that decisions should be risk-based (NRC 2008; Roberts and Abernathy 1996).

Much literature exists concerning factors that influence risk perception: whether the risks are voluntary versus involuntary, natural versus artificial, exotic versus unfamiliar, and the like (MacGregor, Slovic, and Malmfors 1999; Renn 2004). This chapter does not revisit this important aspect, but we do recognize that risk perception plays an important role in public attitudes and regulatory decision making.

Although risk assessments for different purposes (e.g., to analyze the risks associated with airline travel or the risks of adverse effects associated with pharmaceuticals) may differ in the details, all risk assessments share the general framework whereby the risk is a function of magnitude and likelihood. For chemical risks, this translates to hazard (likelihood of health effect) x exposure (duration, frequency, magnitude, and timing). The basic structure of chemical risk assessments for regulatory decision making as a framework involves four steps: hazard identification, dose-response assessment, exposure assessment, and risk characterization (NRC 1983).

More recently, the NRC explained how risk assessment "involves a search for 'causal links' or 'causal chains' verified by 'objective' analytic and experimental techniques," which in the case of chemical risk assessment means quantifying exposure and hazard dose-response data (NRC 2007, 13). The characterization of risk provides important information to help regulatory agencies determine whether action to abate or mitigate risks is necessary or to evaluate available options to achieve those goals. Risk communication is the process of explaining the nature and magnitude of the risks and risk management options to the public.

Hazard identification may be based on the study of the effects of chemicals on laboratory animals or on observations from human exposures, such as those that may occur in an occupational setting. As our understanding of biological processes evolves, scientists have identified new endpoints, prompting debate regarding what type of endpoint constitutes an "adverse effect" for the purposes of risk assessment. Responding to this debate, the NRC noted that "a goal of public health is to control exposures before the occurrence of functional impairment of the whole organism" and that "dividing effects into dichotomous categories as adverse and non-adverse is problematic" (NRC 2007, 56). Instead, it is important to understand the context in which precursor events can be related to overt adverse events.

How Much Is Too Much?

An often-cited basic toxicological principle is that the dose makes the poison, a concept first stated by Paracelsus in the 16th century. Dose-response assessment has recently come under increased intense scrutiny. Although dose response was traditionally seen as either a linear or threshold response, we now understand that more complex dose-response relationships exist. These dose-response relationships for a chemical substance can vary by endpoint. They are also predicated on the timing of the dose, because transient windows of sensitivity exist in biological development given the characteristics of the receptor (which include, among other factors, age, sex, genetic makeup, co-exposures, and co-morbidities). The quantification of such additional stressors and their potential impacts on health outcomes is a challenging, yet critical, component of risk assessment (NRC 2008).

Exposure assessment is the process of describing how a population may be exposed to a chemical substance, including the pathways by which the population may encounter the chemical and the magnitude, frequency, and duration of the exposure. Exposure can be very difficult to predict, and a few notable failures have catapulted chemical nomenclature into the public lexicon. For example, finding bisphenol A, brominated flame retardants, and perfluorooctanoic acid in fetal cord blood – all chemicals found in common consumer products – raised important questions regarding the validity of common assumptions used to predict human exposures (Apelberg et al. 2006; Ikezuki et al. 2002; Mazdai et al. 2003).

Over the past two decades, a variety of community, environmental, and public health groups[1] have raised questions regarding the soundness of using risk assessment as the principal tool for environmental decision making. Two factors have substantially added to the urgency of need for an alternative approach: (1) a growing sense that chronic diseases are increasing at a pace not entirely explained by better tracking or genetic predisposition and (2) delays in regulatory action as risk assessments were required to become more precise and certain (Myers 2006).

At the same time, advances in science are calling into question the protectiveness afforded by a reliance on traditional toxicological endpoints. However, new scientific methods can take a long time to be incorporated into the compendium of accepted test methods, because the process of validation can be rigorous and time consuming. For example, it has taken more than a decade to validate test methods to be used in EPA's endocrine disruptor screening program (EPA 2010). Although some delays may be unavoidable, there appears to be growing agreement that public health can and must be protected while data and evidence continue to be gathered.

Part of the criticism of risk assessments is that they presuppose that some level of harm is acceptable (e.g., one incident of an adverse effect in a million or one in 10,000). Increasingly, with advances in analytical methods, achievement of a level of zero residual, and hence, zero risk, is technically infeasible. Similarly, drugs designed to treat disease have side effects and are therefore not without risk. In most instances, risks associated with the initial introduction of chemicals into commercial or consumer products might be "acceptable," yet consumers are not given an opportunity to address this question.

Drawing the Line

An inherent tension exists between the fluid nature of advances in scientific understanding and the need to draw a "line in the sand" for the purposes of regulatory decision making. On the one hand, we want decision makers to use the best and most

[1] For instance, see critiques of risk assessment by Clean Production Action (2010) and SEHN (1998).

recent scientific information. On the other hand, delays in regulation caused by the collection of more data could lead to unnecessarily prolonged exposures, unnecessarily increased risks to certain populations, or lost market opportunities. This tension has been successfully exploited by those interested in strengthening barriers to chemical regulation (Mooney 2004), particularly with the passage in 2001 of the Data Quality Act (Public Law 106-554), a law that requires the White House's Office of Management and Budget (OMB) to develop and oversee the implementation of policies, principles, standards, and guidelines applicable to the dissemination of public information by federal agencies.

On the surface, this law seems reasonable, because it requires federal agencies to accomplish the following goals by enacting data quality guidelines:

1. Ensure and maximize the quality, objectivity, utility, and integrity of information including statistical information prior to dissemination.
2. Allow affected individuals and/or organizations to seek and obtain correction of information maintained and disseminated by the agency that does not comply with OMB or agency guidelines.
3. Report to OMB regarding the number and nature of complaints received by the agency regarding agency compliance with OMB guidelines. (SKAPP 2010, para. 3)

In fact, data quality challenges have been used to derail or delay implementation of environmental regulations (OMBWatch 2007). A U.S. Government Accountability Office (GAO) investigation determined that it can take up to two years to address these data quality challenges (GAO 2006). Thus, competing interests increase pressure on federal risk assessors.

Several high-profile risk assessments of toxic chemicals have been under development for many years, prompting a congressional hearing on the reasons behind the delays. According to a news report about the hearing, the EPA has not been able to complete assessments for key chemicals of public health concern, including dioxins, formaldehyde, naphthalene, trichloroethylene, and tetrachloroethylene (Lewis 2008). It was revealed during the hearing that a backlog of assessments exists for EPA's Integrated Risk Information System, a database that summarizes chemical hazard information for use in risk assessments. The GAO indicated that these delays were due to a variety of reasons, but noted that 48 risk assessments had been in process for more than five years and 12 of those for longer than nine years (GAO 2008). Some have suggested that this delay is due to political interference through the Office of Management and Budget (Lewis 2008). It is within this context – evolving science, surprising exposures, and challenges to regulatory risk assessment – that the precautionary principle should be viewed.

Integrating the Precautionary Principle and Risk Assessment

In the view of the Environmental Defense Fund, the precautionary principle and risk assessment are not mutually exclusive. In fact, some have argued that many U.S.

environmental laws are inherently precautionary (e.g., the National Environmental Policy Act, which requires consideration of consequences before implementing a federally funded project and consideration of alternatives, including taking no action; Myers 2006).

All risk assessments include uncertainty (because of a lack of complete knowledge) and variability (an inherent characteristic of the population; NRC 2008). Although uncertainty can be reduced with additional information collection, population variability will always be present. The NRC has recommended that risk assessments use qualitative and quantitative approaches to address both uncertainty and variability in each step of the risk assessment process. We suggest that the consideration of uncertainty and variability in the risk assessment represents an implicit application of precaution, whereby significant uncertainty or variability requires a greater use of precaution in the risk assessment.

Integrating Precaution into Risk-Based Approaches

The positive and negative potentials of nanotechnology products provide a good case study for the use of precaution in risk-based assessments. Nanotechnology is a relatively new area of knowledge that promises a dazzling array of opportunities in areas as diverse as manufacturing, energy, health care, and waste treatment. Given the enormous commercial and societal benefits that may potentially come from this technology, nanomaterials and their products and applications will likely be widely produced and used.

In the *Wall Street Journal* on June 14, 2005, DuPont CEO Chad Holliday and EDF President Fred Krupp jointly called for broad collaboration by interested stakeholders to identify and address potential environmental, health, and safety risks of nanotechnology. They noted that "the novel properties that make nanoparticles so promising" could affect human health and ecosystems in unexpected ways. They called for "sound, disciplined research and commercialization guided by thoughtful regulatory standards." Krupp and Holliday also noted that we have reasons to be cautious, citing the previous use of innovative new discoveries – chlorofluorocarbons, DDT, leaded gasoline – to replace problematic materials only to later discover their harmful unintended consequences. They concluded that "an early and open examination of the potential risks of a new product or technology is not just good common sense – it's good business strategy" (Krupp and Holliday 2005, B2).

The EDF and DuPont subsequently entered into a partnership to develop the Nano Risk Framework (NRF). It was subsequently organized by a multidisciplinary team from both organizations, including experts in biochemistry, toxicology, environmental sciences and engineering, medicine, occupational safety and health, environmental law and regulations, product development, and business development.

For two years, the EDF and DuPont collaborated to develop a comprehensive, user-friendly framework for evaluating and addressing the environmental, health, and safety risks of nanomaterials across all stages of a product's lifecycle – from initial sourcing through manufacture, use, and recycling or disposal. The interdisciplinary team also consulted a wide range of stakeholders and pilot-tested the Nano Risk Framework on several materials and applications. Although the NRF builds on the traditional risk assessment paradigm, it does include the following precautionary aspects:

1. The NRF encourages evaluation of potential risks across the value chain, rather than being limited to workers in manufacture or consumer uses.
2. The evaluation of risks, including intrinsic hazard, environmental fate and transport, and potential exposures, should be fully evaluated before taking the nanomaterials to market.
3. The NRF recommends the use of reasonable worst case scenarios when critical information is lacking.
4. The evaluation should be transparent.
5. Recognizing that the accrual of information is fluid, the NRF takes an iterative, rather than static approach, so that information is continually updated and responses to the analytical results adjusted. (Medley and Walsh 2007)

The NRF is not a regulation but it demonstrates that such proactive assessments can help companies make better decisions. Approximately 7,000 copies of it are currently in circulation, including versions in French, Spanish, and Chinese.

Companies have used the NRF in a variety of ways. Some organizations, including DuPont, Nanostellar, and University of Massachusetts Lowell, have implemented it exactly as designed and worked through its principles to make decisions about specific materials and products. GE, Procter & Gamble, Lockheed Martin, and Lloyd's report that they have incorporated elements of the framework into their existing practices, updating their own processes with specific details from the NRF. Finally, other companies including GlaxoSmithKline and Intel have reviewed and referenced the NRF as a benchmark as they develop or update their own corporate policies.

The framework implicitly incorporates a precautionary approach, although it avoids the term "precautionary principle" because of the confusion often associated with this term. Moreover, the NRF demonstrates that risks can be assessed in the absence of complete information and public health can be protected if precaution is used when developing assumptions.

A Matter of Timing

Causation is difficult to prove; prevention is easier to implement. The cost of attaining the requisite burden of proof is incompatible with the needs of a regulatory system that must address tens of thousands of chemicals. We have yet to address the problem of multiple chemical exposures, for which standard risk assessment techniques are lacking.

We need to develop regulatory approaches that allow for more fluid decision making. The integration of precaution into decision making makes the most sense, given the needs of regulators, evolving science, and concern with protecting the most sensitive members of our population.

In the meantime, rather than waiting for regulatory reform or scientific certainty, more companies are choosing to take precautionary action by eliminating or reducing their use of hazardous chemicals in consumer products as well as throughout their value chain. While protecting workers, consumers, and the environment, the implementation of precautionary approaches to risk management can also result in net economic benefits to companies by reducing the cost of handling, storing, transporting, and disposing of hazardous materials (Torrie 2009).

The EDF shares the view that the precautionary principle is a useful risk management concept, particularly when information regarding hazard or exposure is provocative yet incomplete. Implementing a precautionary decision to restrict material use in the face of key data gaps could actually provide an incentive for the prompt generation of additional data by manufacturers in the hope that additional data might result in less restrictive use requirements, or it may spur the development of less hazardous alternatives, creating new markets and economic opportunities. If the additional data "exonerate" the material, then indeed, loosening restrictions may be warranted. In contrast, if the additional data validate initial concerns, then the precautionary action served an important role in limiting or preventing risky exposures.

References

Apelberg, B. J., L. R. Goldman, A. M. Calafat, J. B. Herbstman, Z. Kuklenyik, J. Heidler, L. L. Needham, R. U. Halden, and F. R. Witter. 2006. "Determinants of Fetal Exposure to Polyfluoroalkyl Compounds in Baltimore, Maryland." *Environmental Science & Technology* 41 (11): 3891–7.

Clean Production Action. 2010. "Differences between Precautionary Principle and Risk Assessment." http://www.cleanproduction.org/Steps.Precautionary.Differences.php.

COMEST (World Commission on the Ethics of Scientific Knowledge and Technology). 2005. *The Precautionary Principle*. Paris: UNESCO.

Commission of the European Communities. 2000. *Communication from the Commission on the Precautionary Principle*. COM(2000)1. Brussels: European Commission. http://ec.europa.eu/dgs/health_consumer/library/pub/pub07_en.pdf.

EPA (Environmental Protection Agency). 2010. "Endocrine Disruptor Screening Program (EDSP)." Washington, DC: EPA. http://www.epa.gov/oscpmont/oscpendo/index.htm.

GAO (U.S. Government Accountability Office). 2006. *Information Quality Act: Expanded Oversight and Clearer Guidance by the Office of Management and Budget Could Improve Agencies' Implementation of the Act*. Report to Congressional Requesters GAO-06-765. Washington, DC: GAO. http://www.gao.gov/new.items/d06765.pdf.

______. 2008. *Toxic Chemicals: EPA's New Assessment Process Will Increase Challenges EPA Faces in Evaluating and Regulating Chemicals*. GAO-08-743T. Washington, DC: GAO. http://www.gao.gov/products/GAO-08-743T.

Ikezuki, Y., O. Tsutsumi, Y. Takai, Y. Kamei, and Y. Taketani. 2002. "Determination of Bisphenol A Concentrations in Human Biological Fluids Reveals Significant Early Prenatal Exposure." *Human Reproduction* 17 (11): 2839–41. doi: 10.1093/humrep/17.11.2839.

Krupp, F., and C. Holliday. 2005. "Let's Get Nanotech Right." *Wall Street Journal*, June 14, B2.

Leopold, A. 1966. *A Sand County Almanac: With Essays on Conservation from the Round River*. New York: Ballantine Books.

Lewis, S., ed. 2008. "Congress Probes White House Delay of Chemical Assessments." *Environmental News Service*, June 12. http://www.ens-newswire.com/ens/jun2008/2008-06-12-093.asp.

MacGregor, D. G., P. Slovic, and T. Malmfors. 1999. "How Exposed is Exposed Enough? Lay Inferences about Chemical Exposure." *Risk Analysis* 19 (4): 649–59.

Mazdai, A., N. G. Dodder, M. P. Abernathy, R. A. Hites, and R. M. Bigsby. 2003. "Polybrominated Diphenyl Ethers in Maternal and Fetal Blood Samples." *Environmental Health Perspectives* 111 (9): 1249–52.

Medley, T., and S. Walsh. 2007. *Nano Risk Framework*. Environmental Defense–DuPont Nano Partnership. Wilmington and Washington, DC: DuPont and Environmental Defense Fund. http://www.edf.org/documents/6496_Nano%20Risk%20Framework.pdf.

Miller, H. I., and Conko, G. 2001. "Precaution without Principle." *Nature Biotechnology* 19: 302–3.

Mooney, C. 2004. "Paralysis by Analysis: Jim Tozzi's Regulation to End All Regulation." *The Washington Monthly*, May. http://www.washingtonmonthly.com/features/2004/0405.mooney.html.

Myers, N. J. 2006. Introduction. In *Precautionary Tools for Reshaping Environmental Policy*, edited by N. J. Myers and C. Raffensperger, 1–7. Cambridge: MIT Press.

NRC (National Research Council). 1983. *Risk Assessment in the Federal Government: Managing the Process*. Washington, DC: National Academy Press.

______. 2007. *Scientific Review of the Proposed Risk Assessment Bulletin from the Office of Management and Budget*. Washington, DC: National Academy Press.

______. 2008. *Science and Decisions: Advancing Risk Assessment*. Washington, DC: National Academy Press.

OMBWatch. 2007. "Data Quality Act." http://www.ombwatch.org/node/3479.

Renn, O. 2004. "Perception of Risks." *Toxicology Letters* 149 (1–3): 405–13.

Roberts, W. C., and C. O. Abernathy. 1996. "Risk Assessment: Principles and Methodologies." In *Toxicology and Risk Assessment: Principles, Methods, and Applications*, edited by A. M. Fan and L. W. Chang, 245–70. New York: Marcel Dekker.

SEHN (Science and Environmental Health Network). 1998. "The Wingspread Consensus Statement on the Precautionary Principle." January 26. http://www.sehn.org/wing.html.

SKAPP (Project on Scientific Knowledge and Public Policy). 2010. "Information Quality Act: History and Guidelines." Accessed December 16. http://www.defendingscience.org/public_health_regulations/Information-Quality-Act-History-and-Guidelines.cfm.

Torrie, Y. 2009. *Best Practices in Product Chemicals Management in the Retail Industry: Moving Business towards Safer Alternatives*. With contributions by M. Buczek, G. Morose and J. Tickner. Lowell, MA: Green Chemistry and Commerce Council. http://greenchemistryandcommerce.org/projects.php.

UN (United Nations). 1992. *Annex I: Rio Declaration on Environment and Development*. Report of the United Nations Conference on Environment and Development, Rio de

Janeiro, Brazil, June 3–14. http://www.un.org/documents/ga/conf151/aconf15126-1annex1.htm.

UNESCO (United Nations Educational, Scientific, and Cultural Organization). 2005. *The Precautionary Principle; World Commission on the Ethics of Scientific Knowledge and Technology (COMEST)*. Paris: UNESCO.

13

Risk Assessment Approaches and Implications

José Falck-Zepeda and Anthony J. Cavalieri

The Cartagena Protocol on Biosafety (CPB) and national laws and regulations mandate that GE crops undergo risk assessment before release into the environment. This chapter explores methods for assessing risks of crop biotechnologies and identifies the role of risk management and regulation as cultivars with new traits are developed in the coming years. Extensive experience with GE crop development and risk assessment in developed countries should inform the spread of useful GE traits to farmers in poorer countries. We address these issues in light of two important trends that are affecting the development of a globally sustainable food system: first, the use of biotechnology to develop cultivars containing traits that ameliorate the major constraints to agricultural productivity including abiotic stress resistance, and second, the emergence of the public sector as developers of small market crops and staple crops for the developing world. We consider the impact of the cost of compliance to biosafety and other regulations on investment in biotechnological crop improvement by both the private and public sector and its eventual impacts on a sustainable global food supply. We pay particular attention to the development of risk assessment methods that are scientifically valid, broadly acceptable, and economically feasible.

GE Crops as Part of a Sustainable Food Supply

Higher yields achieved through agricultural intensification represent one of the few approaches for increasing food production and reducing habitat destruction resulting from expanding agriculture. It is essential that intensification does not have negative effects on water quality and biodiversity. GE crops can also make an important contribution to a sustainable food supply (Byerlee et al. 2007; FAO 2004; Pew Initiative on Food and Biotechnology 2004). It is critically important to solve the risk management issues in a cost-effective method that has application in countries with limited scientific capacity and limited resources.

Emergence of Public Sector Research to Develop GE Crops

A small number of large corporations and small biotechnology companies led the initial efforts to develop GE varieties. The major multinational companies involved in crop improvement continue to make large financial investments in research and development of GE crops. As a result, these companies continue to lead in the development of GE products.

In recent years, public sector research organizations have also invested in the development of GE crops for developing countries. These efforts involve important humanitarian crops, which may not justify corporate investments but are critically important to smallholder farmers. In some cases, corporations have provided intellectual property, technical support, and funding to assist the public sector. Corporations have also made their proprietary technology available for poor farmers through public–private partnerships that pair company contributions with foundation and government funding.

Public sector research organizations involved in GE crop cultivar development are now encountering the need to address risk assessment for field-testing and eventual release of GE products. The current regulatory environment has focused on prerelease testing and characterization. The public sector lacks the funding to develop GE crops, and high costs of gaining regulatory approval prevent the public sector from making them available to farmers. In the 14 years since the introduction of GE crops, only two publically developed crops (one papaya variety and two cotton varieties) have reached farmers. Even corporations have limited their efforts to large-scale crops and traits with very large markets, because it is difficult to justify large regulatory costs for products with limited return.

Meeting the Needs for a Sustainable Food Supply: The Importance of Risk Assessment and Regulation of GE Crops

GE crops are regulated products in most countries. Developers of GE crops have to demonstrate to regulatory authorities that their products are safe based on a widely accepted set of risk assessment procedures. These procedures may be quite expensive. A reasonable expectation is that developers will prefer to use unregulated approaches for improving crop productivity over those that involve additional regulatory costs – other things kept equal – unless there is a robust expectation of societal gains from using the product.

A number of recent reviews of the regulatory systems for GE crops have provided useful suggestions for improving these systems (Bradford et al. 2005; NRC 2002; Pew Initiative on Food and Biotechnology 2004; Romeis et al. 2008). Additionally, a number of scientists have proposed approaches for risk assessment of GE crops using biological understanding of the traits, the crops, and their environment (Bradford et al. 2005; Hancock 2003; Romeis et al. 2008). Because the technology underlying the development of GE crops is constantly changing, the question is how to stay current with the technology and incorporate and implement identified procedural improvements and experience over time. In spite of the increasing need for GE products, Jaffe (2006) has found that deregulation of new GE varieties in the United States slowed considerably in the five years from 2000–5.

Use of Global Regulatory Experience with GE Crops

Transgenic corn, soybean, cotton, and canola containing genes for insect and herbicide tolerance have been grown on hundreds of millions of hectares, and the resulting food has been consumed by hundreds of millions of people (James 2007). Additional GE food crops, including cassava, eggplant, papaya, rice, sweet potato, and squash, have been grown by farmers or are in the process of being field-tested (James 2007). GE sugar beet, potato, and wheat cultivars have been tested and deregulated, but have not been commercialized (Chapotin and Wolt 2007). Many more crops and traits are currently in the development process and are slowly entering the regulatory pipeline (Atanassov et al. 2004). The experience with these crops has been largely positive – enabling increased management options for farmers, reduced pesticide use, and in some cases improved yields (Brookes and Barfoot 2007). Farmer adoption has been very rapid in those countries with functioning regulatory systems that deregulate GE crops and in a few countries where unauthorized environmental release preceded regulatory approval. Although there have been a few widely reported regulatory violations, the development and commercialization of GE crops have generally proceeded in an orderly manner (Chapotin and Wolt 2007).

There is now significant experience and familiarity with several insect- and herbicide-resistant GE crops that has supported data sharing and regulatory acceptance of data generated elsewhere; however, crops modified for nutritional and health enhancement, biotic and abiotic stress tolerance, and other agronomic traits (nutrient use efficiency, photosynthetic efficiency, etc.) are still in various stages of development and evaluation, both for efficacy and for risks, which may face additional and/or new regulatory questions and issues during their risk assessment. To what extent is the experience with insect and herbicide resistance applicable to other traits? To what extent is the experience with the four crop species approved for deliberate release or commercialization, including events in corn, soybean, cotton, and canola, relevant to other important crop species? This question is quite relevant for future development of global biosafety regulatory systems.

Technical Improvements in GE: Opportunities for Standardized Approaches

Many of the initial concerns about genetic engineering of crops were related not only to the transgene inserted but also to the various approaches used to accomplish the modification. Advancements over the past 25 years have improved these techniques and eliminated a number of concerns. Transformation efficiencies have improved, and more precise and predictable gene insertion techniques are being introduced (Darbani et al. 2007). Increased genotype independence allows faster and less expensive generation and evaluation of transgenic materials, although many agronomically important genotypes remain difficult or impossible to transform. Antibiotic resistance genes used as selectable markers are not as widely used or are no longer present in finished transgenic varieties. Other improvements that are currently in various stages of development include "targeted" gene insertions using homologous recombination, mini-chromosomes, or other approaches that standardize the results of transformation, minimize possible position effects, and decrease uncertainty about gene insertion.

Bradford et al. (2005) have argued that biosafety regulatory reductions or simplifications are possible. They proposed that antibiotic markers do not pose a threat, and thus regulators should consider allowing their use for GE crop development. They also argued that event-based regulation should be eliminated. In part, they base their arguments on the accumulated knowledge from traditional and mutation breeding results, as well as study of the genomes of domesticated species. When compared to these processes, GE does not introduce unique or problematic changes to the genome.

Although Bradford et al. (2005) convincingly argued that, from a scientific point of view, biosafety systems should no longer regulate the use of antibiotic genes used as selectable markers, opponents of GE crops have been very negative about the use of antibiotic markers. Thus many private GE crop developers have moved to other systems for selecting transgenics or removing or segregating away markers. Consumer concerns and technical abilities may have led to the recognition that antibiotic marker genes are no longer required and do not need to be present in the new generation of GE cultivars. In fact, innovating companies have already adopted alternative approaches that are becoming widely used. Movement away from the use of antibiotic markers should also be a priority in public development programs.

Another possibility to reduce the regulatory burden is reducing or eliminating event-based regulatory approaches, which sufficient experience now makes feasible. It is also possible to eliminate the need for event-specific sequencing and testing by improving methods of gene insertion so that transgene insertion occurs in defined sections of the genome. Several approaches for accomplishing this technique are now available and should be routine in the near future. Ensuring the freedom to operate using these and other research innovations that reduce the regulatory burden to public organizations working on crops of interest to the developing world can facilitate their adoption.

We propose an additional approach for managing risk, increasing acceptance, and reducing costs in the longer term. Continuing improvements in the precision of GE provide the opportunity for GE crop public and private sector developers to agree on a standard portfolio of core development tools. As developers gain experience with the use of core development tools, they may seek reduced regulation from regulators. Of course, new transgenic varieties still need to be characterized (Nickson 2008). However, as methodologies for GE continue to become more standardized, attention can focus on the effects of new transgenes rather than the genetic hardware required for gene insertion and expression.

In the early stages of GE crop development, the level of expertise required and intellectual property protection for the relevant techniques provided both a barrier to entry and a source of competitive advantage for the companies involved; however, this situation has changed. Standardization would move development of GE crops much closer to other types of engineering, for which standards are common for the underlying technology, thereby minimizing the regulation of routine technology. For this approach to work, it would be necessary either to focus on technology already in the public domain or increase freedom to operate for agreed-on technologies. The approach suggested here could evolve with time. New technical approaches could be added to the standard list and others replaced as agreed-on replacements were found by innovators and passed into public use. Of course, many of the concerns about the risks of transgenic crops have to do with the transgenes themselves, rather than the genetic engineering processes. These issues are addressed in the next section.

Gene Discovery, More Valuable Traits, and Regulatory Evaluation

Second-generation traits, like abiotic stress resistance, nutrient use efficiency, and photosynthetic efficiency, all address fundamental and binding crop production problems (Bhatnagar-Mathur, Vadez, and Sharma 2007). The use of rapidly improving genomics tools for gene discovery suggests that the technical hurdles for producing valuable crop varieties using biotechnology will continue to fall (NRC 2008). These developments have enabled the identification of many genes that, at least in early testing, provide crops with drought tolerance, heat tolerance, salinity tolerance, disease resistance, improved nutrient use efficiency, and other traits that will improve or stabilize crop yields across environments (Naylor et al. 2004).

These traits also change the risk–reward balance by increasing the reward of developing and growing GE crops. The biosafety regulatory process still needs to consider risks, but the need also exists to weigh the risks relative to more significant advantages, especially in terms of increasing food security. The development of more relevant and valuable traits could result in broader acceptance and use of GE technologies and products. These traits will be particularly valuable in developing countries where other input options may not be available and populations are growing most rapidly. Infrastructure and the capacity for developing transgenic crops containing these traits and for providing the improved cultivars to farmers are required, as are appropriate regulatory systems.

Nutritional Improvements: Allergenicity, Toxicology, and Other Food Safety Issues

Traits introduced for nutritional enhancement intentionally modify the biochemical makeup of food and feed crops. Nutritional enhancements that may improve crops for human and animal use include the bio-fortification of staple crops, reduction of allergens or other anti-nutritional factors, and modification of oil and protein composition (Newell-McGloughlin 2008). Because allergenicity and toxicity are specific to the gene product and thus are unlikely to be related to the specific environment, food safety evaluations are relatively straightforward (Jaffe 2006). In contrast to GE crops modified for agronomic traits, GE crops modified for nutritional traits are less likely to result in environmental impacts, because they do not confer selective fitness advantages to the crops or to wild relatives with which they might cross. Of course, these nutritionally modified crops, as any other GE crops, still require an appropriate and feasible food safety evaluation. Because nutritional traits have special importance in developing countries where malnutrition is still a major issue, the management of risk assessments in a developing-country setting requires further study.

Most risk assessments for toxicology and allergenicity contemplate a set of distinct procedures for analyzing potential risk attributes. Procedures include examination of the compositional analysis, history of safe use, host crop and agronomic characteristics, homology sequence matching to known allergens using databases such as AllergenOnline or Allermatch, and digestibility studies. In most cases, feeding studies for investigating toxicology are required, as the previously mentioned procedures may not yield adequate information about potential food safety issues. Risk assessors implement current practices for food safety based on FAO/WHO and CODEX guidelines.

Data generated during food safety assessments may be used in other countries, taking into consideration differences in consumption and preparation (cooking) patterns. However, new or unintended responses to specific foods and food components have been reported among distinct ethnic groups. For developing countries, it may be appropriate to develop tiered approaches that assess the potential for food safety issues taking into consideration the data, knowledge, and experience accumulated both in other countries and in that country. This approach will help avoid unnecessary testing and concentrate on those issues that may constitute valid regulatory triggers. The main gaps for food safety assessments in developing countries include development of methods and approaches for assessment and evaluation of potential food risks. Potential approaches for food/feed safety assessment in biosafety regulatory processes that are worth exploring include the following:

1. Development of tiered approaches for risk assessment
2. Development of advanced analytical techniques to predict allergenicity potential (Soeria-Atmadja et al. 2004)

3. Development of the regulators' capacity for assessing evidence presented by developers
4. Exploration of the need for and the development of animal models to support decision making
5. Exploration of the need for new procedures for novel crops and traits

Environmental Assessments and Long-Term Ecological Evaluation (Modeling)

The main objective of the Cartagena Protocol on Biosafety (CPB) is the environmental risk assessment of GE organisms. This risk assessment includes the possibility that GE organisms may themselves become invasive or allow related species to become invasive and thus reduce biodiversity. The potential mechanisms for environmental effects of GE cultivars include the modification of the crop and related species' fitness in ways that may affect their competitive ability. The current focus of biosafety assessments is on biodiversity in an agricultural setting and spatial location. Risk research has only begun to address the broader impacts on overall biodiversity, because it is quite difficult to separate the effects of the specific GE organisms from the impact of agriculture and its practices. In fact, the key issue for risk assessors is to determine whether a particular GE organism will have a deleterious impact on agricultural biodiversity that is different from those of standard agricultural practices while providing an estimation of damage.[1]

Implementation of the CPB in countries that are party to the treaty implies a risk assessment process in which potential impacts are evaluated through confined field trial evaluations, as well as an analysis of other characteristics, such as history of safe use and biology and physiology of the organism. Postrelease monitoring efforts may follow preapproval assessments and regulatory decision making, but in practice, this does not always happen. The task at hand for developing countries is to develop a set of procedures that will examine potential impacts on biodiversity, with the added requirement that the process is feasible and focused on the right questions. In essence, the need exists to fine-tune the accuracy and precision of procedures used to examine risk potential.

In contrast to food safety data, knowledge used in environmental risk assessments tends to be quite specific to a particular agro-ecological niche or area, so there is usually less of a possibility to use data generated elsewhere. However, there are opportunities to reduce the regulatory burden on cultivar developers for developing countries in some cases through tiered approaches, in which risk analysts compare the GM crop's risk profile in a particular area with the risk profile of the same GM

[1] Note that using this assumption is equivalent to assuming that current agricultural practices are "risk-less." Thus, the question remains of what constitutes the appropriate benchmark or counterfactual by which to compare the potential impact of the introduction of a particular GE organism. Furthermore, even if a particular experiment determines there is an impact on a parameter of interest (i.e., change in population numbers for an indicator organism), what does this tell us about impact? In the end the need to connect likelihood of occurrence with damage, implies valuation of biodiversity, which is a contentious issue on its own.

crop in a similar agro-ecological zone. One example is the procedure suggested by Hancock (2003), in which a preliminary assessment of the environmental risk of GE crops compiles data on the geographical range of compatible related species, a determination of the invasiveness of the crop and its congeners, and the description of the modified crop phenotype. A combination of these three pieces of information plus other collateral information can define how and what type of experimentation developers need to conduct before the crop is approved for environmental release. A second approach is the one proposed by Linacre et al. (2006), in which risk analysts combine collateral information (defined by the authors as the related data that inform decision makers about the value of a parameter) from available sources, Bayesian methods, and credibility theory to generate information to support regulatory decision making. In a third approach, Romeis et al. (2008) have proposed a specific science-based method for assessing risks that insect resistance traits will affect nontarget insects. Finally, Nickson (2008) has suggested a systematic science-based approach for evaluating GE stress-tolerant crops. All of these approaches could help regulators and developers focus on the specific questions and critical impact pathways that will help design appropriate experiments that give reasonable and efficient measurements of environmental impact on biodiversity.

Evaluation of environmental risks can be complex and includes indirect effects on biodiversity and land use when formerly fallow and natural land is cultivated. This type of change in land use is only now being considered as part of the discussion of biofuel crops (Chapotin and Wolt 2007). Risks related to gene transfer in natural populations only occur in those areas where there are wild relatives of the crops. The relative contribution of transgenes to fitness of natural populations depends on the environment. For instance, wild populations receiving a transgene for drought tolerance might be more successful in a water-limited environment.

A critical gap in environmental risk assessments is the lack of models and procedures available to examine the long-term ecological consequences of introduction of a particular GE organism into specific environments. Analytical models represent a cost-efficient approach for risk assessment within a biosafety regulatory application. Widely validated models may support decision making while providing the public with a higher degree of confidence in the completeness of the risk assessment procedure and while complying with the goals and objectives of the CPB and its guiding principle, the precautionary approach. Procedures to perform long-term modeling of potential impacts can only buttress the robustness of the risk assessment process in a cost-efficient manner.

The widespread scientific and regulatory experience with initial GE products can also be useful to countries that are only now considering adoption of GE crops containing genes for insect and herbicide resistance. It is critically important to make the most of this experience, particularly as it allows understanding of the longer term environmental impacts of growing transgenic crops, a current weakness

Table 13.1. *Cost of compliance with the National Biosafety Regulatory Framework*

Type of crop (example)	Crop	Country	Event approved in developed countries	Estimated costs of biosafety regulations (US $)
Food Crop	Maize	India	Yes	500,000–1,500,000
	Maize	Kenya	Yes	980,000
	Rice	India	No	1,500,000–2,000,000
	Rice	Costa Rica	No	2,800,000
	Beans	Brazil	No	700,000
	Mustard	India	No & have to seek approval in export markets	4,000,000
	Soybeans	Brazil	Yes	4,000,000
	Potatoes	South Africa	Yes	980,000
	Potatoes	Brazil		980,000
	Papaya	Brazil	Yes	
Nonfood Crop	Cotton	India	Yes	500,000–1,000,000
	Jute	India	No	1,000,000–1,500,000

Source: Bayer, Norton, and Falck-Zepeda (2010).

of regulatory programs. In addition, well-designed experiments that monitor longer term population biology of wild relatives of GE crops can complement long-term monitoring of samples of commercial farmers' fields (Hancock 2003; Romeis et al. 2008).

Cost of Compliance with Biosafety Regulation

Several studies have documented the cost of compliance with biosafety regulations for a diverse set of countries and commodities. Regulatory compliance costs vary across commodities and countries, ranging from $53,556 for *Bt* eggplant in India to $2.25 million for viral-resistant rice in Costa Rica (Table 13.1). The regulatory cost estimates for the technologies described in Table 13.1 are lower than the costs incurred by private companies in the United States. Regulatory costs in the United States vary from $100,000 to $4 million for food and nonfood crops as seen in Table 13.2. Estimates of regulatory costs by those in the private sector range from $20 to $30 million (McElroy 2003). Compliance costs in Table 13.1 are comparable to those incurred by the public sector in China and India (Pray et al. 2006).

Differences in national regulatory processes and level of development largely influence cost levels. Some governments, such as China, have been sensitive to criticisms about the time taken for regulatory assessments and the cost of regulations. These countries have implemented revisions of the regulatory process and modified the

Table 13.2. *Estimated costs per biosafety activities for the United States, India, and China*

Activity	Cost ranges United States (US $)	Cost estimates India (US $)	Cost estimates China (US $)
Molecular characterization	300,000–1,200,000		
Toxicology (90-day rat trial)	250,000–300,000		14,500
Allergenicity (Brown Norwegian rat study)		150,000	
Animal performance and safety studies	300,000–840,000		
Poultry feeding study		5,000	
Goat feeding study – 90 days		55,000	
Cow feeding study		10,000	
Fish feeding study		5,000	
Anti-nutrient			1,200
Gene flow		40,000	11,200
Impact on nontarget organisms			11,600
Baseline and follow-up resistance studies (ea.)		20,000	
Protein production/ characterization	160,000–1,700,000		
Protein safety assessment	190,000–850,000		
Nontarget organism studies	100,000–600,000		
ELISA development, validation, and expression	400,000–600,000		
Composition assessment	750,000–1,500,000		
Agronomic and phenotypic assessment	130,000–460,000	30,000–205,000	
Socioeconomic studies		15,000–30,000	
Facility/management overhead costs	600,000–4,500,000		
Total Cost Approval	3,180,000–12,550.000	195,000	53,000–90,000

Source: Bayer, Norton, and Falck-Zepeda (2010).

process in an attempt to reduce costs. Pray et al. (2006) presented an example where China modified the process to conduct confined field trials.

The cost of compliance may also vary with the type of institution undertaking the regulatory compliance. Pray, Bengali, and Ramaswami (2005) and Pray et al. (2006) found that regulatory costs incurred by private companies are usually higher than those incurred by the public sector. A plausible explanation is that the public sector usually underestimates or subsidizes costs of tests and salaries. According to Pray et al. (2005, 2006) in China, the cost of approval of a new GM field crop event differed between private companies and the government by about $30,000, and the cost per trial for private firms is typically about three times more than for government research institutes.

A high cost of compliance with biosafety regulations can affect public and private crop improvement investments in different ways. The most obvious outcome is that it may reduce investment in the development of regulated products, thereby reducing the flow of potentially valuable products to farmers. A more subtle outcome of high compliance costs is that public and private organizations may require a higher rate of social/private rate of return compared to nonregulated products, because regulated products imply a higher development risk. Furthermore, high compliance costs may force the public sector to shift research to those commodities with more innate private return characteristics compared to crops of public interest; however, this result is not exclusive for the public sector, given that the private sector focused on only four crops for the first generation of products. Crops of little interest to the private sector include orphan, neglected, and underutilized crops that may have a relatively high social rate of return on those investments seeking to eliminate binding productivity constraints in those crops (Naylor et al. 2004).

To date, high compliance costs relate to relatively well-described crops and traits for which science has accumulated a large body of knowledge on which to base regulatory assessments. Scientific and regulatory knowledge has spilled over to other countries, thus reducing the potential cost needed to generate new information. The fact that knowledge can flow between countries provides a rationale for regional efforts to coordinate and harmonize assessments and decision-making processes described later in the chapter. Therefore, current estimates for regulatory compliance costs may represent a lower boundary for the expected costs for new crops and traits that may enter the regulatory pipeline in the near future. The opportunity to learn from existing regulatory processes to improve their efficiency clearly exists, and the cost of performing individual scientific assessments may be declining over time. However, these factors may not be sufficient to balance the expected cost increase for new products/traits entering the regulatory pipeline. Bayer, Norton, and Falck-Zepeda (2008) showed that the costs of compliance with biosafety regulations in the Philippines were similar or even larger than development costs. They also showed that regulatory compliance costs have less impact on the net benefits to society resulting from potential product adoption than short regulatory delays. Short regulatory delays of as little as three years may reduce net benefits to society significantly.

Finally, high regulatory compliance costs may constitute a barrier to entry for smaller organizations and the public sector, because they are sunk costs once investments are made (which are in advance of knowing the outcome). In countries where there is high regulatory uncertainty regarding how, if, and when a product may complete the regulatory process, the public and private sectors may reduce their levels of investment because they are less prepared to manage uncertainty than decision making under risk. Organizations typically assess the cost of compliance, the probability of obtaining an approval for commercialization, and the potential market before making a decision based on the risk-weighted returns. With uncertainty, this decision becomes more difficult and may represent a binding constraint of investment for organizations in fully developing their products. This also applies to public sector institutions, although with a modified public sector objective and decision-making process.

Scientists involved in the private sector development of GE crops originally envisioned a wide range of crops and traits being developed (Charles 2001). As scientists realized that the cost of completing risk analysis and gaining regulatory approval was likely high, newer analyses of the potential return on investment resulted in a narrower range of crops and traits with positive returns to investment. Attempts to manage risks of GE crops have, for diverse reasons, resulted in regulatory systems and stewardship regimes that are expensive and that have limited the application of biotechnology to a small subset of crops and traits. In this environment, companies' investment in research on traits and crops that would have limited markets is not economically viable, because the eventual return would not justify the regulatory cost. The importance of private investment in crop improvement is driving a divergence of crop productivity, based on funding levels. For example, funding for maize improvement exceeds $1 billion per year (primarily from private sources), whereas funding of wheat is less than $200 million (largely from public funds).

Biosafety Regulatory Systems for the Developing World: A Growing Divide?

The initial experience with GE crops has been led by the private sector. The issues related to risk management have been addressed by large, technically sophisticated organizations with funding and scientific capacity for risk management, product development, and stewardship; extensive experience with regulated technologies; and the ability to comply with regulatory processes. The high level of controversy concerning GE crops has added to the expense and technical sophistication necessary to participate in development and distribution of GE crops.

However, the public sector is increasingly becoming involved in developing GE crops, particularly small market crops and crops for the developing world (Byerlee et al. 2007; Cohen 2005; Naylor et al. 2004). GE crop development in and for developing countries continues to advance at a relatively rapid pace (Cohen 2005). How public organizations deal with these challenges is likely to determine the impact of GE technologies on food production in much of the world.

Several issues arise because regulatory costs for many of the food crops now under development for the developing world will be borne by the public sector. First, crop and trait evaluations will include potential return on investment and potential humanitarian impact. Second, in the case of many subsistence crops, regulatory approval will be required in many small countries – not just a few large ones. Finally, if regulatory costs are equal to those in large developed countries for large commercial crops like maize and cotton, very few of these potentially useful crops will end up being available to smallholders and resource-limited farmers.

Chapotin and Wolt (2007) have shown that many of the GE traits developed for crops grown in the developed world could address binding constraints affecting developing countries and resource-poor smallholder farmers. Farmers in developing countries would deem traits for insect and disease resistance, drought tolerance, salinity tolerance, heat tolerance, and nutrient use efficiencies highly valuable because they may lack access to other inputs. The ultimate value of these traits to resource-poor smallholder farmers will depend on whether countries can invest in pro-poor GM crop improvements, develop regimes to evaluate and manage risks, and provide infrastructure that will allow access to the improved seeds and adequate stewardship to ensure efficacy and monitoring.

Unfortunately, developing countries have chronically underinvested in science, technology, and innovation (Pardey et al. 2006). In fact, in Sub-Saharan Africa, investments stagnated during the 1980s and 1990s (Beintema and Stads 2006). Cohen (2005) found that the public sector underinvested in agricultural R&D generally and in biotechnology specifically during the same time period. The implication of such serious underinvestments is that investments in biotechnology and biosafety, especially by the public sector, may be insufficient to address pressing needs, especially when focused on resolving national constraints. Although the public sector in many developing countries has invested in some agricultural biotechnology research (Atanassov et al. 2004; Falck-Zepeda et al. 2008), few of its technologies have progressed to commercialization stages (Cohen 2005).

Many developing countries, particularly those in Sub-Saharan Africa and Southeast Asia, lack the minimal infrastructure and scientific capacity to conduct risk assessments and implement biosafety regulations. Johnston et al. (2008) concluded that as many as 100 developing countries lack the necessary technical and management capacity to implement effective biosafety regulatory regimes, but a growing number of countries are investing in the development of national biosafety frameworks. These countries are also working diligently to draft instruments that will allow the adoption of effective and protective biosafety systems. Implementation of functional biosafety systems becomes even more pressing when novel traits, crops, and transformation protocols enter the regulatory pipeline for risk assessments.

Taking into consideration the limited resource base and binding capacity constraints, developing countries need to explore alternatives to assess, manage, and communicate risk from GE crops, while carefully balancing their potential benefits to resource-poor farmers. Use of regional biotechnology regulatory systems offer significant advantages in both developed and developing countries, although there is a limited precedent for such regional systems. Birner and Linacre (2008) have explored many of the policy and governance issues related to these systems, particularly in West Africa. Regional efforts to define food safety and environmental risk assessment issues are likely to produce more efficient and cost-effective regional systems, particularly for developing countries.

A relatively easy and cost-effective approach is enabling and supporting the sharing of regional data and experience of risk assessment efforts. There are two distinct approaches to sharing data and experience among countries. The first approach is the creation of mechanisms that facilitate knowledge exchange. One example is the Biosafety Clearinghouse at the CPB. It is a depository of regulatory decisions and other regulatory-related knowledge that facilitates data sharing among countries that are parties to the Protocol. Unfortunately, the data available at the Clearinghouse are uneven and incomplete, but countries are taking steps to address this limitation. The second approach is the creation of bilateral and multilateral mechanisms to facilitate data sharing and the potential for countries to pool human and financial resources and expertise, as well as gain access to data and knowledge generated by partner countries and others. An example of this type of arrangement is the ongoing negotiations for the implementation of the Regional Approaches to Biosafety and Biotechnology Regulations by the Common Market in East and Southern Africa. As described in *Nature* (2010), this is a multilateral approach that seeks to address binding regulatory constraints while taking advantage of existing strengths and capacity of some countries in the region.

Conclusion

The introduction of new traits and crops in a wider array of countries may expand benefits to producers in these countries. As second-generation GE crops address binding constraints of potentially higher value than first-generation crops, their development will be more critical to developing countries. Developing countries need to devise innovative and novel approaches for the risk assessment of these and other GE crops without compromising safety. Innovative approaches exist that allow risk assessment of GE crops in a cost-efficient manner. Developing countries will require the political will to implement these innovative approaches, because they may be novel and somewhat different from those in place in developed countries.

The need exists for smart and efficient biosafety regulations, but not more regulations per se. Additional regulations, unnecessary procedures, and regulatory time delays tend to increase the cost of development and compliance with biosafety regulations. Additional, unnecessary costs reduce the present value of GE crops and may even prevent the release of the technology. However, in most cases, costs do not affect the present value of benefits as much as do regulatory time delays. Cost becomes a barrier to entry for private companies and especially the public sector.

References

Atanassov, A., A. Bahieldin, J. Brink, M. Burachik, J. I. Cohen, V. Dhawan, R. V. Ebora et al. 2004. *To Reach The Poor: Results from the ISNAR-IFPRI Next Harvest Study on Genetically Modified Crops, Public Research, and Policy Implications*. EPTD (Environment and Production Technology Division) Discussion Paper 116. Washington, DC: IFPRI (International Food Policy Research Institute). http://www.ifpri.org/sites/default/files/publications/eptdp116.pdf.

Bayer, J. C., Norton, G. W., and J. B. Falck-Zepeda. 2008. "The Cost of Biotechnology Regulation in the Philippines." Paper presented at the Annual Meeting of the American Agricultural Economics Association, Orlando, FL, July 27–9. http://purl.umn.edu/6507.

______. 2010. "Cost of Compliance with Biotechnology Regulation in the Philippines: Implications for Developing Countries. *AgBioForum* 13(1): 53–62.

Beintema, N. M., and G. J. Stads. 2006. *Agricultural R&D in Sub–Saharan Africa: An Era of Stagnation*. Washington, DC: IFPRI (International Food Policy Research Institute). http://harvestchoice.org/maps-etc/Beintema%20%20Stads%202006%20ASTI%20–%20Agricultural%20RD%20in%20Sub-Saharan%20Africa.pdf.

Bhatnagar-Mathur, P., V. Vadez, and K. K. Sharma. 2007. "Transgenic Approaches for Abiotic Stress Tolerance in Plants: Retrospect and Prospects." *Plant Cell Reports* 27 (3): 411–24. doi:10.1007/s00299-007-0474-9.

Birner, R., and N. Linacre. 2008. *Regional Biotechnology Regulations: Design Options and Implications for Good Governance*. IFPRI (International Food Policy Research Institute) Discussion Paper 753. Washington, DC: IFPRI. http://www.ifpri.org/sites/default/files/publications/ifpridp00753.pdf.

Bradford, K. J., A. V. Deynze, N. Gutterson, W. Parrott, and S. H. Strauss. 2005. "Regulating Transgenic Crops Sensibly: Lessons from Plant Breeding Biotechnology and Genomics." *Nature Biotechnology* 23: 439–44.

Brookes, G., and P. Barfoot. 2007. "Global Impact of GM Crops: Socio-Economic and Environmental Effects in the First Ten Years of Commercial Use." *AgBioForum* 9 (3): 139–51.

Byerlee, D., A. de Janvry, E. Sadoulet, R. Townsend, and I. Klytchnikova. 2007. *World Development Report 2008: Agriculture for Development*. Washington, DC: World Bank. doi: 10.1596/978-0-8213-7233-3.

Chapotin, S. M., and J. D. Wolt. 2007. "Genetically Modified Crops for the Bioeconomy: Meeting the Public and Regulatory Expectations." *Transgenic Research* 16 (6): 675–88.

Charles, D. 2001. *Lords of the Harvest*. Cambridge: Perseus Publishing.

Cohen, J. I. 2005. "Poorer Nations Turn to Publicly Developed GM Crops." *Nature Biotechnology* 23: 27–33. doi:10.1038/nbt0105-27.

Darbani, B., A. Eimanifar, C. N. Stewart, Jr., and W. N. Camargo. 2007. "Methods to Produce Marker-Free Transgenic Plants." *Biotechnology Journal* 2: 83–90. doi:10.1002/biot.200600182.

Falck-Zepeda, J. B., C. Falconi, M. J. Sampaio-Amstalden, J. L. Solleiro, E. Trigo, and J. V. Lazo. 2008. "A Quantitative Assessment of Agricultural Biotechnology Capacity and Innovation in 18 Latin American and the Caribbean Countries." Presentation at the 12th ICABR (International Consortium on Agriculture Biotechnology Research) Conference, "The Future of Agricultural Biotechnology: Creative Destruction, Adoption, or Irrelevance?" Ravello, Italy, June 14.

FAO (Food and Agriculture Organization of the United Nations). 2004. *The State of Food and Agriculture 2003–2004*. Rome: United Nations.

Hancock, J. F. 2003. "A Frame Work for Assessing the Risk of Transgenic Crops." *Bioscience* 53 (5): 512–19.

Jaffe, G. 2006. "Regulatory Slowdown on GM Crop Decisions." *Nature Biotechnology* 24: 748–9. doi: 10.1038/nbt0706-748.
James, C. 2007. *Global Status of Commercialized Biotech/GM Crops: 2007*. ISAAA (International Service for the Acquisition of Agribiotech Applications) Brief 37. Ithaca, NY: ISAAA.
Johnston, S., C. Monagle, J. Green, and R. Mackenzie. 2008. *Internationally Funded Training in Biosafety and Biotechnology – Is it Bridging the Biotech Divide?* With R. Mackenzie. Nishi-ku: UNU – IAS (United Nations University – Institute of Advanced Studies).
Linacre, N., J. B. Falck-Zepeda, J. Komen, and D. MacLaren. 2006. *Risk Assessment and Management of Genetically Modified Organisms under Australia's Gene Technology Act*. EPT (Environment and Production Technology) Discussion Paper 157. Washington, DC: IFPRI (International Food Policy Research Institute).
McElroy, D. 2003. "Sustaining Agbiotechnology through Lean Times." *Nature Biotechnology* 21 (9): 996–1002.
Nature. 2010. "Transgenic Harvest." Editorial. *Nature* 467: 633–4. doi: 10.1038/467633b
Naylor, R. L., W. P. Falcon, R. M. Goodman, M. M. Jahn, T. Sengooba, H. Tefera, and R. J. Nelson. 2004. "Biotechnology in the Developing World: A Case for Increased Investment in Orphan Crops." *Food Policy* 29: 15–44.
Newell-McGloughlin, M. 2008. "Nutritionally Improved Agricultural Crops." *Plant Physiology* 147: 939–53.
Nickson, T. E. 2008. "Planning Environmental Risk Assessment for Genetically Modified Crops: Problem Formulation for Stress-Tolerant Crops." *Plant Physiology* 147: 494–502.
NRC (National Research Council). 2002. *Environmental Effects of Transgenic Plants: The Scope and Adequacy of Regulation*. Washington, DC: National Academy Press.
———. 2008. *Achievements of the National Plant Genome Initiative and New Horizons in Plant Biology*. Washington, DC: National Academy Press. http://www.nsf.gov/bio/pubs/reports/nrc_plant_genome_report_in_brief.pdf
Pardey, P. G., N. M. Beintema, S. Dehmer, and S. Wood. 2006. *Agricultural Research: A Growing Global Divide?* Food Policy Report 17. Washington, DC: IFPRI (International Food Policy Research Institute).
Pew Initiative on Food and Biotechnology. 2004. *Feeding the World: A Look at Biotechnology and World Hunger*. Washington, DC: Pew Initiative on Food and Biotechnology. http://www.pewtrusts.org/uploadedFiles/wwwpewtrustsorg/Reports/Food_and_Biotechnology/pew_agbiotech_feed_world_030304.pdf.
Pray, C. E., P. Bengali, and B. Ramaswami. 2005. "The Cost of Regulation: The Indian Experience." *Quarterly Journal of International Agriculture* 44 (3): 267–89.
———, B. Ramaswami, J. Huang, R. Hu, P. Bengali, and H. Zhang. 2006. "Costs and Enforcement of Biosafety Regulations in India and China." *International Journal of Technology and Globalization* 2 (1/2): 137–57.
Romeis, J., D. Bartsch, F. Bigler, M. P. Candolfi, M. M. C. Gielkens, S. E. Hartley, R. L. Hellmich, et al. 2008. "Assessment of Risk of Insect-Resistant Transgenic Crops to Nontarget Arthropods." *Nature Biotechnology* 26 (2): 203–8.
Soeria-Atmadja, D., A. Zorzet, M. G. Gustafsson, and U. Hammerling. 2004. "Statistical Evaluation of Local Alignment Features Predicting Allergenicity Using Supervised Classification Algorithms." *International Archives of Allergy and Immunology* 133 (2): 101–12. doi:10.1159/000076382.

14

The Context for Biotechnology in Sustainable Agriculture

Marty D. Matlock

> It is true that the tide of the battle against hunger has changed for the better during the past three years. But tides have a way of flowing and then ebbing again. We may be at high tide now, but ebb tide could soon set in if we become complacent and relax our efforts. For we are dealing with two opposing forces, the scientific power of food production and the biologic power of human reproduction.
>
> *Borlaug (1970)*

"Everything is connected. Everything is changing." These axioms are fundamental principles of systems ecology first described by Odum (1988) and are the foundation of ecosystem design. They are critical for understanding and conceptualizing solutions to the challenges of developing sustainable agriculture strategies. The third axiom, "we are all in this together," is a normative claim that connects ecosystem theory with sustainability. The World Commission on Environment and Development defined sustainability as "development that meets the needs of the present without compromising the ability of the future generations to meet their own needs" (WCED 1987).

Yet the ethics of sustainability are difficult to define beyond this general framework. Nieto (1997) identified principles of sustainable development that included a voice for all people in an open and transparent process, a respect for the rights of future generations, a redefinition of the relationship between the human species and the ecosystems on which we depend, a science-based understanding of the limits of ecosystem services, an understanding of the interconnected impacts of activities throughout the supply chain and across spatial scales, enhanced self-sufficiency at the community level, and a pragmatic implementation of practices to test, revise, and adapt to changing conditions. Although these principles are very aspirational, they can also guide formulation of specific goals for sustainability within a sector of the global economy. Agriculture represents the core of all economic systems and has the largest global and ecological footprint. If sustainable agricultural practices can be adopted, much progress can be made in moving the entire supply chain of goods and services across the economy to a sustainable footing.

Sustainable goals for agriculture can be formulated to respond to these definitions and principles of sustainability. In response to these aspirations and the very real challenges facing humanity in the next 40 years, the goals of sustainable agriculture must include the following:

1. Provide secure and safe food, feed, fiber, and fuel supply chains globally to feed 9.25 billion people by 2050.
2. Increase global production with no loss of habitat.
3. Increase global production with reduced impacts on water resources.
4. Increase global production with fewer inputs (fertilizer, energy, pesticides, labor).
5. Increase global production while increasing ecological services.
6. Increase global prosperity through equitable resource allocation in global markets.

The role of biotechnology in meeting these objectives must be evaluated in the context of the pressures on Earth's critical ecosystem services. These pressures are a function of current global population dynamics, increasing demands for agricultural products, land use pressures associated with those demands, and water resource demands. The context of biotechnology in reducing these stresses should include an assessment of the implications of global climate change on agricultural production and the opportunities demonstrated by biotechnology to address these challenges.

Global Population Dynamics and Agricultural Production

In 2010 the human population on Earth reached 6.8 billion. By 2050, Earth's population will likely reach 9.25 billion (UN 2009b). The population added to Earth in the next 40 years will exceed the total population in 1950. These projections are based on median estimates of population growth by country, using median fertility and mortality estimates (Figure 14.1). This growth is occurring despite declines in global fertility rates (UN 2009b). In 1970–5 world fertility rates (WFRs) were 4.5 children per woman, with WFRs in least developed countries as high as 6.6, less developed at 5.2, and more developed regions at 2.1. In 2000–5 the WFR was 2.6, with least developed regions at 5.0, less developed at 2.6, and more developed at 1.6. By 2045–50 those rates are expected to decline even further, with least developed regions reaching 2.4 and less developed countries decreasing to 2.1 children per woman.

These reductions in fertility are a direct consequence of the dramatic reductions in abject poverty around the world, largely driven by increased agricultural productivity within least developed regions. Without these reductions, at the current fertility rate the population would exceed 12 billion by 2050, with 9.8 billion in less developed regions (Figure 14.1; UN 2009b). The rate of decrease of fertility throughout the

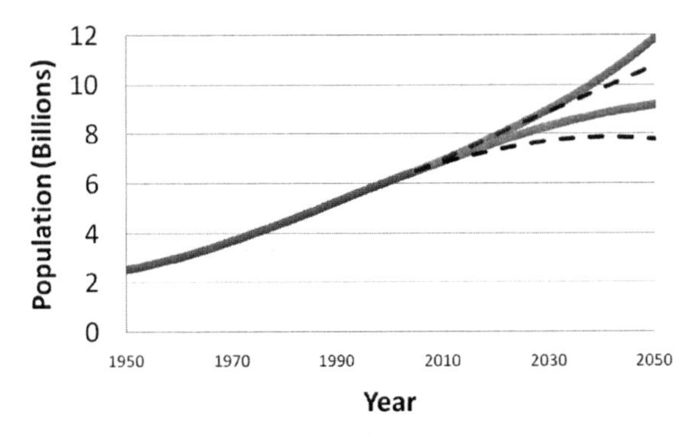

Figure 14.1. United Nations Estimates of Global Population Growth through 2050. The red line represents no decrease from 2005 fertility rates. The green line represents the median estimate of fertility, bounded by the upper and lower estimates. *Source:* UN (2009b).

three categories (least, less, and more developed) is expected to decline as birth rates approach replacement rates (approximately 2.1, dependent largely on child mortality).

Declining fertility means that the median age of human populations will increase. The median age of humans was 28 in 2009 and will likely reach 38 in the next 40 years (UN 2009b). Current populations in less developed regions are very young, in large part because of the devastating impact of human immunodeficiency virus (HIV; UN 2009b). Almost 50 percent of people in less developed countries are younger than 24 years old; almost 30 percent are less than 15 years old. In least developed countries more than 40 percent are less than 15 years old, largely because of the ravages of HIV on the mid-age population.

The global percentage of population over 60 will double by 2050, and the number of people over 60 in less developed and least developed countries will exceed 1.6 billion (Figure 14.2). The number of elders on Earth is projected to pass the number of children in 2047 (UN 2009a). The implication for future economic prosperity is significant: Economic growth and labor markets, family composition, living arrangements, childhood education support, health care services, epidemiology, and almost every other facet of economic, social, and political domains will be affected by this age shift. The potential support ratio (PSR) is the ratio of people between the ages of 15 and 65 to those over 65 and represents the potential workers to support the aging sector. The PSR declined from 12:1 to 9:1 from 1950–2007; by 2050 the PSR will likely reach 4:1 (UN 2009a). The ability to produce, process, and distribute agricultural products using nonmechanized practices is dependent on a vibrant and relatively young workforce. That workforce will be in short supply in 40 years, forcing a shift to labor-efficient and mechanized forms of production.

Population density distributions are not even, nor are they distributed relative to the ecological carrying capacity of the region. This inequity will be increased in the coming decades. The added population will emerge entirely in developing countries

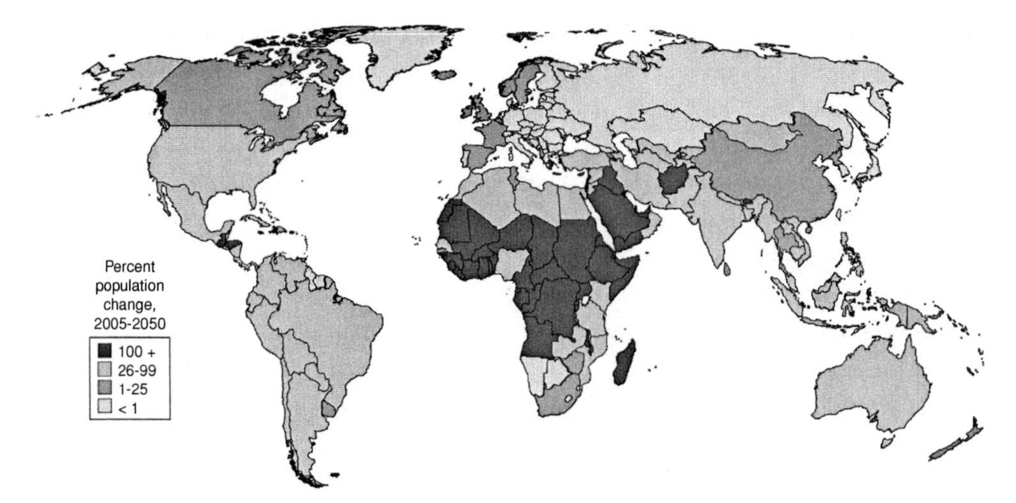

Figure 14.2. Percent Population Change, Projected from 2005 to 2050. *Source:* PRB (2005).

(Figure 14.2). When immigration is accounted for, populations of more developed regions will increase from 1.23 to 1.28 billion (less than 5 percent); without immigration this population would have decreased almost 7 percent (UN 2009a). Currently 1.5 billion people live in deep poverty (< $1 per day). This disparity motivates increasing numbers of people to seek to increase their prosperity (security of food and water, security of person, security of opportunity, especially education). Migrations of large human populations across eco-political regions have been central to the human experience. Increasing resource pressures are resulting in reduced tolerance and opportunity for those immigrants, leading to social and political strife that often results in systemic violence. From 1990 to 2005 the number of international immigrants increased almost 25 percent (UN 2009a). The immigrant population in 2005 was 3 percent of the global population. One in four lived in North America, and one in three lived in Europe. Immigration is a measure of the degree to which political and ecological pressures are disrupting economic and social communities.

Increasing Demands for Agricultural Products

The first Green Revolution, an effort to reduce poverty, resulted in increases in food production to unprecedented levels. Dr. Norman Borlaug was honored with the 1970 Nobel Peace Prize for his work in directing research that resulted in increased crop production efficiency globally. Borlaug and his colleagues used genetically improved crops (predominantly dwarf wheat and rice) combined with irrigation, fertilizers, and mechanization to improve yields in Asia and the Americas. Through integrated pest control, cultivation, and crop genetics research the agricultural science community tripled wheat production from 2 to 6 Mg/hectare over the period (IFPRI 2002). The development and adoption of high-yield variety (HYV) rice, sorghum, millet, maize, cassava, and beans soon followed. Rice yields have more than doubled globally since the introduction of IR8 in 1966 (Cantrell and Hettel 2004). The doubling of production of cereal grains was achieved with less than 5 percent increase in cultivated area (Tride 1994). A series of droughts in India in the early 1960s had threatened to plunge the country into mass starvation, but the increases in production beginning in 1966 ameliorated the impact of the drought. Rural economies flourished with increased production; per capita incomes in Asia doubled between 1970 and 1995 (IFPRI 2002). The number of poor in rural India dropped from 65 percent in 1965 to 34 percent in 1993.

For 21st-century citizens to realize the significance of these innovations it is important to reflect on the state of civilization in the aftermath of World War II. In 1950 an estimated 70 percent of the human population was chronically malnourished. Hunger was the human condition. Even in the United States hunger was prevalent in urban and rural areas. The notion that Earth could produce enough food for 3 billion people was outside the realm of imagination. Many farmers were still using beasts of burden (predominantly draft horses) to till the soil in poor rural areas. Farms were relatively small as a consequence, because a man and a team of horses could not cultivate much more than 130 ha in a season. Mechanization changed the equation. With the advent of petrochemical-powered agricultural tillage and harvesting machinery, a single farmer could cultivate thousands of acres in a season. In addition, soil fertility and pest management improved dramatically, predominantly through exogenous fertilization and chemical pest control. Within a generation, the human condition in the United States and Europe had changed. An army of committed men and women who spent their lives improving the prosperity of the poorest on the planet revolutionized the human condition in the middle of the 20th century. Less than 50 percent of Earth's human population was chronically malnourished in 1970. These people were predominantly in Asia, Africa, and Central and South America.

Within 20 years of Norman Borlaug's acceptance of the Nobel Peace Prize for leading the Green Revolution, humanity was being fed. By 1990 only 16 percent of humanity was chronically malnourished. Yet percentages do not tell the entire story; in 1990 the population was 5.8 billion and growing at what seemed an unstoppable rate. The chronically malnourished numbered 842 million people. Currently (2010) we are feeding more than 5.6 billion people; this number is equivalent to the entire world population of 1994 (UN 2009b). The notion that Earth could provide food for almost 6 billion people was outside the realm of confidence for agricultural experts in 1970. Yet we are now meeting that goal. Clearly, the Green Revolution delivered more people from hunger than ever before in the history of humanity.

Humanity has come very close to eradicating chronic hunger. The estimated global nourishment deficit in 2005 was roughly 8.75×10^{13} kcal. Global production of rice was almost 632 Tg (FAOSTAT 2009). Assuming rice provides an average of 360 kcal

Table 14.1. *Global grain production, 2007*

Grain	Area cultivated (ha)	Production (Mg)	Yield (Mg/ha)
Maize	142,331,335	637,444,480	3.41
Wheat	204,614,529	549,433,727	2.75
Rice	153,324,898	588,563,933	3.37
Total	500,270,762	1,775,442,140	–

Source: FAOSTAT (2009).

per 100 grams and conversion efficiency from harvest to plate of 70 percent, the 2005 nutritional deficit could be eradicated with 35 Tg of rice, or about 100 grams of rice per person per day. This amount represents less than 6 percent of global production and could be met by reducing spoilage in rural production areas through the use of flat-bed dryers (Berthelsen 2008). Chronic malnourishment in 2009 is not an agronomic problem; it is an economic problem. Chronic malnourishment is more connected to stable food prices and supplies than to small landholder production efficiency.

Global food production is increasing both in terms of efficiency and effectiveness (Table 14.1). In 2004 global cereal production was 2.27 Pg, almost 9 percent higher than 2000 (FAOSTAT 2009). During that same period meat production rose almost 11 percent to 260 Tg. Production of fruits and vegetables increased more than 14 percent to 1.38 Pg. Fertilizer production increased 12 percent from 2000–4 to almost 148 Tg globally. Overall, per capita agricultural production was up 5.3 percent from 2000 to 2004, with cow milk as the highest value commodity (FAOSTAT 2009).

However, although modern agro-economies have been feeding almost 86 percent of humanity, 923 million people or 18 percent of the population were chronically undernourished in 2007 and 963 million in 2008 (FAOSTAT 2009). The World Food Summit, convened in 1996, established 1990 (842 million people chronically undernourished) as the baseline for measuring hunger (FAOSTAT 2009). It set as a goal the reduction of hunger by half by 2015; as Borlaug anticipated 25 years ago and for the first time since 1950 we are losing ground in reaching that goal (Figure 14.3).

Many critics of agriculture cite this as evidence of the failure of the Green Revolution. Yet the data simply do not support this conclusion: The Green Revolution worked well beyond the imaginations of agricultural leaders. In 2008, modern agriculture fed six billion people, the total world population of 1996. Instead, high food prices share much of the blame. The most rapid increase in chronic hunger in recent years occurred between 2003–5 and 2007, during periods of corn, soybean, and rice price spikes. Hunger in the opening decade of the 21st century is a prosperity problem.

The UN Comprehensive Framework for Action (CFA) presents two sets of actions to promote a comprehensive response to the global food crisis (UN 2010). The first set focuses on meeting the immediate needs of vulnerable populations, a triage intervention. The second set builds resilience and contributes to global food and

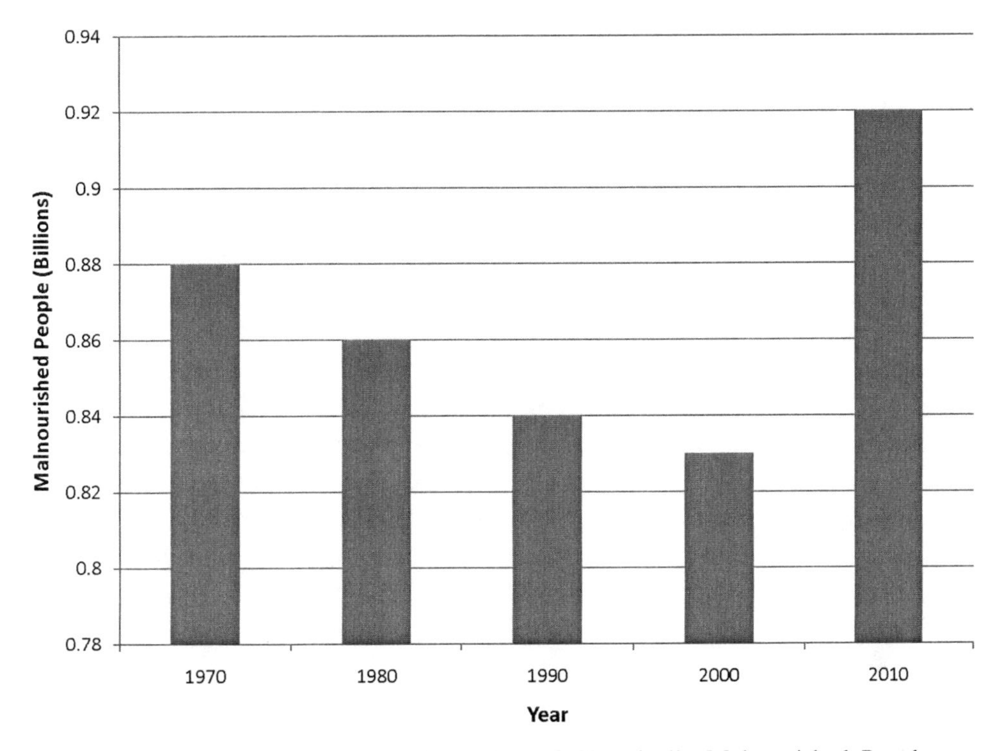

Figure 14.3. Proportion and Total Number of Chronically Malnourished People. *Source:* UN (2009b).

nutrition security through infrastructure and education. To support these two sets of actions, the CFA suggests strengthening coordination, assessments, monitoring, and surveillance systems. The CFA also recommends four key outcomes: (1) enhancing and making emergency food assistance, nutrition interventions, and safety nets more accessible; (2) boosting smallholder farmer food production; (3) adjusting trade and tax policies; and (4) managing macroeconomic implications.

The intent of the Green Revolution was not just to alleviate hunger; rather it was to enhance prosperity. The process of enhancing prosperity would in turn reduce fertility rates, decreasing the rate of increase in human demand for food and other resources. The objective was to buy time so that governments and societies could adopt policies that stabilized exploding global consumption rates. The agricultural scientists who devoted their careers to this effort succeeded in their mission; the political and social leaders of the world did not. In 1984 Borlaug said he had "watched his life's work devoured by the population monster" (pers. comm., 1984).

Prosperity is directly correlated with fertility (Figure 14.4; PRB 2007). "Development is the best contraceptive," Karan Singh, Indian Ambassador to the United States said in 1974 (Singh 2008). Fertility rates are declining globally to below 4 children per woman. As Borlaug extolled, human prosperity is directly correlated with population. In this context the Green Revolution worked; what failed was the global effort

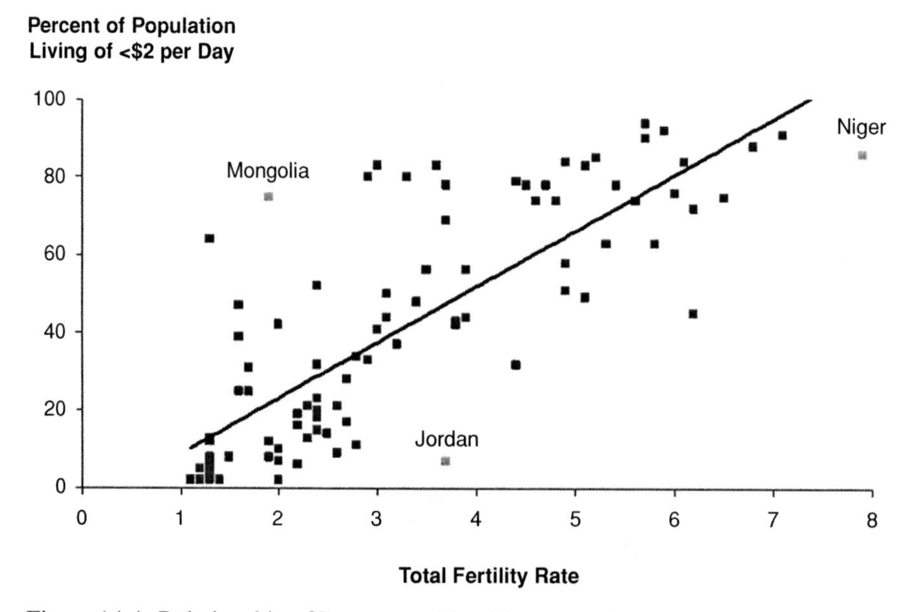

Figure 14.4. Relationship of Poverty and Fertility Rates Globally. *Source:* PRB (2007).

to reduce population growth. Current levels of prosperity cannot be maintained with increasing growth pressures. More than 3 billion more people will inhabit Earth in 2050 than do today; the demand on resources exerted daily by 9.25 billion people is hard to imagine. Fertility rate declines that will result in a stable population in 2050 suggest that humanity can control its appetite and begin reducing its consumptive impact on natural resources. Feeding this population will require an increase in agricultural production of 50 to 100 percent.

The increased demands for food, feed, fiber, and fuel made by 9.25 billion people will be extraordinary, likely creating resource compression and supply chain constriction. Food production will need to increase by at least 50 percent in the next 40 years to meet imminent demand. This increase is not outside the reach of modern agricultural capacity. The increases in production in the first part of this decade suggest that the trajectory of growth is in the appropriate scale to meet demand. However, compounding challenges of competition between food and biofuels for land and water, as well as the uncertain impacts of climate change, make developing strategies for meeting future demands more difficult to assure.

Land Use Pressures

Humans have changed the biomes of Earth. Agricultural land use (crop, pasture, and rangeland) occupies an estimated 40 percent of all of Earth's non-ice-covered surface area (Foley et al. 2005). Earth's largest terrestrial biome is now agriculture. Human activities currently appropriate between 35 and 50 percent of primary productivity

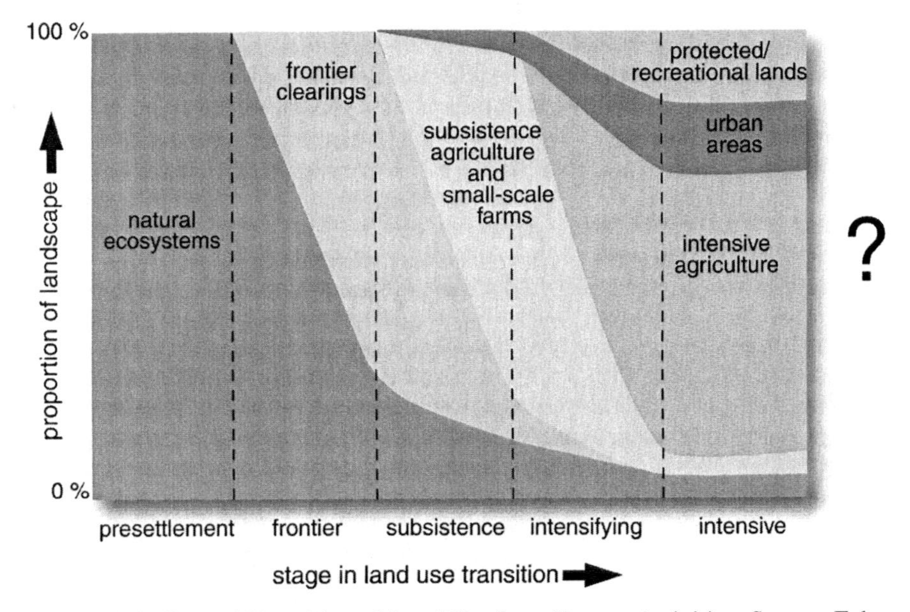

Figure 14.5. General Transition of Land Use from Human Activities. *Source:* Foley et al. (2005).

from the landscape (Vitousek 1994). Ecosystem services are the benefits obtained from ecosystems including provisioning services (e.g., food and water), regulating services (e.g., flood and disease control), cultural services (e.g., spiritual, recreational, and cultural benefits), and supporting services (e.g., nutrient cycling) that maintain the conditions for life on Earth (Millennium Ecosystem Assessment 2005).

Ecosystem services are in decline globally (Millennium Ecosystem Assessment 2005), with more than 30 percent of ecosystem services in serious decline, largely because of habitat loss associated with land use change. Land use pressures are driven by the market for products of primary productivity, especially food, biofuels, timber, and fiber.

Land use changes have historically occurred along a predictable transition line from forested to intensive agricultural production (Figure 14.5). More than 10 million km^2 of forest biomes have been destroyed by human activities through land use conversion (Foley et al. 2005). That process took centuries in preindustrialized society. Tropical forest land use change is now occurring at an increasing rate (Lepers et al. 2005). Forest land use change occurred predominantly in the tropics in the analysis period between 1990 and 2000, with deforestation hotspots concentrated in the Amazon basin and Southeast Asia (Lepers et al. 2005).

Dryland areas in Asia were also identified as hotspots of land use change. This land use change was associated with expansion of croplands, which was noted across all continents (Lepers et al. 2005). Croplands increased most intensively during the last decade of the 20th century in Southeast Asia, but also in Bangladesh, the Indus

Table 14.2. *Annual and daily deaths and DALYs attributed to water-borne disease*

Disease	Deaths	DALYs
Schistosomiasis	14,000	1,932,000
Trachoma	0	1,239,000
Ascariasis	3,000	505,000
Trichuriasis	2,000	481,000
Hookworm disease	7,000	1,699,000
Total disease	26,000	5,856,000
Daily disease	71	16,044
Diarrhea	2,219,000	72,776,516
Daily diarrhea	6,079	199,388
Daily childhood diarrhea	4,343	178,663

Sources: Mathers et al. (2004); Prüss et al. (2002); WHO (1999, 2004).

Valley, the Great Lakes region of Eastern Africa, and in the Amazon basin. Cropland area decreased in North America (predominantly the southeastern United States) and eastern China (Lepers et al. 2005).

Increased demand for agricultural products for feed, food, fuel, and fiber will result in increased pressure on land use change. Land use change is not the product of poverty; rather, it is the vehicle for prosperity for most economies that are primary productivity based. Land use change occurs in response to economic opportunities, with some constraints from governmental and other institutional factors (Lambin et al. 2001). These are global drivers, moderated by local factors. Land use change will likely have the largest effect in decreasing biodiversity for terrestrial ecosystems in this century (Sala et al. 2000); this impact on biodiversity will likely be larger than that of climate change. Human prosperity in the 21st century may come at the cost of infinitely valuable and irreplaceable species and ecosystems. Increasing productivity on current agricultural lands will reduce the pressures of land use change and thus will conserve biodiversity.

Water Resource Demands

Water is the first need for human survival. An estimated 30 percent of people currently live in areas of chronic water stress (Millennium Ecosystem Assessment 2005; Vörösmarty et al. 2000). Water resource demands have two major facets: water quality and water quantity. Water quality issues relate predominantly to pathogens and mineral (salt) content. The disease burden from water, sanitation, and hygiene is estimated to be 4 percent of all deaths and 5.7 percent of the total disease burden in disability-adjusted life years (DALYs) occurring worldwide (Table 14.2).

The rate at which water is cycled across the landscape affects the concentration of pollutants. As water resources become scarcer, a given volume of water is cycled more frequently as it moves through the hydrologic cycle. Salinity of many water resources in arid systems is rising because of diversion of flows for irrigation and other uses (Reynolds et al. 2007). The irrigated water collects salts as it flows across tilled soils, evaporating (leaving more salts behind) and infiltrating into groundwater (Williams 1999).

Water resource allocation will be an increasingly contentious issue in the 21st century. More than 70 percent of freshwater globally is appropriated for irrigation for agricultural production. In less developed countries as much as 90 percent of freshwater resources are used for agriculture. Freshwater consumption worldwide has more than doubled since World War II and is expected to rise another 25 to 40 percent by 2030 (Foley et al. 2005). Hoekstra and Chapagain (2008) suggest that minimum water rights should be elevated to a human right to potable water before any other allocation is made. More than 2.5 billion people live in arid and semi-arid areas (mean annual rainfall between 25 and 500 mm); these regions will become increasingly stressed as populations increase and pressures on finite water resources continue to grow.

Meeting the water demands for 9.25 billion people will require allocation of freshwater for basic human consumption, agricultural production, biofuel production, municipal sanitation, industrial use, and other applications. There is a growing concern that humanity has passed peak water, or the point of maximum production/utilization of water resources (Gleick 2009). Humanity uses 26 percent of evapotranspiration and 56 percent of accessible terrestrial runoff (Postel et al. 1996). Globally 20 percent of freshwater fish are in danger of extinction or are already extinct, and 47 percent of all listed endangered species in the United States are freshwater species (Jackson et al. 2001). Increasing water consumption will decrease biodiversity.

Conclusion

Sustainable agricultural production will require land, water, energy, and labor. All four of these elements will be in decreasing supply over the next four decades. The goal of sustainable agriculture should be to meet the food, feed, fiber, and fuel needs of 9.25 billion people without one hectare more of land, without one drop more of water. Clearly achieving this goal will require every tool available, including biotechnologically enhanced crops. In 150 to 200 years humanity will have a population approaching 6 billion again. This future generation will be faced with meeting the nutritional demands of the population with depleted petrochemical energy supplies. Efficiency in energy, water, and land use for crop production is critical to sustaining humanity.

To meet this challenge, agricultural producers need to develop better metrics for sustainable agriculture at the field, farm, region, and national scales. These metrics must be easy to implement, outcomes-based, and scalable. They should also align and coordinate with indices that assess progress toward sustainable management goals. The agricultural community must also develop cohesive and timely reporting protocols to inform policy at local, national, and global scales so management and policy strategies can be adapted to changing conditions and situations.

There can be no permanent progress in the battle against hunger until the agencies that fight for increased food production and those that fight for population control unite in a common effort. Fighting alone, they may win temporary skirmishes, but united they can win a decisive and lasting victory to provide food and other amenities of a progressive civilization for the benefit of all mankind.

Borlaug (1970)

References

Berthelsen, J. 2008. "Anatomy of a Rice Crisis." *Global Asia* 3 (2): 26–31.

Borlaug, N. 1970. Acceptance Speech for Nobel Peace Prize. Oslo, Sweden, December 10. http://nobelprize.org/nobel_prizes/peace/laureates/1970/borlaug-acceptance.html.

Cantrell, R. P., and G. P. Hettel. 2004. "New Challenges and Technological Opportunities for Rice-Based Production Systems for Food Security and Poverty Alleviation in Asia and the Pacific." Presented at the FAO Rice Conference, Rome, Italy, February 12–13.

FAOSTAT. 2009. *Food Security Statistics* (accessed January 10, 2010). http://www.fao.org/faostat/foodsecurity/index_en.htm.

Foley, J. A., R. DeFries, G. P. Asner, C. Barford, G. Bonan, S. R. Carpenter, F. S. Chapin, et al. 2005. "Global Consequences of Land Use." *Science* 309: 570–4. doi: 10.1126/science.1111772.

Gleick, P. 2009. *The World's Water: 2008–2009*, vol. 6. New York: Island Press.

Hoekstra, A., and A. Chapagain. 2008. *Globalization of Water*. Malden, MA: Blackwell.

IFPRI (International Food Policy Research Institute). 2002. *Achieving Sustainable Food Security for All by 2020: Priorities and Responsibilities*. Washington, DC: IFPRI. http://www.ifpri.org/sites/default/files/publications/actionshort.pdf.

Jackson, R. B., S. R. Carpenter, C. N. Dahm, D. M. McKnight, R. J. Naiman, S. L. Postel, and S. W. Running. 2001. "Water in a Changing World." *Ecological Applications* 11 (4): 1027–45.

Lambin, E. F., B. L. Turner, H. J. Geist, S. B. Agbola, A. Angelsen, J. W. Bruce, O. T. Coomes, et al. 2001. "The Causes of Land-Use and Land-Cover Change: Moving beyond the Myths." *Global Environmental Change* 11 (4): 261–9. doi: 10.1016/S0959-3780(01)00007-3.

Lepers, E., E. Lambin, A. Janetos, R. DeFries, F. Achard, N. Ramankutty, and R. Scholes. 2005. "A Synthesis of Information on Rapid Land Cover Change for the Period 1981–2000." *Bioscience* 55 (2): 115–24.

Mathers, C. D., C. Bernard, K. M. Ilburg, M. Inoue, D. M. Fat, K. Shibuya, C. Stein, N. Tomijima, and H. Xu. 2004. *Global Burden of Disease in Data Sources, Methods, and Results*. Global Programme on Evidence for Health Policy discussion paper 54. Geneva: WHO (World Health Organization). http://www.who.int/healthinfo/paper54.pdf.

Millennium Ecosystem Assessment. 2005. *Ecosystems and Human Well-Being: General Synthesis*. Washington, DC: Island Press. http://www.maweb.org/documents/document.356.aspx.pdf.

Nieto, C. C. 1997. "Toward a Holistic Approach to the Ideal of Sustainability." *Society for Philosophy and Technology*. 2 (2): 41–8.

Odum, H. T. 1988. "Self-Organization, Transformity, and Information." *Science* 242 (4882): 1132–9. doi: 10.1126/science.242.4882.1132.

Postel, S. L., G. C. Daily, and P. R. Ehrlich. 1996. "Human Appropriation of Renewable Fresh Water." *Science* 271 (5250): 785–8. doi: 10.1126/science.271.5250.785.

PRB (Population Reference Bureau). 2005. *2005 World Population Data Sheet.* http://www.prb.org/pdf05/05WorldDataSheet_Eng.pdf.

______. 2007. *Population & Economic Development Linkages 2007 Data Sheet.* http://www.prb.org/pdf07/PopEconDevDS.pdf.

Prüss, A., D. Kay, L. Fewtrell, and J. Bartram. 2002. "Estimating the Burden of Disease from Water, Sanitation, and Hygiene at a Global Level." *Environmental Health Perspectives* 110 (5): 537–42.

Reynolds, J. F., D. M. Stafford Smith, E. F. Lambin, B. L. Turner, II, M. Mortimore, S. P. J. Batterbury, T. E. Downing, et al. 2007. "Global Desertification: Building a Science for Dryland Development." *Science* 316 (5826): 847–51. doi: 10.1126/science.1131634.

Sala, O. E., F. S. Chapin, III, J. J. Armesto, E. Berlow, J. Bloomfield, R. Dirzo, E. Huber-Sanwald, et al. 2000. "Global Biodiversity Scenarios for the Year 2100." *Science* 287 (5459): 1770–4. doi: 10.1126/science.287.5459.1770.

Singh, K. 2008. "My Life: Chronology." http://www.karansingh.com/index.php?action=chronology.

Tride, D. 1994. *Feeding and Greening the World.* Wollingford: CABI (Centre for Agricultural Bioscience International).

UN (United Nations). 2009a. *International Migration Report 2006: A Global Assessment.* New York: UN. http://www.un.org/esa/population/publications/2006_MigrationRep/fullreport.zip.

______. 2009b. *World Population Prospects: The 2008 Revision Population Database.* (Population for world for each year from 1950–2050, accessed Jan 8, 2010). http://esa.un.org/unpp/.

______. 2010. *United Nations Millennium Development Goals.* http://www.un.org/millenniumgoals/pdf/MDG%20Report%202010%20En%20r15%20-low%20res%2020100615%20-.pdf#page=13.

Vitousek, P. M. 1994. "Beyond Global Warming: Ecology and Global Change." *Ecology* 75 (7): 1861–76. doi:10.2307/1941591.

Vörösmarty, C. J., P. Green, J. Salisbury, and R. B. Lammers. 2000. "Global Water Resources: Vulnerability from Climate Change and Population Growth." *Science* 289 (5477): 284–8. doi: 10.1126/science.289.5477.284.

WCED (World Commission on Environment and Development). 1987. *Our Common Future.* New York: Oxford University Press.

Williams, W. D. 1999. "Salinisation: A Major Threat to Water Resources in the Arid and Semi-Arid Regions of the World." *Lakes and Reservoirs: Research and Management* 4 (3–4): 85–91. doi: 10.1046/j.1440-1770.1999.00089.x.

WHO (World Health Organization). 1999. *World Health Report: Making a Difference.* Geneva: WHO. http://www.who.int/whr/1999/en/.

______. 2004. *Health Statistics and Information Systems: Global Burden of Disease.* (Disease and Injury Regional: Standard DALYs; WHO regions, world estimates; Accessed November 18, 2010.) http://www.who.int/healthinfo/global_burden_disease/estimates_regional/en/index.html.

15

Agricultural Biotechnology

Equity and Prosperity

Gregory D. Graff and David Zilberman

The global increase in agricultural productivity achieved during the 20th century may be one of the most significant technological achievements of the human race (Edgerton 2008). To a large extent, it resulted from the repeated application of the basic principles of Mendelian genetics in crop breeding. Numerous incremental innovative steps have translated into major yield gains over time. Agricultural biotechnology has introduced a discrete leap in the same process, a radical innovation in new tools that allows for more precise modification of the genetic makeup of crops. Whereas the Green Revolution of the 20th century resulted in a few super-varieties that combined wide arrays of desired traits, the emerging "Gene Revolution" of the 21st century will allow for precise genetic improvements of a much larger number of locally adapted crop varieties with a much larger array of improved characteristics – one gene (or genetic system) at a time. However, this new wave of innovation in agricultural genetics is more scientifically intensive and continues to evolve as our scientific knowledge grows. The tools of biotechnology – combined with our knowledge of biological systems and the right economic, policy, and institutional frameworks – will be able to provide a foundation of economic prosperity and environmental sustainability for global agriculture in the decades to come (see Figure 15.1).

Agricultural biotechnology is still in its infancy relative to crop breeding and agrochemical technologies that have preceded it. In the 1980s and early 1990s, fundamental innovations in the use of *Agrobacterium* to transfer DNA into plants were developed, and a legal and regulatory framework was established in the United States and other leading adopter countries for the regulation of this new capability. Initial improvements included long shelf-life tomatoes and herbicide-tolerant soybeans. Although a considerable range of research and product development opportunities were pursued – including crop plants capable of nitrogen fixation, endowed with disease resistance, and able to grow in saline soils – the first generation of commercially successful transgenic crops were those that embodied pest control traits; in particular, insect-resistant varieties of corn and cotton (particularly those with various types of *Bt* genes) and

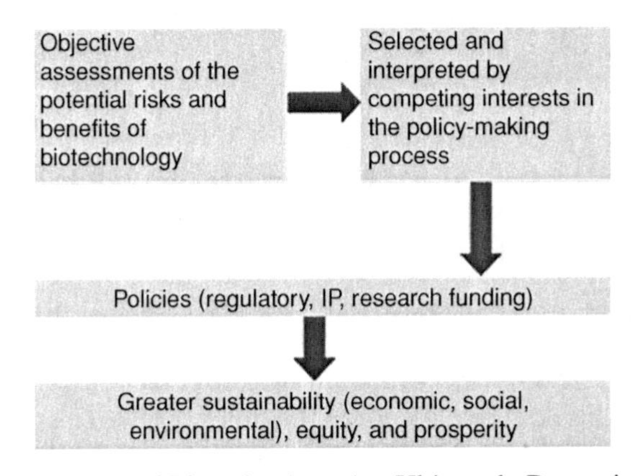

Figure 15.1. The Impacts of Biotechnology Are Ultimately Determined by the Characteristics of the Technology Itself, but Only as Selected and Interpreted through the Policy Process.

herbicide-tolerant soybean and canola (particularly RR varieties that withstand the herbicide glyphosate).

The high rates of adoption of first-generation commercial traits, spread across tens of millions of acres, have established a foundation for detailed statistical analysis of their impacts – whether those impacts be on yields, costs, biodiversity, or other economic or environmental parameters that these traits might affect. We examine the impacts and implications of first-generation traits in the next section, followed by a discussion of the future as a new generation of product-quality traits approaches commercialization and as agriculture assumes a new major role in supplying energy inputs to the economy.

We recognize, however, that the capacity of farmers, industry, and ultimately consumers to benefit from the positive impacts of agricultural biotechnology depends on national policies and institutional setting, including product approvals and intellectual property rights. These issues are addressed in the section, "Shaping Policies for Sustainability." Finally, we provide a summary and conclusions.

The Impacts of Transgenic Crops: Lessons of First-Generation Traits

Transgenic traits have been introduced on a large scale in four major field crops – corn, soybean, cotton, and canola – and on a very limited scale in some specialty crops, such as tomatoes, squash, and papaya. The introduction and cultivation of this first generation of transgenic crops have provided significant experience and opportunities for evaluation. Of these first-generation crops, more countries approved and adopted transgenic varieties of cotton than any other, and its performance across these different contexts provides several important lessons about the impacts of first-generation pest control traits.

Impacts Depend on Preexisting Practices and Conditions

The impacts of a transgenic variety are path-dependent; that is, they depend on the prior conditions where it is adopted. In situations when *Bt* cotton replaces effective pesticides, the impact is mostly reduced pesticide use and, as some evidence has shown, improved human and environmental health (R. Bennett et al. 2004; Huang et al. 2002, 2003). Adoption of *Bt* cotton can have very substantial yield effects when it introduces control of a pest that previously was not effectively controlled by pesticides or other means. Thus, reported yield effects of introducing *Bt* cotton can range from 0 to as high as 80 percent (Qaim, Yarkin, and Zilberman 2005; Qaim and Zilberman 2003).

Impacts Are Stochastic in Nature

The impacts of adopting a transgenic variety at a certain location are stochastic and vary in response to weather and other conditions. For example, a pest control trait will have significant impact during years of severe pest infestations and minimal impacts in those years when pest pressure is minimal. If the trait is inserted into a generic variety, the overall yield effect of the transgenic variety is likely to be positive and significant when infestation occurs but may actually be negative when there is no pest pressure. In general, transgenic pest control traits have value as risk management tools. They affect not only the mean of profits facing farmers but also the entire distribution of profits. In addition to their impacts on average profits, transgenic traits also reduce risk whether measured by variance or other methods. This quality provides value to risk-averse producers. Another lesson, which derives from those described earlier, is that assessing the impacts of agricultural biotechnology based on a one-year experience may result in misleading conclusions. The overall impact of transgenic traits depends on the effectiveness of the traits per se, the quality of the crop varieties into which they are introduced, and the specific climatic and pest conditions that occur (Zilberman, Ameden, and Qaim 2007).

Impacts Can Be Nonpecuniary – Convenience and Time Savings

There is a large body of literature showing that the primary reasons for adoption of transgenic varieties are the yield effects and reductions in cost and risk. Two other significant reasons that have come to light are nonpecuniary benefits, such as convenience and time savings (Piggott and Marra 2008; Zilberman et al. 2007). Chemical and mechanical pest control frequently requires significant allocation of time and effort to monitoring the pest situation. Both pest-resistant and herbicide-tolerant varieties constitute preventive measures that both save time and reduce anxieties typically associated with the available alternatives for pest control, such as chemical pesticides and integrated pest management.

Impacts Can Be Indirect – Reduced Fusarium Infestation

There is evidence that transgenic pest resistance traits can have secondary benefits by reducing secondary pest infestations or infections typically caused by initial pest damage. For example, analysis of *Bt* corn indicates that it has reduced postharvest levels of mycotoxins, which are often introduced at the site of insect damage to corn, and increased the quality of stored grain and silage, especially in humid climates (Wu 2006; Wu, Miller, and Casman 2004).

Impacts Are Scale-Neutral

Although many agricultural technologies have features that make them more attractive to larger farm operations, this is usually not the case with transgenic varieties. Seeds have traditionally been considered a technology with constant returns to scale (Dorfman, Samuelson, and Solow 1987). Their adoption does not require investment in new and expensive capital. For example, adoption of *Bt* corn does not require investment in expensive application equipment or protective clothing, as would be necessary for chemical pesticides. Smaller farms may lack the expertise required for effective management of chemical pesticides and integrated pest management and may have inferior storage facilities relative to larger farms. These qualities of smaller farms make transgenic varieties that reduce pest damage both on the field and in storage relatively more attractive. However, smaller or poorer farmers may not have the access to funds that allow them to buy the more expensive transgenic varieties. When funds are a constraint, a major policy priority is the establishment of policies to provide the financial resources to purchase transgenic varieties.

No Evidence of Harm Even after Extensive Use

Although "absence of evidence" does not constitute "evidence of absence," there is, as of yet, no compelling evidence of damage to the environment or to human health from growing and consuming transgenic crops, despite the hundreds of millions of acres planted and hundreds of millions of tons eaten over the last decade (CAST 2005). We emphasize that this lack of compelling evidence is particularly notable because clearly there is a strong opposition to GM agriculture that would be quick to seize on and widely publicize any incriminating evidence. Much concern has been raised about the potential environmental effects of transgenic varieties. These effects include gene flow to nearby wild species or landraces closely related to a transgenic crop species, the emergence of resistant weeds, the reduction of crop biodiversity as a few transgenic super-varieties cover large portions of cultivated lands, and resistant pest populations (Ellstrand 2003; Letourneau and Burrows 2002). Although there have been some highly publicized cases of such impacts – transgenic varieties mixing with nontransgenic varieties or human exposure to transgenic varieties only approved for animal consumption[1] – in these cases no human life was at risk. Of course accidents do happen, and the impacts of transgenic varieties on the environment and human health should be monitored. However, no substantial environmental catastrophes have yet occurred. Moreover, based on accumulated evidence of the behavior of transgenic crops within agro-ecological settings and the human diet, the likelihood of such a major problem is greatly diminished.

Incentives to Avoid Yield Drag Will Help Maintain Crop Biodiversity

One lesson has been the importance of inserting transgenic traits into many diverse varieties to prevent yield drag. For example, by 2002 in the United States, more than 1,100 local varieties of soybean contained the Roundup Ready trait (Qaim et al. 2005). Over the years, varietal experimentation and selection by farmers have resulted in patterns of varietal choices that optimize yield and other product properties specific to local conditions. When a transgenic trait is not available in a local variety, a farmer must switch to a generic variety in which the trait is available. This switch is likely to result in yield and other losses. Farmers trade off gains from the trait with losses from the generic variety.

Several studies have shown that, under plausible conditions, the concern about reduced crop biodiversity can be addressed. If farmers are likely to face yield drag by planting a generic transgenic variety that is different from their locally optimized variety, then seed suppliers are likely to gain if they are able to sell farmers a transgenic version of their local variety. Seed suppliers will add the transgenic traits to local varieties if they have access to the genetics and if the cost of producing the transgenic local variety can be recaptured. Thus, we should expect that a significant share of locally optimized varieties will become transgenic in cases where the necessary genetic materials are available at low transaction costs and where there is sufficient technical capacity to backcross or modify local varieties at relatively low cost. Indeed, there is evidence that crop biodiversity is well maintained with the introduction of transgenic varieties in countries where these conditions are met (e.g., the United States, China, and India; Zilberman et al. 2007). Concern regarding lack of technical capacity may lead to the introduction of only a few generic transgenic varieties in Africa, unless that capacity is upgraded.

[1] For example, see the Starlink™ corn case (Bucchini and Goldman 2002).

Managing Resistance

Concerns about the development and spread of resistance within target pest populations led to introduction of formal refuge requirements in the use of *Bt* varieties. Thus far, it seems that the emergence of resistance to *Bt* toxins has been mostly contained (Bates et al. 2005). However, Nichols et al. (2009) reported emergence of glyphosate-resistant weeds. Key to the development process of transgenic pest control, just as with chemical pest control, is to continually develop traits and management strategies that enable farmers to stay ahead in the race against evolutionary forces resulting in resistance.

Some Positive Environmental Impacts

Despite the environmental concerns about agricultural biotechnology there have been significant reductions in the environmental footprint of farming due to the use of GM pest control traits. These reductions are largely due to the substitution away from chemicals and the promotion of no-till practices (Barfoot and Brookes 2007; Johnson, Strom, and Grillo 2008; Phipps and Park 2002). Studies in China have documented that adoption of *Bt* varieties has reduced the use of chemical pesticides and improved the health of farm workers (Huang et al. 2002, 2003; Sydorovych and Marra 2007). The adoption of herbicide-resistant varieties has promoted the adoption of low- and no-tillage practices and slowed soil erosion. Low- and no-tillage practices have sequestered carbon in the soil, reducing the amount of greenhouse gases in the atmosphere (Barfoot and Brookes 2007; CAST 2004). Finally, increases in yield and farm productivity associated with the use of transgenic varieties have improved food supply and reduced food prices, thereby reducing incentives to expand the agricultural land base. One of the great achievements of agricultural technologies in the second half of the 20th century was more than doubling agricultural output with only a 20 percent increase in area farmed (Edgerton 2008; Sunding and Zilberman 2001). Agricultural biotechnology is contributing to the trend of meeting growing food demand with lower impact on uncultivated lands and on the biodiversity they contain.

The Unrealized Potential of First-Generation Traits

Despite the achievements, the overall contributions to prosperity by these first-generation pest control traits and the equity of distribution could have been greater. Preliminary comparisons of global supply and demand trends of major commodity crops – those for which biotech traits have been developed and those without – suggest that the technology has made a positive impact on supply of and demand for those crops in which it has been deployed (Sexton et al. 2008, 2009). Changes in annual consumption of corn (which has developed biotech traits) between 2004 and

2008 compared to changes in annual consumption of wheat (which has not developed biotech traits) over the same years shows that consumption of corn grew over these seasons while consumption of wheat was largely stagnant. Although there are clearly numerous factors involved in the differences in consumption trends between these two commodities, one potentially significant difference was the wide availability of transgenic corn varieties versus the lack of transgenic wheat varieties. Production trends show similar differences.

The fundamental distributional question of social equity and prosperity that emerges from this discussion is, What would the impact have been on commodity prices, global food security, and, by extension, human welfare if these same traits that have already been commercialized in corn, soybean, canola, and cotton had also been made available in sorghum, rice, and wheat? What would the impact on global equity and prosperity have been if these existing, proven traits were made as widely available in Sub-Saharan African or Southeast Asian countries as they were in the United States, Canada, and Argentina? Even in the United States, which leads in biotech adoption, what increases in productivity and decreases in global food prices – along with decreases in pesticide use and other environmental impacts – could have been achieved if first-generation traits like insect resistance and herbicide tolerance were deployed in other major crops? In sum, the first generation of transgenic crops, consisting largely of pest control traits, has contributed to agricultural productivity by increasing yields, as well as reducing some of the costs and risks of farm production. In so doing, it has produced tangible gains in human welfare while delivering environmental benefits. The technology has demonstrated objective possibilities for achieving some of the goals of sustainability.

Reassessment of Biotechnology for Future Applications

A fundamental reassessment of biotechnology and its potential for the economics and the sustainability of agriculture is underway, fueled by the large increases in food and energy prices experienced in 2007 and 2008. In fact, the National Research Council (NRC) has assembled a committee to investigate this issue (NRC 2008). Although one reason for such analysis is to consider the role of biotechnology in light of rising energy prices, the other is to assess its potential as second-generation innovations are developed, which enhance the quality and efficiency of agricultural products in their entire range of uses for food, feed, fiber, and fuel.

Agricultural Biotechnology and Energy Prices

In the agricultural sector, increased energy prices are likely to drive up prices of energy-intensive inputs such as fertilizers, mechanized cultivation, and even irrigation (fuel to run irrigation pumps). If, for the sake of argument, rising input prices are not

Table 15.1. *Ten categories of product quality improvements that have been worked on in the plant biotechnology R&D pipeline*

1. Protein and amino acid content	6. Esthetics and convenience (color, taste, smell, size, seedless, etc.)
2. Oils and fatty acid content	7. Ripening control and shelf life
3. Carbohydrate and sugars content	8. Fiber/biomass composition
4. Micronutrient (vitamin, mineral, and functional component) content	9. Environmental traits (bioremediation)
5. Reduced toxin and allergen content	10. Multiple traits combined

Source: Graff, Zilberman, and Bennett (2009).

accompanied by increases in output prices, agricultural production would suffer, and transgenic varieties that substitute for these energy-intensive inputs – whether by reducing crop damage (insect resistance), water requirements (drought tolerance), or fertilizer use (nitrogen use efficiency) – would have extra adoption benefits. Such transgenic traits would compensate for lost production due to rising input prices and thereby would increase food supply.

However, it is more likely that we will continue to see increases in energy prices accompanied by higher food prices. In recent years, as we observed earlier, the rise in food prices and the decline in food inventories were greater in those crops without transgenic traits – such as rice and wheat – than in crops with transgenic traits like corn, cotton, and soybean (Sexton et al. 2008, 2009). Increased energy prices led to the diversion of about 20 percent of the U.S. corn harvest to ethanol production. This diversion contributed to lower fuel prices by 1 to 3 percent but to higher commodity prices by 16 to 32 percent (Sexton et al. 2009). The two means by which biotechnology is most likely to reduce pressure on the food sector are (1) to increase primary adoption of first-generation transgenic varieties that increase primary crop productivity and thus reduce pressure on commodity prices and (2) to introduce second-generation traits that enhance the efficiency-of-use of agricultural commodities, meaning fewer quantities are needed to achieve the same benefits.

Second-Generation Traits

We distinguish this forthcoming second generation of transgenic traits by the fact that they exhibit fundamentally different economic impacts: They improve the quality of agricultural output rather than reduce the costs or risks associated with agricultural production. Examples include improved nutrition for human consumption, optimized animal feed rations, and more easily degraded woody biomass for feed, pulping, and biofuels. Observation of the agbiotech R&D pipeline (Graff, Zilberman, and Bennett 2009) reveals ten broad categories of second-generation quality traits (Table 15.1).

These characteristics generally benefit the users rather than the producer of the agricultural output. They can be expected to increase efficiencies-of-use of the output from agricultural and agro-forestry systems, improve environmental health, and improve human health, especially among those in the lowest socioeconomic strata.

Increased efficiencies-of-use of agricultural commodities occur in downstream processes in the agricultural value chain, such as feeding animals or paper pulping. Less of the commodity, or less of other inputs, is needed for the process to achieve comparable results. For example, if the conversion efficiency of corn fed to cattle can be improved 3 percent by genetically balancing the amino acid content and increasing protein or oil content, then feed rations can be decreased by about 3 percent and weight gain will not change. Ultimately, this means 3 percent less land, fertilizer, water, and other inputs required to produce corn for cattle feed. Another example would be improved fiber characteristics in the wood of trees grown for paper milling. By reducing the need for mechanical and chemical processing, this could save pulping costs on the order of \$10 per m^3 (Sedjo 2007). Similarly, by genetically improving the shelf life of vegetables, waste could be reduced: For every ton of tomatoes harvested, a greater percentage would reach the end consumers, improving returns and easing logistics for distributors and retailers. Moreover, consider the efficiency impacts of a shelf-life trait in Africa, where lack of roads and refrigeration means that average wastage of fresh vegetables is more than twice as high as in the United States or Europe (FAOSTAT 2010).

Environmental health can be improved both directly and indirectly by quality enhancements. Consider the indirect environmental health impacts of the examples in the previous paragraph, whether from reductions in land and water necessary to feed cattle or from reductions in energy use and chemicals for pulping. An example of a direct environmental health improvement is properly balanced amino acids and bioavailable phosphorus content in animal feed crops. These qualities would reduce excess nitrogen and phosphorus excreted in animal waste and then polluting waterways.

Improved human health would result from increased supplies – thus decreased costs – of essential nutrients. Although access to basic nutrients is not a constraint in high-income countries, there are still ways that biotechnology can be used to reduce costs of supplying high-value nutrients or complying with food health regulations. For example, genetic modification can make soy oil rich in omega-3 precursors, potentially making omega-3 fatty acids ubiquitous in the U.S. diet. Other soy oil quality modifications can reduce the need for hydrogenation of vegetable oils to achieve desired baking qualities, thus eliminating the occurrence of trans fats in the U.S. diet at competitive costs. Yet, it is the mass of humanity living on less than \$1 a day who stand to benefit more from greater supplies and lower costs of essential nutrients. The debate over the last decade has revolved around a single strategy – bio-fortification – of a single crop, rice, with a single nutrient, beta carotene. In theory and practice, however, low- and middle-class consumers in developing countries could benefit from an entire range of strategies, crops, and nutrients to make balanced nutrition more prevalent and less costly, including greater availability of fresh produce, functional foods, micronutrient bio-fortification, healthier oils, and higher protein contents.

Shaping Policies for Sustainability

The extent to which economies and ecosystems have been able to benefit from the technology has depended on governance and institutions, most significantly national regulatory frameworks and intellectual property rights. Economic analysis of innovation suggests that investment in developing a new technology and the adoption of that technology fundamentally depend on two things: profit incentives and regulatory constraints. To a significant extent, agricultural biotechnology has been a product of the "educational-industrial complex," which consists of university and public sector researchers identifying new technological paradigms and a wide array of companies, including both start-ups and major corporations, investing in the development of new products using those new paradigms. Private investment is incentivized, particularly in biotechnology, by intellectual property rights. Investment by companies and adoption by farmers depend on their abilities to earn a profit subject to regulatory constraints and access to intellectual property rights.

Regulatory Requirements

The regulatory environment governing transgenic varieties has significantly shaped the evolution of R&D investments and the patterns of commercialization of this technology. When European policy effectively obstructed the registration process and then established costly labeling and traceability requirements followed by strict GM-content and gene-flow rules, it decreased not only the introduction of new transgenic products into Europe but also the expected profitability of new transgenic products introduced everywhere in the world. It reduced incentives on the margin for private innovators to invest in developing new products with both first- and second-generation traits. Yet Europe was not alone. Even though China has developed a large line of transgenic crops, it has only allowed the introduction of a few of them. The regulatory framework in other developing countries continues to maintain much more strict control than the regulatory frameworks in high-adoption countries like the United States, Canada, and Argentina.

Economic theory suggests that a socially optimal regulatory framework will consider the expected discounted net benefits of a new technology relative to the alternative technology most likely to be used. In reality, the regulatory process evaluates separately and individually each new technology – whether a trait or a chemical compound – and allows those that meet some established, objective benchmark of minimal or zero impact on the environment or health. The logic of this process compares each new technology against an absolute standard, rather than comparing them against one another or against actual technologies that are already in practice – many of which were grandfathered in under current regulatory standards because they were already in the marketplace before standards were adopted. Such a system does not encourage the development of new technologies that could replace some of the more problematic ones already in the marketplace. A reform of policy processes to allow consideration of the costs and benefits of proposed transgenic varieties relative to current agricultural technologies, rather than relative to a hypothetical practice with zero impact, would go a long way toward achieving the social optimum.

The costs of regulatory compliance also affect the product development strategies pursued by companies and therefore the adoption patterns among farmers. For example, if a company has to navigate costly regulatory filings for every insertion of a transgenic trait into every variety of a crop species, the firm will only introduce the trait into the most widely used varieties. As discussed earlier, this would likely lead to yield and crop biodiversity losses, as farmers trade off yield against the benefits gained from the transgenic trait by shifting to a transgenic variety that is not optimized for local conditions. However, if a single filing can approve the use of a transgenic trait in multiple varieties without duplicating the costs for each and every variety, then the company is likely to insert the trait into many varieties. This will have the effect of increasing productivity and preserving biodiversity in that crop.

This example can extend to a case in which, if a trait is found effective in a family of closely related crops, registration requirements that extend approval across the family of crops, rather than being limited to each individual crop, may be optimal. Similarly, in some regions with many small countries or sparse populations, such as Africa or Central America, it may not make sense for every small nation to develop and manage its own independent regulatory process independently. Instead, a regulatory framework implemented and managed on a regional level could reduce costs and enhance the introduction of new technologies across the whole region. In general, governments should follow a regulatory approach that avoids redundancy; review of new transgenic varieties or crops should be required only in those cases where additional testing can provide necessary safety information not available from previously approved varieties.

Intellectual Property Rights

Because many agricultural applications of biotech exhibit constant returns to scale, they can be adopted just as readily by small subsistence farmers as by large industrial farmers. Yet to do so, small farmers need access to (1) information about the trait and its management and (2) seeds containing the trait. The dissemination of information about biotech crops is an important role for agricultural extension. The dissemination of seed or other genetic materials that contain biotech traits can, at least in theory, be a role for either public or private channels. Whether the private or public sector or both take on this role depends on the level of funding that public agricultural R&D programs receive and on the disposition of intellectual property rights over traits, crop varieties, and methods of GE. Proprietary claims over all three of these – genes, germplasm, and tools – have important bearing on the incentives and constraints for private innovation in agricultural genetics and on the ability of public R&D to develop biotechnologies and transfer them to farmers.

To the extent that property rights over a technology are secure, an owner is able to exclude others from using it. Legally constructed excludability gives an owner the opportunity to sell a technology, thus creating incentives to innovate and disseminate new technologies. If the benefits generated by using a new technology are greater than the price charged by the owner of that technology, users are better off. Conversely, if the price charged by the owner exceeds the benefits generated by the technology, then users will not adopt the new technology.

Agricultural genetics, however, often has significant aspects of a "commons." Everyone benefits, to some degree, from the unrestricted exchange of genetic materials. Low-income farmers, in particular, tend to benefit from low-cost or free access to seed that is saved and shared within local communities. Smaller regional seed companies around the world thrive on a business model of breeding traits from elite germplasm into localized varieties. Ultimately, any crop variety – whether engineered or not – represents a cumulative integration of numerous genetic inputs and development processes.

Introducing private assets into such a commons-based system could efficiently be executed, in theory, if the owners of the privately held components were compensated each time that each component was used to the benefit of a producer or consumer. However, such a scheme is difficult to implement in agriculture because it introduces transaction costs that exceed the incremental values of the genetics being transacted. In particular, the scheme is unworkable for the "long tail" of low-income and subsistence growers who each privately benefit, but at a level so low that the profit margin of selling proprietary seeds to them could not cover a private company's costs of doing business.

The solutions that therefore most often appear to be workable are those that manage agricultural genetics, either as public goods or as private assets. When provided as a public good in the public domain, costs of exchange stay extremely low, and transaction costs to try to account for ownership are unnecessary. To the extent that private seed vendors do enter the market, they compete on costs of multiplying and disseminating public domain genetics, guaranteeing seed quality, or providing complementary inputs and services. When agricultural genetics are regarded as private assets, private firms consolidate all of the necessary inputs within their respective trait development and breeding programs. They then sell only to those portions of the market that have a willingness to pay that exceeds their marginal costs at the point of sale.

Market-based hybrid solutions such as intellectual property clearinghouses or patent pools may eventually bridge the two systems (Delmer et al. 2003; Graff and Zilberman 2001). They do so by decreasing the transaction costs of exchanging agricultural genetics, making it feasible to offer privately owned traits in a wider array of both public and privately owned germplasm. They also can provide feasible mechanisms for what is, in effect, a price differentiation strategy between growers who have a positive net willingness to pay and those who do not. To date, two variants of this sort of solution have been attempted in agriculture: the Public Intellectual Property Resource for Agriculture (PIPRA) and the African Agricultural Technology Foundation (AATF). PIPRA is a consortium of universities and agricultural research institutes developing jointly owned GE tools that will be widely available for licensing both by private and public developers of agricultural genetics (Atkinson et al. 2003; A. B. Bennett et al. 2008; Graff et al. 2003). AATF is engaged in licensing technology globally from across an array of private companies and universities to be deployed in transgenic crops for subsistence-level farmers in Sub-Saharan Africa (AATF 2005).

Conclusion

Biotechnology has radically expanded the capacity of scientists to develop genetic materials for farming, and applications have increased agricultural productivity. Like any other radical technology, agricultural biotechnology can arouse uncertainty and fear among the public. It also affects the relative value of existing assets and technologies (e.g., increasing the value of seeds and decreasing the value of agro-chemicals), which may lead to attempts to manipulate regulatory and intellectual property policies to control and delay the introduction of these technologies (Graff and Zilberman 2004).

The short experience we have had with transgenic varieties suggests that they can significantly increase agricultural supply, improve food quality, and reduce environmental side effects and health risks of agricultural production. Furthermore, they reduce the efforts and sophistication required for farming and do not have any inherent increasing returns to scale – making adoption by small farmers equally likely. A forthcoming second generation of agricultural biotechnology can enhance the nutritional content of food and improve food quality. Enhanced agricultural capacity will also likely overcome climatic constraints. These capabilities that are likely to be unleashed by transgenic varieties are especially valuable as we foresee a future in which agriculture is increasingly becoming a supplier of energy and society is concerned by the challenge of climate change.

However, unleashing the potential of agricultural biotechnology is an institutional and policy challenge. First, mechanisms must be established to assure that, if or when agricultural biotechnology results in negative side effects, they will be identified and controlled immediately. Concurrently, a regulatory system should be cost effective and reasonable to allow the technology to flourish and provide incentives to innovate and adopt the technology to enhance social welfare.

References

AATF (African Agricultural Technology Foundation). 2005. *A New Bridge to Sustainable Agricultural Development in Africa*. Nairobi: AATF.

Atkinson, R. C., R. N. Beachy, G. Conway, F. A. Cordova, M. A. Fox, K. A. Holbrook, D. F. Klessig, et al. 2003. "Public Sector Collaboration for Agricultural IP Management." *Science* 301 (5630): 174–5.

Barfoot, P., and G. Brookes. 2007. "Global Impact of Biotech Crops: Socio-Economic and Environmental Effects, 1996–2006." *AgBioForum* 11 (1): 21–38.

Bates, S. L., J. Z. Zhao, R. T. Roush, and A. M. Shelton. 2005. "Insect Resistance Management in GM Crops: Past, Present and Future." *Nature Biotechnology* 23: 57–62.

Bennett, A. B., C. Chi-Ham, G. Graff, and S. Boettiger. 2008. "Intellectual Property in Agricultural Biotechnology: Strategies for Open Access." In *Plant Biotechnology and Genetics: Principles, Techniques, and Applications*, edited by N. Stewart, 325–42. New York: J. Wiley & Sons.

Bennett, R., R. Phipps, A. Strange, and P. Grey. 2004. "Environmental and Human Health Impacts of Growing Genetically Modified Herbicide-Tolerant Sugar Beet: A Life-Cycle Assessment." *Plant Biotechnology Journal* 2: 273–8.

Bucchini, L., and L. R. Goldman. 2002. "Starlink Corn: A Risk Analysis." *Environmental Health Perspectives* 110 (1): 5–13.

CAST (Council for Agricultural Science and Technology). 2004. *Climate Change and Greenhouse Gas Mitigations: Challenges and Opportunities for Agriculture*. Task Force Report 141. Ames, IA: CAST.

———. 2005. *Crop Biotechnology and the Future of Food: A Scientific Assessment*. CAST Commentary QTA 2005-2. Ames, IA: CAST.

Delmer, D., C. Nottenburg, G. Graff, and A. B. Bennett. 2003. "Intellectual Property Resources for International Development in Agriculture." *Plant Physiology* 133: 1666–70.

Dorfman, R., P. A. Samuelson, and R. M. Solow. 1987. *Linear Programming and Economic Analysis*. New York: Dover Publishing.

Edgerton, D. 2008. "The Charge of Technology." *Nature* 455: 1030–1.

Ellstrand, N. C. 2003. *Dangerous Liaisons? When Cultivated Plants Mate with Their Wild Relatives*. Baltimore: Johns Hopkins University Press.

FAOSTAT. 2010. *SUA/FBS (Supply Utilization Accounts and Food Balances) Database* (accessed November 18). http://faostat.fao.org/site/354/default.aspx.

Graff, G., S. Cullen, K. Bradford, D. Zilberman, and A. B. Bennett. 2003. "The Public-Private Structure of Intellectual Property Ownership in Agricultural Biotechnology." *Nature Biotechnology* 21 (9): 989–95.

———, and D. Zilberman. 2001. "An Intellectual Property Clearinghouse for Agricultural Biotechnology." *Nature Biotechnology* 19: 1179–80.

———. 2004. "Explaining Europe's Resistance to Agricultural Biotechnology." *Agricultural and Resource Economics Update* 7 (5): 1–4.

———, D. Zilberman, and A. B. Bennett. 2009. "The Contraction of Agbiotech Product Quality Innovation." *Nature Biotechnology* 27 (8): 702–4.

Huang, J., R. Hu, C. Pray, F. Qiao, and S. Rozelle. 2003. "Biotechnology as an Alternative to Chemical Pesticides: A Case Study of *Bt* Cotton in China." *Agricultural Economics* 29 (1): 55–67.

———, S. Rozelle, C. Pray, and Q. Wang. 2002. "Plant Biotechnology in China." *Science* 295 (5555): 674–6.

Johnson, S. R., S. Strom, and K. Grillo. 2008. *Quantification of the Impacts on US Agriculture of Biotechnology-Derived Crops Planted in 2006*. Washington, DC: National Center for Food and Agricultural Policy.

Letourneau, D. K., and B. E. Burrows, eds. 2002. *Genetically Engineered Organisms: Assessing Environmental and Human Health Effects*. Boca Raton, FL: CRC Press.

Nichols, R. L., J. Bond, A. S. Culpepper, D. Dodds, V. Nandula, C. L. Main, M. W. Marshall, et al. 2009. "Glyphosate-Resistant Palmer Amaranth (*Amaranthus palmeri*) Spreads in the Southern United States." *Resistant Pest Management Newsletter* 18 (2): 8–9.

NRC (National Research Council). 2008. *The Impact of Biotechnology on Farm Economics and Sustainability*. Washington, DC: National Academy Press. http://www8.nationalacademies.org/cp/projectview.aspx?key=48978.

Phipps, R. H., and J. R. Park. 2002. "Environmental Benefits of Genetically Modified Crops: Global and European Perspectives on Their Ability to Reduce Pesticide Use." *Journal of Animal and Feed Sciences* 11 (1): 1–18.

Piggott, N. E., and M.C. Marra. 2008. "Biotechnology Adoption over Time in the Presence of Non-Pecuniary Characteristics That Directly Affect Utility: A Derived Demand Approach." *AgBioForum* 11 (1): 58–70.

Qaim, M., C. Yarkin, and D. Zilberman. 2005. "Impact of Biotechnology on Crop Genetic Diversity." In *Agricultural Biodiversity and Biotechnology in Economic Development*, edited by J. Cooper, L. M. Lipper, and D. Zilberman, 283–307. New York: Springer.

———, and D. Zilberman. 2003. "Yield Effects of Genetically Modified Crops in Developing Countries." *Science* 299 (5608): 900–02.

Sedjo, R. A., 2007. "Regulation of Biotechnology for Forestry Products." In *Regulating Agricultural Biotechnology: Economics and Policy*, edited by R. E. Just, J. M. Alston, and D. Zilberman, 663–682. New York: Springer.

Sexton, S., G. Hochman, D. Rajagopal, and D. Zilberman. 2008. "The Role of Biotechnology in a Sustainable Biofuel Future." *AgBioForum* 12 (1): 130–40.

———, D. Rajagopal, G. Hochman, D. Zilberman, and D. Roland-Holst. 2009. "Biofuel Policy Must Evaluate Environmental, Food Security, and Energy Goals to Maximize Net Benefits." *California Agriculture* 63 (4): 191–8.

Sunding, D., and D. Zilberman. 2001. "The Agricultural Innovation Process: Research and Technology Adoption in a Changing Agricultural Sector." In *Handbook of Agricultural and Resource Economics*, edited by B. L. Gardner and G. C. Rausser, 207–61. Amsterdam: Elsevier Science.

Sydorovych, O., and M.C. Marra. 2007. "A Genetically Engineered Crop's Impact on Pesticide Use: A Revealed-Preference Index Approach," *Journal of Agricultural and Resource Economics*, 32(3): 476–91.

Wu, F. 2006. "*Bt* Corn's Reduction of Mycotoxins: Regulatory Decisions and Public Opinion." In *Regulating Agricultural Biotechnology: Economics and Policy*, edited by R. E. Just, J. M. Alston, and D. Zilberman, 179–200. New York: Springer.

———, J. D. Miller, and E. A. Casman. 2004. "The Economic Impacts of *Bt* Corn Resulting from Mycotoxin Reduction." *Journal of Toxicology* 23 (2&3): 397–424.

Zilberman, D., H. Ameden, and M. Qaim. 2007. "The Impact of Agricultural Biotechnology on Yields, Risks, and Biodiversity in Low-Income Countries." *Journal of Development Studies* 43 (1): 63–78.

Index